Demographic Methods

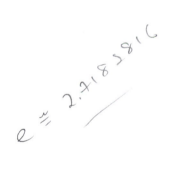

Demographic Methods

Andrew Hinde

Senior Lecturer in Population Studies, University of Southampton

A member of the Hodder Headline Group
LONDON NEW YORK SYDNEY AUCKLAND

First published in Great Britain in 1998 by
Arnold, a member of the Hodder Headline Group
338 Euston Road, London NW1 3BH
http://www.arnoldpublishers.com

Copublished in the United States of America by
Oxford University Press Inc.,
198 Madison Avenue,
New York, NY 10016

British Library Cataloguing in Publication Data
A catalogue entry for this book is available from the British Library

Library of Congress Cataloging-in-Publication Data
A catalog entry for this book is available from the Library of Congress

ISBN 0 340 71891 9 (hb)
ISBN 0 340 71892 7 (pb)

Production Editor: James Rabson
Production Controller: Helen Whitehorn
Cover designer: Terry Griffiths

Typeset in 10/12pt Times by Academic & Technical, Bristol
Printed and bound in Great Britain by MPG Books, Bodmin, Cornwall

Contents

Preface

This book arose out of the need for a single textbook to cover the material taught in a final-year undergraduate course in demographic methods at the University of Southampton. Although a limited number of textbooks dealing with demographic methods and demographic analysis are in existence, none of these was found to meet the requirements. The course in question is designed for students with some previous study of mathematics and/or statistics, and the existing books either tend to be written for non-mathematicians or to assume a considerable amount of experience with calculus and matrix algebra. In this book, I have tried to steer a middle course, using some calculus where appropriate, but not assuming too much familiarity with this subject.

The course, which I have taught for the past four years, tries to dispel the belief that demographic analysis is a 'hard' subject. It often seems to be difficult, because the computations involved can be extensive, and because the nature of the data with which demographers often have to work means that the calculations can become very 'fiddly'. The fundamental principles involved, however, are in fact rather simple; moreover, they carry across from the analysis of mortality to the analysis of marriage, fertility (and even – though less obviously – migration).

To take two examples, the calculation of demographic rates, and the need to ensure correspondence between the events in the numerator and the population exposed to the risk of experiencing the events, are common to the analysis of mortality, marriage, fertility and migration. The life table, which is the demographer's most important and most versatile tool, is now used routinely to analyse almost all demographic processes, especially now that its extension in the form of survival analysis has become an integral part of the demographer's tool-kit.

In this book, therefore, I have tried to focus attention on basic principles, and to stress how these may be applied to analyse the main demographic processes. Because it is the simplest of these processes, mortality is covered first (Chapters 2–6). Marriage is the next simplest, and is treated second (Chapter 7). The analysis of fertility, which poses a number of additional challenges, is covered in Chapters 8–11.

The book also deals with the analysis of population dynamics in general, in Chapters 12–14. These chapters aim to integrate the analysis of mortality and fertility to show how they act together to determine population growth and structure. For simplicity, migration is ignored in these chapters. It is considered separately in Chapter 15. Finally, Chapters 16–18 cover population projection, which requires the integration of all the preceding material.

I said earlier that the book assumes some (but I hope not too much) mathematical background. Although calculus is used in some chapters, I have tried to structure the text in such a way that students without any calculus can benefit from it. Sections which require calculus are indicated in the list of contents with an asterisk. They may be omitted if desired.

Most of the chapters include exercises. A few of these are pure drill, but most are designed either to test understanding or to provide practice in tackling the kinds of challenge which arise when real data are confronted. The exercises are also designed to double as worked examples, and to that end a comprehensive set of solutions is included, showing how solutions are arrived at. The solutions to the exercises are, therefore, an integral part of the text.

Spreadsheet files (for Quattro® Pro v. 6.0 and Microsoft® Excel) with the data for all the numerical exercises, together with a number of additional data files, have been placed on the World Wide Web at **http://www.arnoldpublishers.com/download/hinde.htm**. Further details of how to use them are included in an Appendix.

Inevitably, the book is somewhat selective, both in what it covers and in the depth in which different techniques are studied. The selection has been made on a number of criteria. Some topics have been omitted, or given only a cursory treatment, because they are not absolutely essential and there are excellent introductory texts available. The most obvious of these are multi-regional methods and the practical application of survival analysis using widely available statistical software (such as SPSS or SAS). For the same reason, many indirect methods of demographic estimation are not described here. The recommendations for further reading at the end of relevant chapters point interested readers in the direction of suitable specialist texts on these topics.

Other topics have been given what may seem to some an over-extended treatment. Some of these are topics which I felt were missing from my own demographic training (for example, the maximum likelihood estimation of a hazard; or force of mortality; or any reasonably rigorous explanation of exactly why a population with constant fertility and mortality develops a constant proportional age structure and ultimately grows in size at a constant rate).

The analysis of migration is treated much less fully than the analysis of fertility and mortality. To some extent this reflects its traditional 'Cinderella' position within demography. This is hardly an adequate excuse. Its relative neglect is also a consequence of the constraints imposed by the length of the book. Chapter 15 provides a basic introduction to the topic, but it really only scratches the surface.

Finally, I have chosen to concentrate on demographic *methods*, rather than demographic *models*. For this reason, model life tables are not given an extensive treatment, and model schedules of nuptiality and fertility have been omitted. I appreciate that these omissions will not please some, but in a book of this length, I felt it desirable not to try to cover too much, and thereby risk the text being too terse for many students.

During over ten years of teaching demographic methods, I have been indebted to many individuals. Richard Smith (now director of the Cambridge Group for the History of Population and Social Structure) first kindled my interest in demography. The influence of the justly famous series of lectures on population dynamics which William Brass gave for many years at the Centre for Population Studies at the London School of Hygiene and Tropical Medicine may be recognized, especially in Chapter 13. My colleagues at Southampton, especially Philip Cooper, Ian Diamond, Mac McDonald and Máire Ní Bhrolcháin, have assisted me greatly during the last eight years. Others who have helped me clarify my understanding of specific issues include John Ermisch, Heather Joshi, Angus Macdonald, Bob Woods and Robert Wright, and there are many others who I hope will forgive me for not mentioning them by name. Successive generations of students at Southampton have pointed out errors in the material – especially the solutions to the exercises (!) – and, by their determined and pointed questioning, have forced me to think hard about how to explain particular concepts. Marge Fauvelle helped type the exercises into spreadsheet files. Of course, any errors and omissions which this book contains are my responsibility alone.

Finally, thanks are due to my wife, Jane, and my three sons, Luke, Dominic and Joel, who have been very patient with me during the evenings and weekends when I have been closeted in a study rather than fulfilling functions which to them seemed more important.

Acknowledgements

The author and the publisher would like to thank the following for permission to use copyright material in this book.

Institut National d'Études Démographiques: Figure 1 from 'Vingt et unième rapport sur la situation démographique de la France', *Population*, **47** (1992), p. 1114. Reprinted by permission.

Centre for Economic Policy Research, London, for an extract from a table in H. Joshi (ed.) *The Changing Population of Britain*, Oxford, Blackwell, © 1989 Centre for Economic Policy Research. Reprinted by permission.

J. Bongaarts and R. Potter: adaptation of Figure 4.1 in *Fertility, Biology and Behaviour: an Analysis of the Proximate Determinants*, London, Academic Press, © 1983 Academic Press. Reprinted by permission.

A.J. Coale and P. Demeny (eds): extract from two tables in *Regional Model Life Tables and Stable Populations*, 2nd edition, New York, Academic Press, © 1983 Academic Press. Reprinted by permission.

Office of Population Censuses and Surveys: extract from a table in *English Life Table No. 14: the Report Prepared by the Government Actuary for the Registrar General for England and Wales*, London, Her Majesty's Stationary Office, © Crown Copyright 1987. Reprinted by permission of the Office for National Statistics.

Office of Population Censuses and Surveys: two tables in *Marriage and Divorce Statistics*, series FM2, no. 14, London, Her Majesty's Stationary Office, © Crown Copyright 1989. Reprinted by permission of the Office for National Statistics.

C. Daykin: a figure from 'Projecting the population of the United Kingdom', *Population Trends*, **44** (1986), pp. 28–33, © Crown Copyright 1986. Reprinted by permission of the Office for National Statistics.

C. Shaw: a figure from '1991-based national population projections for the United Kingdom and constituent countries', *Population Trends*, **72** (1993), pp. 45–50, © Crown Copyright 1993. Reprinted by permission of the Office for National Statistics.

Macro International Inc. for a table from S. Coulibaly, F. Dicko, S.M. Traoré, O. Sidibé, M. Seroussi and B. Barrère, *Enquête Démographique et de Santé, Mali 1995–96*, Bamako, Mali, Cellule de Planification et de Statistique, Ministère de la Santé, de la Solidarité et des Personnes Agées, Direction Nationale de la Statistique et de l'Informatique, and Calverton, MD, Macro International Inc. Reprinted by permission.

1

Some Demographic Fundamentals

1.1 Introduction

Demography is the study of population structure and change. With the increasing complexity of modern society, it is becoming ever more important to be able to measure accurately all aspects of change in the size and composition of the population, and to be able to make estimates of what the future size and composition of the population might be. Demographers are the professionals who carry out this task.

This book attempts to describe and explain the methods which demographers use to achieve their aim. It considers the particular processes which are within the purview of demography. It shows how these processes may be measured, and how their operation in different populations, and in the same population over time, may be compared.

The subject matter of this book is sometimes referred to as *formal demography*, to distinguish it from the broader field of *population studies*. The latter field involves not only the measurement of demographic processes but also the study of their relationships to economic, social, cultural and biological processes. Readers who are interested in this broader field will find a good introduction in Daugherty and Kammeyer (1995).

This introductory chapter sets the scene for what is to follow. In Section 1.2 the fundamental demographic processes are listed. In Section 1.3 we show how the operation of demographic processes can be viewed in terms of people making transitions between a relatively small number of definable states, or conditions. Section 1.4 describes how demographers typically measure the speed at which people are making these transitions. In Section 1.5 the idea of population heterogeneity is introduced.

The analysis of demographic change relies on the availability of accurate data about the relevant population characteristics and processes. In Section 1.6 the main sources of data are briefly described. We shall not devote a lot of space to a general description of data sources. There are a number of good introductions available elsewhere (see the list of further reading towards the end of this chapter). In later chapters, however, the characteristics and limitations of particular sources in the context of specific applications of demographic analysis will be discussed.

1.2 The basic demographic equation

One of the fundamental facts about population change is that populations only change because of a limited, countable, number of events. For example, consider the population of a country. Suppose that this country at some time t contains P_t persons, and that 1 year later it contains P_{t+1} persons. Then we can write down the following equation:

$$P_{t+1} = P_t + B_t - D_t + I_t - E_t, \tag{1.1}$$

where B_t and D_t are respectively the number of births and deaths occurring in the population between times t and $t + 1$, and I_t and E_t are respectively the number of immigrants to and emigrants from the country during the same period.

The quantity $B_t - D_t$ is known as the *natural increase* (if the number of deaths exceeds the number of births, then we have $D_t > B_t$, which implies negative natural increase, or natural decrease). The quantity $I_t - E_t$ is known as the *net migration*.

Equation (1.1) is often referred to as the *basic demographic equation*, or sometimes as the demographic balancing or accounting equation. It says that a country's population size can only change because of three types of event: births, deaths and migration. These three events are known as *components of population change*.

The process by which a population bears children is known as its *fertility*, and the process by which the members of the population are reduced by death is known as *mortality*. Fertility, mortality and migration are, therefore, the three fundamental demographic processes. Chapters 2–6 of this book consider the analysis of mortality, Chapters 8–11 are concerned with the analysis of fertility, and Chapter 15 is an introduction to the analysis of migration. Fertility, mortality and migration are, however, not the only processes of interest to demographers. Other processes which are studied include, for example, marriage and divorce, which we consider in Chapter 7.

1.3   Demographic processes as transitions between states

One way of representing the components of population change is to view them as a set of transitions made by individuals between various states. The basic demographic equation may, in this way, be represented by four states: 'alive, and in the population'; 'alive, but in another population'; 'not yet born'; and 'dead'.

The components of demographic change are then represented by transitions between these states (Figure 1.1). Notice that in some cases, transitions between two states can take place in both directions, whereas in other cases a transition in only one direction is possible. A state which people can never leave (for example, 'dead') is called an *absorbing state*.

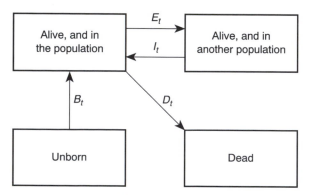

Figure 1.1 Multiple-state representation of the basic demographic equation; B_t, D_t, E_t and I_t represent transitions (see text for definition of symbols)

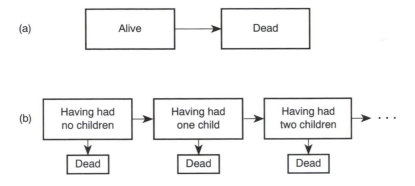

Figure 1.2 Two further examples of multiple-state representations

This way of viewing demographic processes may be called a *multiple-state representation*. Multiple-state representations are increasingly popular in demography, and often help demographers to understand complex processes.

Two further examples, drawn respectively from the analysis of mortality and the analysis of fertility, illustrate their usefulness in enabling demographers to conceptualize the processes they wish to study. First, mortality may be viewed as a single transition from the state 'alive' to the state 'dead' (Figure 1.2a). Second, women who are in the process of bearing children can be viewed as moving successively through the states 'having had no children', 'having had one child', 'having had two children', and so on (Figure 1.2b). Multiple-state representations will be used from time to time during this book.

1.4   Demographic rates

Understanding population change involves measuring and analysing its components: in the multiple-state representation it involves measuring the 'speed' with which the population is making the transitions between various states.

The simplest measure of any transition is the number of events which occur in a given time period. However, this is of little use for practical purposes since it is heavily influenced by the number of people who are around to experience those events. Clearly, the more people there are, the more births and deaths there are likely to be. For comparative work, what is needed is a way of measuring the number of the transitions in relation to the population size.

Demographers therefore measure events in terms of *rates*. A demographic rate is normally defined as

$$\text{rate} = \frac{\text{number of events of a specific type in a given time period}}{\text{number of people at risk of experiencing that type of event in the given time period}}.$$

It is sometimes referred to as an *occurrence/exposure ratio*, because the numerator is the number of occurrences of an event (within a given period), and the denominator measures the population exposed to the risk of experiencing that event. Reflecting this, the denominator of a rate is often called the *exposed-to-risk*, and we shall use that term from time to time during this book.

One commonly used example of a demographic rate is the crude death rate, which is a widely used measure of mortality (see Section 2.2). It is defined as follows:

$$\text{crude death rate} = \frac{\text{total number of deaths in a given year}}{\text{total population in that year}}.$$

Notice that the numerator of the crude death rate is based on the number of deaths within a given time period (in this case one year). It is normal for a demographic rate to relate to a specific time period in this way (although it is not essential for the period to be one year).

THE PRINCIPLE OF CORRESPONDENCE

When we attempt to calculate or estimate rates, it is important to ensure that the events in the numerator correspond with the exposed-to-risk in the denominator. By 'correspondence' we mean that, if a person is included in the exposed-to-risk in the denominator, and he/she experiences the event during the relevant time period, then the particular event which involves him/her must be included in the numerator. Conversely, if a person experiences the event during the relevant time period, and the particular event which involves him/her is recorded in the numerator, then he/she must be included in the exposed-to-risk in the denominator.

This may seem an obvious principle, but ensuring that it applies in practice is not always easy. If it were straightforward to ensure that the principle of correspondence applies when we try to measure fertility, mortality and migration in real populations, then demography would be a much easier subject than it is.

1.5   Population structure

One of the most important aspects of populations is their heterogeneity. People are not all the same. For a start, almost all populations contain people of different ages, and most contain both males and females.

Age and sex are, demographically, the most important ways in which people differ. That is why demographers often present analyses of population characteristics 'broken down by age and sex'. However, there are many other ways which are important in particular situations. These include educational level, occupation, marital status, the physical environment in which people live, life-style (for example, the level and type of sexual activity), income and nutrition.

Many of these factors influence the rates at which the components of demographic change operate. To take two examples: birth and death rates vary greatly with age; and death rates are higher among single people than married people. Because of this, demographers commonly use specific rates to measure population change. *Specific rates* are rates which apply only to specific subgroups within the population. By far the most common form of specific rates are *age-specific rates*, but others are sometimes used (for example, rates specific to occupation or marital status).

1.6   Data sources

In order to calculate a rate, data are required on both the number of events occurring within the given time interval, and the population exposed to the risk of experiencing those events.

How are these data normally obtained? There are three main sources: population censuses; vital registration; and surveys. In this section we describe these briefly.

POPULATION CENSUSES

These are the most widely used source of data about the exposed-to-risk. Most countries have regular censuses (typically taken every 10 years), in which everyone resident in the country on a particular night is counted, and asked to reply to various questions about age, sex, occupation, marital status, and so on. These answers allow demographers to make an accurate calculation of the population structure on the night of the census, and provide sufficient data to enable demographers to calculate the exposed-to-risk of most (though not all) rates of interest.

Almost every country has had at least one census, and, even in developing countries, regular censuses are now quite usual (for example, Tanzania, one of the world's poorest countries, has had three censuses – in 1967, 1978 and 1988).

The range of pieces of information typically asked in censuses may be illustrated using the 1991 population census of England and Wales. This required each person to give details of their name, sex, date and country of birth, marital status, relationship to the head of the household in which they live, usual address, usual address one year ago, ethnic group, occupation, place of work, and some information about their educational qualifications (Dale and Marsh, 1993). In developing countries, in particular, censuses often also ask questions of women about the number of children they have ever had, and the number of children they have had during the past 12 months.

VITAL REGISTRATION

This provides data about the events themselves. In developed countries, it is usually a legal requirement to register the birth of every child, all marriages, and each death. At the time of registration, other details may be collected (see Table 1.1 for a list of those collected in

Table 1.1 Information collected in England and Wales by the vital registration system

Births	Marriages	Deaths
Date of birth	Date of marriage	Date of death
Place of birth	Place of marriage	Place of death
Name of child	Names of bride and groom	Name of deceased
Sex of child		Sex of deceased
	Occupations of bride and groom	Occupation of deceased
	Previous marital status of bride and groom	
	Ages of bride and groom	Age of deceased at death
		Cause of death (up to three causes)
Names of child's parents	Names of parents of bride and groom	
Occupations of child's parents	Occupations of fathers of bride and groom	
	Form of ceremony	
Description of informant		Description of informant

England and Wales). For example, when the birth of a child is registered, details of the occupation of the mother and father are sought. When a death is registered, the age of the deceased is asked for.

The registration of migration is much less widespread. Some countries – for example, the Netherlands and Sweden – have systems of *continuous registration* in which all changes of permanent residence are registered, as well as births, marriages and deaths. Using the notation in equation (1.1), this implies knowledge of B_t, D_t, I_t and E_t for all t, and thus, once the population size, P_t, is known at one point in time, it does away with the need for subsequent censuses (at least in principle). In most countries, however, migration is not registered. Net migration in these countries may be estimated by rearranging equation (1.1) to give

$$I_t - E_t = P_{t+1} - P_t - B_t + D_t.$$

This method of estimating net migration is considered more fully in Chapter 15.

In developing countries, systems of vital registration are rare, and comprehensive systems even rarer (this is the main reason why questions about fertility are included in censuses in those countries). Until recently, this meant that estimates of demographic rates for these countries used to rely almost entirely on one of two methods: the use of *population models*, such as the stable population model; and *indirect estimation*, using a variety of ingenious techniques devised by demographers. Both of these are still widely used. The use of population models is covered in Chapters 13 and 14. Indirect estimation is, for the most part, beyond the scope of this book. Readers who are interested in it should consult the excellent descriptions of indirect methods in Brass (1975) or United Nations (1983).

SURVEYS

Although census and vital registration data can provide much of the information which demographers need to calculate the demographic rates of interest, there are occasions when additional, more detailed, information is required. In England and Wales, for example, nowhere in censuses or vital registration systems are questions routinely asked about how many children a woman has already had, which is a quantity of interest in many analyses of fertility.

Gaps in the data provided by censuses and vital registration can be filled by carrying out special surveys to elicit the particular information required. Surveys providing data of interest to demographers may be divided into two types.

In *prospective studies*, a defined group of people is followed for a number of years, and the dates at which events of interest occur to them noted. A British example is the Office for National Statistics Longitudinal Study of a 1% sample of the population, which has been running since 1971.

In *retrospective surveys*, a sample of people is interviewed at a single point in time, and asked questions about their lives so far, including the dates at which events of interest to demographers happened to them.

In the last two or three decades, large-scale retrospective sample surveys have been carried out in many developing countries, and these are now widely used to estimate rates. The most important sets of such surveys are the World Fertility Survey, which was mainly carried out during the 1970s, and the Demographic and Health Survey, which began during the 1980s and which is still going on. The Demographic and Health Survey now covers more than 40 developing countries, including many without efficient vital registration systems.

These surveys are carried out by interviewing the members of a randomly selected sample of several thousand households in each country. The questions asked include many of those typically asked in population censuses, but the interview is much more detailed than the census form, and includes questions on other topics as well.

In various places in this book, we shall need to consider the analysis of both prospective and retrospective survey data in more detail. It will be seen that both types of survey have advantages and disadvantages, but that in the majority of cases, the disadvantages may be overcome by the use of appropriate demographic methods.

Further reading

General introductions to population studies can be found in Daugherty and Kammeyer (1995), Lucas and Meyer (1994), Weeks (1989) and Yaukey (1990).

Exercises

1.1 Demographers are often interested in changes in the structure of the population classified by marital status. Consider the four marital statuses: 'single' (that is, never married), 'married', 'widowed' and 'divorced'. Draw a multiple-state representation showing these states and the possible transitions between them.

1.2 A lecturer is giving a final-year course on the analysis of mortality to a group of undergraduate students. The lectures are very boring, and the number of students attending them gradually falls as the course progresses, even though many of those who stop attending the lectures remain registered for the course and intend to sit the examination. Of course, there are also some students who stop attending the lectures because they die, suspend their registration because of illness, or change courses.

Assuming that, once a student has ceased to attend lectures for whatever reason, he or she never resumes attending them, draw a multiple-state representation of this process.

2

The Measurement of Mortality

2.1 Introduction

This chapter introduces the measurement of mortality by considering in detail the various kinds of mortality rate used by demographers. In Section 2.2 the crude death rate is described, and in Section 2.3 the calculation of age-specific death rates is illustrated. Section 2.4 then explains the rationale behind, and the difference between, two types of mortality rate commonly used by demographers: initial rates and central rates. These two types of rate are shown to be manifestations of two different approaches to analysing demographic data: that based on time periods; and that based on birth cohorts. Section 2.5 introduces the Lexis chart as a means of representing and illustrating the difference between initial and central rates, and thereby between the period and cohort approaches. In Section 2.6 the formula which is commonly used to convert age-specific death rates of one type into the other is derived, with the aid of a Lexis chart. Finally, Section 2.7 summarizes the advantages and disadvantages of the two types of mortality rate.

2.2 The crude death rate

The simplest measure of mortality is the number of deaths. However, this is not of much use for practical purposes since it is heavily influenced by the number of people who are at risk of dying.

Because of this, as we saw in Section 1.4, demographers typically measure mortality using *rates*. A death rate is defined as

$$\text{death rate} = \frac{\text{number of deaths in a specified time period}}{\substack{\text{number of people exposed to the risk of dying} \\ \text{during that time period}}}.$$

Thus, in order to measure mortality, data are required about the number of deaths, and about the number of people exposed to the risk of dying. Data on the number of deaths are usually obtained from death registers, and data on the number of people exposed to the risk of dying are typically obtained from a population census. Of course, survey data may also be used, especially in countries where death registration is deficient, or the quality of census data is suspect.

The simplest conceivable death rate is probably the total number of deaths in a given time period divided by the total population. This measure is called the *crude death rate*. The time period used is typically one calendar year. Thus

$$\text{crude death rate} = \frac{\text{total number of deaths in a given year}}{\text{total population}}.$$

An immediate issue arises with the measurement of the total population. During any year, the population will usually change. At what point in the year, therefore, should it be measured? Conventionally, the point chosen is half-way through the year (30 June). The population on 30 June is called the *mid-year population*. Using this definition of the population exposed to the risk of dying, therefore,

$$\text{crude death rate} = \frac{\text{total number of deaths in a given year}}{\text{total mid-year population}}.$$

true rate.

Denoting the crude death rate in year t by the symbol d_t, the total number of deaths in year t by θ_t, and the total population on 30 June in year t by P_t, we can write

$$d_t = \frac{\theta_t}{P_t}.$$

meter

pop at begining of year but , estimate

Now, for simplicity, the subscripts t are usually omitted because, unless otherwise stated, the period of time over which the crude death rate is measured may be assumed to be a single calendar year. Thus

$$d = \frac{\theta}{P}.$$

Since death is a relatively rare event in most populations, the crude death rate is often small. For this reason, it is often expressed as *the number of deaths per thousand of the population*, or

$$d = \frac{\theta}{P} \times 1000.$$

Thus, for example, the population of Peru on 30 June 1989 has been estimated to be 21 113 000 (excluding some Indian people in remote areas). It is estimated that there were 200 468 deaths in Peru in 1989. The crude death rate in Peru in 1989 is therefore equal to 200 468/21 113 000, which is 0.00950, or, multiplying by 1000, 9.5 per thousand.

2.3 Age-specific death rates

The crude death rate does not provide a great deal of information about mortality. In particular, the risk of dying varies greatly with age, and the crude death rate indicates nothing about this variation. Because of this, demographers often find it useful to use *age-specific death rates*. The age-specific death rate at age x years is defined as

$$\text{age-specific death rate at age } x \text{ years} = \frac{\text{number of deaths of people aged } x \text{ years}}{\text{population aged } x \text{ years}},$$

in a given calendar year. When we refer to 'age x years', we mean 'aged x last birthday'. The denominator, as before, is the mid-year population.

Denoting the age-specific death rate at age x years last birthday by the symbol m_x, the number of deaths of people aged x years last birthday by θ_x, and the population aged x

years last birthday by P_x, we can write

$$m_x = \frac{\theta_x}{P_x},$$

or, if preferred,

$$m_x = \frac{\theta_x}{P_x} \times 1000.$$

Note that the subscripts x denote years of age, not calendar years.

Age-specific death rates can be calculated for single years of age, or for age groups, such as 5–9 years last birthday, 10–14 years last birthday, and so on. Because mortality is also known to vary by sex, age-specific death rates are usually calculated separately for males and females. When age-specific rates are calculated for age groups, a special notation is used to denote the precise age group under consideration. The symbol $_n\theta_x$ denotes the number of deaths to people between the exact ages x and $x + n$ years. The symbol $_nP_x$ is used to denote the mid-year population of people between the exact ages x and $x + n$ years, and the symbol $_nm_x$ denotes the age-specific death rate between exact ages x and $x + n$ years. Thus, for example, the age-specific death rate at ages 5–9 years last birthday, $_5m_5$, is calculated using the formula

$$_5m_5 = \frac{_5\theta_5}{_5P_5}.$$

To take an example, the male population aged 35–44 years last birthday in England and Wales on 30 June 1995 is estimated to have been 3 333 000. The number of deaths reported in England and Wales of males in this age group during the calendar year 1995 was 5860. The age-specific death rate in 1995 for males aged 35–44 years last birthday was, therefore, 5860/3 333 000, or 0.00176. Multiplying this by 1000 gives a rate of 1.76 per thousand.

There is one (and only one) age group for which a different method of calculating age-specific death rates is employed. This is the age group 'under 1 year', or '0 last birthday'. For this age group, the denominator is taken to be the number of live births in the calendar year in question, rather than the mid-year population aged under 1 year. For example, in England and Wales in 1995 there were 648 100 live births, and 3970 deaths to infants under 1 year. The infant mortality rate is therefore equal to 3970/648 100, which is 0.00613 or 6.13 per thousand births. Notice that this rate refers to both sexes. To measure infant mortality, unlike that of other age groups, demographers quite often use a rate referring to both sexes combined.

2.4   The two types of mortality rate

So far, we have been looking at rates in which the denominator is a mid-year population, and the numerator is the number of deaths during the whole of the relevant calendar year.

This procedure violates the principle of correspondence, described in Section 1.4. Why? Two important reasons are as follows:

1   Someone who dies in the relevant year, but before 30 June, will not be alive on that date, and will not be included in the mid-year population, yet that person's death will be included in the numerator.

2   Consider someone whose birthday is on 7 September, and who dies on 9 October. Suppose this person is aged x years last birthday on 30 June. Then when he dies he

will be aged $x + 1$ years last birthday. He will be included in the denominator of the age-specific rate at age x last birthday, but his death will be included in the numerator of the age-specific rate at age $x + 1$ last birthday.

A similar problem affects the calculation of age-specific death rates for infants aged under 1 year when the number of births during the entire year is the denominator. Consider an infant who died on 31 March 1997 aged nine months. This child was born on 30 June 1996. His/her death is included in the numerator for the age-specific death rate at age 0 last birthday in 1997, but his/her birth is included in the denominator for the age-specific death rate at age 0 last birthday in 1996.

What can be done about this? Can rates be obtained in which the numerator and denominator correspond exactly? Yes, they can, but they require additional data. What we really need is to know the exact period of exposure at each age during the given year for each person at risk of dying. Thus, for example, the person whose $(x + 1)$th birthday was on 7 September, and who died on 9 October, would be regarded as contributing $250/365$ of a year's exposure during that year at age x last birthday (since there are 250 days between 1 January and 7 September), and $32/365$ of a year's exposure during that year at age $x + 1$ last birthday (since there are 32 days between 7 September and 9 October). Summing these fractions of a year over the whole population under investigation for each year of age, and using the result in the denominator, would give a rate in which the numerator and denominator corresponded exactly.

In practice, such detailed information is not usually available except from special (and expensive) investigations designed to elicit it. Therefore demographers rely on mid-year populations as an *approximation* to the correct exposed-to-risk. The approximation is usually quite close in large populations.

There is, however, another approach to measuring mortality rates, which does not lead to violations of the principle of correspondence (at least at the 'person level'). In this second approach, what is done is to calculate the number of people who have their xth birthday during a given period, and then follow them up until either they celebrate their next birthday, or they die (whichever happens first). Dividing the number who die by the original number having their xth birthday gives us an age-specific death rate at age x.

This kind of age-specific death rate is often called a *q-type rate*, and is given the symbol q_x, to distinguish it from the first kind of age-specific rate, m_x, which is called an *m-type rate*. Demographers also use the terms *initial rates* for q-type rates and *central rates* for m-type rates. This is because in q-type rates the exposed-to-risk is defined at the start, or initiation, of the year of age under investigation (that is, when the members of the exposed-to-risk celebrate their xth birthday), whereas in m-type rates the exposed-to-risk is an estimate of the number of persons aged x last birthday at the time the events took place. The average age of these persons is $x + \frac{1}{2}$ years: they are half-way through (in the 'centre' of) the year of age in question.

Strictly speaking, q-type rates do not lead to an exposed-to-risk which is exactly right. Those people who die between exact ages x and $x + 1$ years are actually only 'at risk' of dying for the period between their xth birthday and the point at which they die (since once they have died, they are no longer at risk). The q-type rate assumes that such people are at risk for the entire year between exact ages x and $x + 1$. Thus q-type rates over-estimate the length of time exposed to risk for every person who dies. Nevertheless, they do get the right number of people in the denominator, which m-type rates calculated using mid-year populations cannot be relied upon to do. That is what is meant by saying that q-type rates maintain the principle of correspondence at the 'person level'.

THE DIFFERENCE BETWEEN THE TWO TYPES OF RATE

The two types of mortality rate are examples of two quite different approaches to measuring the components of population change. One approach calculates rates based on a specific calendar time period (m-type rates). This is known as the *period approach*. The other approach calculates rates based on the experience of a specific group of people born during a specific calendar period (q-type rates). Since q-type rates are based on a group of people who celebrate their xth birthday during a given period, it follows that they must all have been born during a period of the same length x years earlier. Such a group of people is known as a *birth cohort*, and this approach is called the *cohort approach*.

2.5   The Lexis chart

The difference between the period and cohort approaches, and hence between m-type and q-type rates, may be illustrated using a diagram called a *Lexis chart*. A Lexis chart has a vertical axis which represents age, and a horizontal axis which represents calendar time. Since people get older as time goes on, the life of any person can be represented on a Lexis chart by a diagonal line running from the horizontal axis until a point which corresponds to the person's age at death measured on the vertical axis (Figure 2.1).

On a Lexis chart, the population alive and aged x last birthday, P_x, at a particular point in time is represented by a vertical line. In Figure 2.2, the line AB represents those alive aged 30 years last birthday on 30 June 1997. Vertical lines like line AB thus represent the denominators of m-type mortality rates.

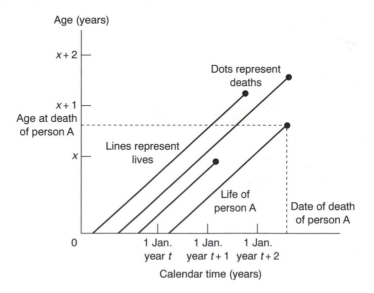

Figure 2.1 Principle behind the Lexis chart. Individual lives are represented by diagonal lines running from the bottom left towards the top right of the chart. As individuals grow older, they move up their 'life lines'. The coordinates of the upper end of each line denote the time of death and the age of the person when he/she died

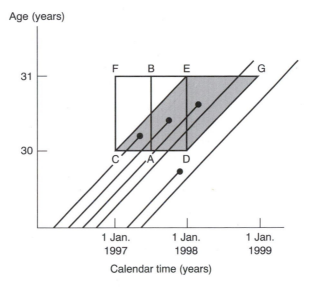

Figure 2.2 A Lexis chart

A set of people who celebrate their xth birthday during a particular time period (which means that they were all born during a particular time period and thus constitute a birth cohort) are represented by a horizontal line. In Figure 2.2, the horizontal line CD represents all the people who celebrated their 30th birthday during the year 1997 (and who were, therefore, all born during 1967). Horizontal lines like line CD represent birth cohorts, or the denominators of q-type mortality rates.

The deaths of people aged x last birthday on their date of death, and who died in a particular time period, are represented by squares. The square CDEF in Figure 2.2 represents all the people who died during the calendar year 1997 who were aged 30 years last birthday when they died. Squares like this represent the numerators of m-type mortality rates.

The deaths of people aged x last birthday on their date of death, and who all celebrated their xth birthday during a particular time period, are represented by parallelograms. In Figure 2.2, the shaded parallelogram CDGE represents all the people who died aged 30 years last birthday when they died, and who celebrated their 30th birthday during the calendar year 1997. Parallelograms like this represent the numerators of q-type mortality rates.

2.6   The relationship between the two types of mortality rate

In principle, there is no necessary relationship between the two types of mortality rate. However, by making a number of assumptions, a theoretical relationship can be derived. Since the assumptions are not too unreasonable, the theoretical relationship works quite well in most practical situations.

Consider the Lexis chart in Figure 2.3. The deaths representing the numerator of the m-type mortality rate at age x last birthday in calendar year t are in the square PQRS. Suppose that there are θ_x of these deaths. The deaths representing the numerator of the q-type mortality rate at age x which most closely overlaps with the numerator of the m-type rate for year t are in the shaded parallelogram TMWN.

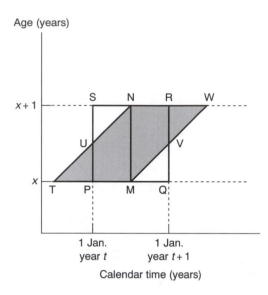

Figure 2.3 Lexis chart illustrating the relationship between the two types of mortality rate

We now make two assumptions. We assume that:

1 mortality only varies with age, and not with calendar time;
2 deaths are evenly distributed across each single year of age.

These two assumptions are made repeatedly in the demographic analysis of mortality. The assumption that mortality only varies with age is easy enough to understand, but the second assumption may need clarification. The assumption that deaths are evenly distributed across each year of age implies that, say, the average age at death of those dying between their 75th and 76th birthdays is 75 years 6 months. It implies that the number of deaths between 75 years exactly and 75 years 6 months is the same as the number of deaths between ages 75 years 6 months and 76 years exactly.

Once we have made these two assumptions, then, on the Lexis chart, within any horizontal band representing the ages between x and $x + 1$, deaths are evenly distributed. This means that the number of deaths is proportional to the area of any portion of the chart. Two sections of the chart which both lie within the horizontal band representing ages between x and $x + 1$ exactly and which have the same area will represent the same number of deaths.

Since the area of the parallelogram TMWN in Figure 2.3 is equal to the area of the square PQRS, then the number of deaths represented by the shaded parallelogram TMWN will also be θ_x.

The *m*-type mortality rate, m_x, is then given by

$$m_x = \frac{\theta_x}{\text{population represented by the vertical line MN}},$$ (2.1)

and the *q*-type mortality rate, q_x, is given by

$$q_x = \frac{\theta_x}{\text{population represented by the horizontal line TM}}.$$

But, using the assumptions above, the number of deaths in the triangle TMN must be $\frac{1}{2}\theta_x$, and, since all the lives which cross the line TM also cross the line MN, we have

$$q_x = \frac{\theta_x}{\text{population represented by the vertical line MN} + \frac{1}{2}\theta_x}. \qquad (2.2)$$

But, from equation (2.1) above,

$$\text{population represented by the vertical line MN} = \frac{\theta_x}{m_x}. \qquad (2.3)$$

Thus, substituting from equation (2.3) into equation (2.2), we have

$$q_x = \frac{\theta_x}{\theta_x/m_x + \frac{1}{2}\theta_x},$$

and the θ_x cancel to leave

$$q_x = \frac{1}{1/m_x + \frac{1}{2}} = \frac{m_x}{1 + \frac{1}{2}m_x},$$

or, as it is often written,

$$q_x = \frac{2m_x}{2 + m_x}.$$

This result is dependent upon the two assumptions we have made. In many practical situations, however, the approximation is satisfactory.

There are certain age groups, though, in which the assumption that deaths are evenly distributed is not valid. This is particularly true of the first year of life. Most deaths to infants during the first year of life take place during the first few weeks of that year. Indeed, in low-mortality populations, it is usual for more than half of all the deaths to infants under the age of 1 year to occur during the first month of life (see Exercise 2.8). Deaths to infants during the first four weeks of life are known as *neonatal deaths*. Neonatal deaths may be measured using the *neonatal death rate*, defined as

$$\text{neonatal death rate} = \frac{\begin{array}{c}\text{number of deaths in a given year to}\\\text{infants aged 28 days or under}\end{array}}{\text{number of births in the given year}}.$$

For example, in the United Kingdom in 1995, there were 732 000 live births, and 3070 neonatal deaths. The neonatal death rate was therefore equal to 3070/732 000, which is 0.0042, or 4.2 per thousand births.

2.7   Advantages and disadvantages of the two types of mortality rate

The two types of mortality rate have their advantages and disadvantages (Table 2.1). Generally speaking, m-type rates have the advantage of being straightforward to calculate from routinely available data. Their disadvantage is that they do not reflect the experience of 'real' people, and, if calculated using mid-year populations, violate the principle of correspondence. The advantages and disadvantages of q-type rates are in a sense 'mirror-images' of those of m-type rates.

Table 2.1 Advantages and disadvantages of the two types of mortality rate

Type	Advantages	Disadvantages
m-type	Data are readily available Easy to calculate Can be applied to a specific calendar time period	Violate the principle of correspondence if based on mid-year populations Do not reflect the experience of a real group of people
q-type	Reflect the experience of real people Do not violate the principle of correspondence (at least at the 'person level')	Do not apply to a particular calendar year Data are not readily available Awkward to calculate

Exercises

2.1 Table 2E.1 gives the total number of deaths in certain years, together with the estimated mid-year populations for those years, for certain countries in Latin America. Use them to calculate the crude death rate for each of these countries.

2.2 Table 2E.2 gives the estimated mid-year population in certain age groups, together with the number of deaths to people in those age groups, for males and females in Argentina in 1986. Use them to calculate age-specific death rates for the two sexes.

2.3 Table 2E.3 gives the estimated mid-year population in certain age groups, together with the number of deaths to persons in these age groups, in England and Wales in 1995. Use the data in the table to calculate age-specific death rates for the relevant age groups. Comment briefly on your results.

Table 2E.1

Country	Year	Estimated mid-year population	Number of deaths
Argentina	1990	32 322 000	295 796
Brazil	1989	147 404 000	1 164 452
Colombia	1990	32 987 000	201 166
Costa Rica	1991	3 064 000	12 452
Mexico	1991	87 836 000	500 615

Source: Wilkie *et al.* (1996, pp. 101, 102, 167).

Table 2E.2

Age group	Males		Females	
	Mid-year population (thousands)	Number of deaths	Mid-year population (thousands)	Number of deaths
1–4	1 422	1 637	1 380	1 325
5–14	3 062	1 390	2 968	920
15–24	2 430	2 816	2 318	1 437
25–44	4 101	9 690	4 023	5 942
45–64	2 755	36 581	2 753	18 535

Source: Wilkie *et al.* (1996, pp. 179–180).

Table 2E.3

Age group	Mid-year population (thousands)		Number of deaths (thousands)	
	Males	Females	Males	Females
1–4	1403	1335	0.40	0.34
5–14	3394	3219	0.61	0.42
15–24	3348	3172	2.45	0.91
25–34	4252	4076	4.10	1.84
35–44	3523	3480	5.86	3.64
45–64	5630	5900	44.20	27.79
65–74	2078	2477	74.50	52.70
75–84	1032	1702	91.60	96.40
85 and over	240	708	46.60	107.50

Source: *Population Trends* 87 (1997), pp. 47 and 55.

2.4 Table 2E.4 gives the numbers of births, and deaths of infants aged under 1 year, classified by sex, in England and Wales in certain recent calendar years.
  (a) Calculate sex-specific infant mortality rates for the years in question.
  (b) Calculate the infant mortality rates for both sexes combined for the years in question.
  (c) Comment briefly on your results.

2.5 Draw a Lexis chart with axes like the one in Figure 2.1. On the chart draw lines to represent the following groups of people:
  (a) people alive aged x last birthday on 30 June in year t;
  (b) people alive aged $x + 1$ last birthday on 1 January in year $t + 1$;
  (c) people who celebrated their xth birthday during year t;
  (d) people who celebrated their xth birthday before the beginning of year t.

2.6 On a Lexis chart with axes like the one in Figure 2.1, mark areas representing the following deaths:
  (a) deaths of people who died in year t aged x last birthday when they died;
  (b) deaths of people who died in year $t + 1$ aged $x + 1$ last birthday when they died;

Table 2E.4

Year	Number of births		Number of deaths of infants aged under 1 year	
	Males	Females	Males	Females
1971	402 500	380 800	7970	5750
1976	301 200	283 600	4880	3460
1981	327 000	308 500	4120	2900
1986	338 200	323 800	3720	2590
1991	357 830	342 200	2970	2190
1993	344 300	328 600	2410	1840
1994	343 500	324 100	2370	1750
1995	331 900	317 000	2290	1680

Source: *Population Trends* 87 (1997), p. 55.

Table 2E.5

Year	Number of births	Number of deaths	
		at ages under 1 year	at ages under 28 days
1971	901 600	16 200	10 800
1976	675 500	9 790	6 680
1981	730 800	8 160	4 930
1986	755 000	7 180	4 000
1991	792 500	5 820	3 460
1995	732 000	4 520	3 070

Source: *Population Trends* 87 (1997), p. 50.

(c)   deaths of people who celebrated their xth birthday in year t and who were aged x last birthday when they died;

(d)   deaths of people who celebrated their xth birthday in year $t + 1$, who died in year $t + 1$, and who were aged x last birthday when they died.

2.7   The equation

$$q_x = \frac{2m_x}{2 + m_x}$$

shows how the *m*-type and *q*-type mortality rates are related to one another (under certain assumptions) over a single year of age. Derive a similar equation for the more general case of an age group of width n years.

2.8   Table 2E.5 gives the numbers of births, deaths of infants aged under 1 year, and deaths of infants aged under 28 days, in the United Kingdom in selected recent calendar years.

(a)   Calculate the percentage of infant deaths in each year which were neonatal deaths.

(b)   Calculate the infant and neonatal mortality rates for each year.

2.9   Someone proposes calculating an infant mortality rate using the number of births in a given calendar year t in the denominator, and the average number of deaths of persons aged under 1 year in the two calendar years t and $t + 1$ in the numerator, arguing that this would better reflect the mortality experience of this birth cohort than the conventional method of calculating an *m*-type infant mortality rate.

(a)   Use a Lexis chart to illustrate the rationale behind this argument.

(b)   Why might the suggestion not work as well in practice as in theory?

(c)   Suggest a modification to the proposal which should lead to an infant mortality rate which better reflects the experience of the births occurring in year t.

3

Comparing Mortality Experiences

3.1 Introduction

Mortality varies greatly with characteristics such as age, sex, occupation, marital status, region and so on. One of the tasks of demography is to try to measure and understand this variation.

In order to measure mortality differences, or to compare the mortality experiences of different subgroups within the population, we need to be able, first, to identify and measure mortality differences; and, second, to ascribe those differences to a particular characteristic (in other words, to be sure that, if the mortality rates of, say, two occupations differ, the difference is really something to do with occupation, and not due to some other factor).

The comparison of the mortality experiences of subgroups within large populations is normally carried out using m-type mortality rates, because official statistics lend themselves much more readily to the use of m-type rates than q-type rates. Thus, for the remainder of this chapter, we shall be considering only m-type mortality rates.

When comparing many populations, it would clearly be useful to have a single number reflecting the mortality experience of each population. However, in Section 3.2 it is shown that some such *single-figure indices*, notably the crude death rate, also have disadvantages. Sections 3.3 and 3.4 describe some single-figure indices which manage to avoid one of the principal disadvantages, that of the confounding of the comparison by age compositional differences among the populations being compared. These single-figure indices are based on a widely used procedure known as *standardization*. Although indices based on standardization are widely used in practice to compare mortality experiences, it is shown in Sections 3.5 and 3.6 that they cannot overcome all the potential difficulties.

3.2 Single-figure indices

A set of m-type age-specific death rates provides quite a complete picture of the mortality experienced by a population in a given period. However, for the purposes of comparing the mortality of subgroups within a population, or for comparing many different populations, looking at the complete set of age-specific death rates is very tedious, since there is a separate rate for each age x for each subgroup. Since people may live to ages well in excess of 100 years, we should have to compare more than 100 figures for each population.

Table 3.1 Comparison of the mortality of two hypothetical populations

Age group	Population 1		Population 2	
	Population (mid-year)	Deaths in year	Population (mid-year)	Deaths in year
0–9	2000	20	1000	10
10–19	1500	5	1000	3
20–29	1000	2	1000	2
30–39	800	2	950	2
40–49	700	2	900	2
50–59	600	4	800	5
60–69	500	10	800	16
70–79	400	30	700	52
80–89	300	50	600	100
90 +	100	50	150	75
Totals	7900	175	7900	267

One way of reducing the number of figures to be looked at is to group the ages using five-year or ten-year age groups. While this reduces the scale of the problem somewhat, there are still about 20 figures to look at for each population using five-year age groups. It would be easier to use a single-figure index of mortality for each population – that is, an index which is one number for each population or subgroup being compared. Such an index would make it easy to compare the mortality experiences of many populations: this could be done by simply ranking the values of the index.

THE CRUDE DEATH RATE AS A SINGLE-FIGURE INDEX OF MORTALITY

The crude death rate, defined in Section 2.2, is, of course, just such a single-figure index. Why not use it? The idea seems attractive in principle, but it has quite serious drawbacks in practice. Consider the two populations shown in Table 3.1. The age-specific death rates in each age group are worked out in Table 3.2.

Table 3.2 Age-specific death rates for the two hypothetical populations in Table 3.1

Age group	Age-specific death rate (m-type)	
	Population 1	Population 2
0–9	0.0100	0.0100
10–19	0.0033	0.0030
20–29	0.0020	0.0020
30–39	0.0025	0.0021
40–49	0.0029	0.0022
50–59	0.0067	0.0063
60–69	0.0200	0.0200
70–79	0.0750	0.0740
80–89	0.1670	0.1670
90 +	0.5000	0.5000

The age-specific death rates in populations 1 and 2 are very close to one another at all ages. The mortality experiences of the two populations are very similar. The crude death rate in population 1, however, is equal to 175/7900, which is 0.022, or 22 per thousand, whereas the crude death rate in population 2 is equal to 267/7900, which is 0.034, or 34 per thousand.

Why is the crude death rate in population 2 more than 50% higher than that in population 1, when all the age-specific death rates are very similar in the two populations? The answer lies in the age distributions of the two populations. Population 2 has a much larger percentage of older people than population 1, and a correspondingly smaller percentage of younger people. Because old people have a higher risk of dying than younger people, the number of deaths in population 2 is greater than that in population 1, even though the death rates at each age are approximately the same in the two populations. Because the crude death rate only takes into account the total number of deaths, it, too, is increased. The differences in the age composition are confounding the comparison of the mortality experiences. Relative values of the crude death rate, therefore, are not a good guide to the mortality experiences of two populations where the age composition of the two populations differs.

Fortunately, there are a number of ways in which single-figure indices which are not so prone to confounding as the crude death rate can be devised. The two most widely used of these are described in the next two sections.

3.3   The standardized death rate

One obvious possibility is to assess the impact of a set of age-specific death rates on a standard population age structure. If we use the same population age structure when comparing different populations we will avoid the problem of confounding. This procedure is known as *direct standardization*. In direct standardization we compare two or more sets of age-specific rates by examining their impact on the same standard age structure.

The standard age structure is usually chosen so that it reflects some kind of 'average' of those of the populations being compared (for example, the age structure of the whole population when comparing subgroups of that population).

The resulting single-figure index is known as the *standardized death rate*. Before introducing the formula used to calculate this measure, let us describe the notation we shall use in the rest of this section. Suppose we are interested in comparing the mortality experience of several populations. We can denote these populations by the letters A, B, etc. Let the m-type age-specific death rates at age x last birthday in population A be denoted by the symbol $^A m_x$. The population exposed to the risk of death is drawn from some standard population, S; let the population exposed to the risk of death at age x last birthday in this standard population be denoted by the symbol $^S P_x$.

With this notation, the standardized death rate for population A is given by the formula

$$\text{standardized death rate for population } A = \frac{\sum_x (^A m_x \, ^S P_x)}{\sum_x \, ^S P_x}, \tag{3.1}$$

where the summations are over all ages x. The standardized death rate is therefore obtained by dividing the total expected number of deaths in the standard age structure by the total standard population. Alternatively, it may be seen as a weighted average of population A's age-specific rates using the standard population structure as weights.

Perhaps a better way of looking at what the standardized death rate does is to multiply both the numerator and denominator of equation (3.1) by the quantity $\sum_x (^S m_x \, ^S P_x)$, which is the total number of deaths in the standard population. This gives

$$\text{standardized death rate for population } A = \frac{\sum_x (^A m_x {}^S P_x)}{\sum_x {}^S P_x} \cdot \frac{\sum_x (^S m_x {}^S P_x)}{\sum_x (^S m_x {}^S P_x)},$$

which may be rearranged to give

$$\text{standardized death rate for population } A = \frac{\sum_x (^S m_x {}^S P_x)}{\sum_x {}^S P_x} \cdot \frac{\sum_x (^A m_x {}^S P_x)}{\sum_x (^S m_x {}^S P_x)}.$$

But $\sum_x (^S m_x {}^S P_x)/\sum_x {}^S P_x$ is just the crude death rate in the standard population, so we can see that the standardized death rate for population A is obtained by multiplying the crude death rate in the standard population by a factor $\sum_x (^A m_x {}^S P_x)/\sum_x (^S m_x {}^S P_x)$, specific to population A. This factor is called the *comparative mortality factor* for population A, or sometimes the *comparative mortality figure* for population A. Either way, it may be abbreviated to CMF. It is the ratio of the expected number of deaths which the standard population's age structure would experience if it had population A's age-specific death rates to the actual number of deaths in the standard population.

Finally, by multiplying and dividing each component of the summation in the numerator of the CMF by $^S m_x$, we obtain

$$\text{CMF for population } A = \frac{\sum_x [^S m_x {}^S P_x (^A m_x / {}^S m_x)]}{\sum_x (^S m_x {}^S P_x)}. \tag{3.2}$$

Thus the CMF for population A is a weighted average of the ratios between the age-specific death rates for populations A and S, using the deaths in the standard population at each age as weights.

Note that the standardized death rate and the CMF may be calculated using either single years of age, or age groups.

3.4   The standardized mortality ratio

A major practical limitation of the standardized death rate and the CMF is that their calculation requires us to know the age-specific death rates in each population to be compared. It is often the case in practice that these are unknown, or, if they are known, they are subject to large errors because of a small population at risk in certain age groups. To overcome this problem, we can take equation (3.2) and replace $^S P_x$ by $^A P_x$, the population exposed to the risk of death at each age in population A. This gives the expression

$$\frac{\sum_x [^S m_x {}^A P_x (^A m_x / {}^S m_x)]}{\sum_x (^S m_x {}^A P_x)},$$

which may be simplified to

$$\frac{\sum_x (^A m_x {}^A P_x)}{\sum_x (^S m_x {}^A P_x)}.$$

How does this help? Well, the numerator of this expression is just the total number of deaths in population A, a quantity which is both more likely to be known, and less likely to be subject to fluctuations caused by small numbers.

The expression $\sum_x (^A m_x {}^A P_x)/\sum_x (^S m_x {}^A P_x)$ is called the *standardized mortality ratio* (SMR) for population A. The SMR is another single-figure index which compares mortality without the problem of confounding. Because it is obtained without a knowledge of the age-specific death rates in the populations to be compared, comparing mortality experiences using the SMR is known as *indirect standardization*. Note that for indirect standardization, age-specific death rates are required for the standard population. We can, however, choose a standard population for which these rates are available and reliable.

The SMR is the ratio of the actual number of deaths in population A to the number of deaths that would be expected in population A if it experienced the age-specific death rates of the standard population. As we have seen, it may be written

$$\frac{\sum_x[{}^Sm_x\,{}^AP_x({}^Am_x/{}^Sm_x)]}{\sum_x({}^Sm_x\,{}^AP_x)}.$$

Thus it is a weighted average of the ratios between the age-specific death rates for populations A and S, using the expected deaths in population A under the standard set of age-specific death rates as weights.

3.5   The limits of standardization

Neither direct nor indirect standardization provides single-figure indices without shortcomings. Both the standardized death rate and the SMR are improvements on the crude death rate only in respect of the fact that they can control for confounding factors, notably differences in the age composition among the populations being compared. There are other limitations of single-figure indices which standardization cannot overcome.

Consider again the two expressions derived in Sections 3.3 and 3.4 for the CMF and the SMR:

$$\text{CMF for population } A = \frac{\sum_x[{}^Sm_x\,{}^SP_x({}^Am_x/{}^Sm_x)]}{\sum_x({}^Sm_x\,{}^SP_x)}$$

and

$$\text{SMR for population } A = \frac{\sum_x[{}^Sm_x\,{}^AP_x({}^Am_x/{}^Sm_x)]}{\sum_x({}^Sm_x\,{}^AP_x)}.$$

Both the CMF and the SMR are weighted averages of the ratios between the age-specific death rates in the populations being compared. Therefore, the values of both the CMF and the SMR will be affected by the magnitudes of the weights used at each age, except when the ratio between the age-specific death rate in population A and the standard population, ${}^Am_x/{}^Sm_x$, is the same for all ages x.

This can be seen clearly by looking at a hypothetical example. Table 3.3 shows the population at risk and the age-specific death rates for two occupations, R and T, together with the corresponding data for all occupations, which can be treated as the standard population, S.

The standardized death rate for occupation R is 0.00908 and that for occupation T is 0.00717. The crude death rate for all occupations is 0.00800. The SMR for occupation R is 1.150, and that for occupation T is 0.860 (zealous readers might care to check these).

This seems to indicate that occupation R has higher mortality than occupation T, and that the mortality for all occupations lies somewhere between the two, after controlling for

Table 3.3 Comparison of the mortality experiences of two occupations

Age group	Occupation R		Occupation T		All occupations	
	$^R P_x$	$^R m_x$	$^T P_x$	$^T m_x$	$^S P_x$	$^S m_x$
16–34	6 000	0.001	3 000	0.003	200 000	0.002
35–44	12 000	0.003	12 000	0.005	300 000	0.004
45–54	12 000	0.005	24 000	0.005	400 000	0.005
55–64	12 000	0.026	18 000	0.015	300 000	0.020
All ages	42 000		57 000		1 200 000	

the different age compositions of the occupations. Yet it is clear from Table 3.3 that in the 16–34 age group, occupation T has mortality three times that of occupation R, and one and a half times that of all occupations. Similarly, in the 35–44 age group, occupation R has lower mortality than occupation T. Conversely, in the 55–64 age group, occupation R has higher mortality than all occupations, and mortality nearly 75% higher than that in the same age group in occupation T.

The reason why the standardized death rate and the SMR take the values they do can be seen by looking at the ratios $^A m_x / ^S m_x$, and the weights which are applied to them in calculating the CMF and the SMR. These are summarized in Table 3.4.

The weights for the two younger age groups are a lot smaller than the weights for the older age groups. Thus, the ratios between the mortality rates in the older age groups dominate the mortality comparison, and the values of the single-figure indices reflect this.

The situation in which standardization works well, therefore, is when $^A m_x / ^S m_x$ is similar for all ages x, for in this case the relative magnitudes of the weights become less relevant. In such a case, we can say that the age pattern of the mortality rates in the two populations is the same, although the level of mortality may differ. The standardized death rate or the SMR will then give a good indication of the relative levels of mortality in the two populations.

In fact, when both the level of mortality and the age pattern of mortality in the two populations differ, it is clear that a single-figure index is *never* going to be adequate to compare the two experiences, since a single number cannot simultaneously measure differences in the level of mortality and differences in the age pattern. In such cases, the differences between the two experiences are too complex to be expressed well by any single-figure index.

Table 3.4 Weights applied when comparing the mortality experiences of two occupations from Table 3.3

Age group	Occupation R			Occupation T		
	$^R m_x / ^S m_x$	Weights		$^T m_x / ^S m_x$	Weights	
		CMF	SMR		CMF	SMR
16–34	0.50	400	12	1.50	400	6
35–44	0.75	1200	48	1.25	1200	48
45–54	1.00	2000	60	1.00	2000	120
55–64	1.30	6000	240	0.75	6000	360

3.6   Other problems commonly encountered when comparing mortality experiences

Standardization is a useful method for controlling the influence of confounding factors when comparing mortality experiences. However, it cannot overcome all difficulties. One of the major practical problems derives from the nature of the data we have to use.

Typically, when calculating mortality rates, data on deaths come from death registers, and data on the exposed-to-risk come from a population census. If such data are used to compare mortality for, say, two occupations, it is important that the classification of occupations is the same in the two sources. This applies regardless of whether we are using crude rates or standardized rates.

There are two difficulties. We can describe them in the context of an investigation into occupational differentials in mortality. They apply, however, to any sort of differentials (for example, by educational level or marital status).

First, people might change jobs between the date of the census and their date of death. This will mean that they appear in the exposed-to-risk for one occupation and in the deaths for another occupation. The rates for both occupations will therefore violate the principle of correspondence. This problem is particularly acute when the change of occupation is for health-related reasons: for in this case persons who have a high risk of death tend to move out of some subgroups being compared, and into others. Such processes are said to *select* people on the basis of their potential mortality. For example, persons engaged in certain very strenuous occupations might be compelled to retire on ill-health grounds prior to death, and take up a less strenuous occupation. This might lead to especially high death rates being recorded in those less strenuous occupations which have little to do with the inherent danger of such occupations. Conversely, low death rates would tend to be recorded in more strenuous (and possibly inherently more dangerous) occupations.

Second, even where a person does not change jobs, his/her occupation might be described differently in the two sources. Sometimes, for example, the status of the occupation of a deceased person is exaggerated by that person's spouse, or other relative, who reports his/ her death. In the 1970–72 investigation of occupational mortality in England and Wales, the exposed-to-risk for each occupation was worked out from information supplied in the 1971 census. This was probably quite accurate. The deaths were classified according to the occupations stated on death certificates for the years 1970–72. These were often vague, or exaggerated the status of the occupations of dead people (for example, deceased electricians were described as 'electrical engineers'). Again, this led to rates which violated the principle of correspondence – there were more dead 'electrical engineers' than there should have been, and fewer dead 'electricians' (Benjamin and Pollard, 1993).

Note that these difficulties arise because different sources are used to obtain the deaths in the numerator and the exposed-to-risk in the denominator of the death rates. An alternative approach is to use a prospective mortality investigation to collect the data. In England and Wales, for example, social and occupational mortality differentials can be assessed using the Office for National Statistics Longitudinal Study (see, for example, Fox *et al.*, 1985). This is a prospective study in which a 1% sample of people is followed. The study incorporates information from both censuses and vital registration. Because the individual study members are identified by their names and other information collected both in the censuses and on death registers, deaths can be exactly matched to census information, so avoiding both the problems mentioned above.

HETEROGENEITY

Another issue which arises in comparing mortality experiences is that of heterogeneity within the populations being compared. Standardization is able to control for the confounding effect of different age compositions, but there may be many other confounding effects present. One of these is the proportion of the population living in institutions (Bulusu, 1985). It is known that the mortality of people living in institutions is much higher than that of other people. It is, of course, possible to extend standardization to cope with the influence of more than one confounding factor, but the data to do this are not always available.

Further reading

There are a number of useful articles dealing with the comparison of the mortality experiences of different subgroups in England and Wales. See especially Britton (1989), Bulusu (1985), Fox *et al.* (1985) and Harding (1995). Benjamin and Pollard (1993, pp. 426–471) provide a much fuller treatment of the difficulties inherent in comparing mortality experiences, especially with respect to occupational differentials. Finally, Cox (1976) describes a number of alternative single-figure indices of mortality which have at various times been proposed.

Exercises

3.1 The standardized death rate for the town of Burnley in Lancashire was 1.23, when the population of England and Wales as a whole was used as the standard. What does this tell you about mortality in Burnley relative to that in England and Wales as a whole?

3.2 The data in Table 3E.1 refers to the male populations of Argentina, Colombia and Panama in the mid-1980s.
   (a)   Calculate the crude death rates for each country.
   (b)   Using the population of Argentina as the standard, calculate standardized death rates for Colombia and Panama.
   (c)   Comment on your results.

Table 3E.1

Age group	Argentina 1986		Colombia 1984		Panama 1987	
	Pop. (thousands)	Number of deaths	Pop. (thousands)	Number of deaths	Pop. (thousands)	Number of deaths
0–4	1 767	11 832	1 857	5 179	150	860
5–14	3 062	1 390	3 372	2 300	286	132
15–24	2 430	2 816	3 123	6 646	243	322
25–44	4 101	9 690	3 724	12 702	294	614
45–64	2 755	36 581	1 587	15 441	134	925
65+	1 129	70 138	478	27 034	51	2 343

Source: Wilkie *et al.* (1996, pp. 179–180).

Table 3E.2

Age group	Mid year population (thousands)			Number of deaths in England and Wales
	England and Wales	Scotland	Northern Ireland	
0–4	3 006	317	131	8 200
5–24	14 958	1 655	552	6 280
25–44	13 082	1 326	375	14 730
45–64	11 040	1 140	296	101 500
65–74	4 619	459	116	155 000
75–84	2 388	232	56	190 400
85+	541	49	13	102 400

Source: *Population Trends* 87 (1997), pp. 47, 48, 50 and 55.

3.3 Using the data in Exercise 3.2, calculate standardized mortality ratios for Colombia and Panama using Argentina as the standard population. Comment on your results.

3.4 Table 3E.2 shows the population by age group in England and Wales, Scotland and Northern Ireland in 1981. The numbers of deaths in England and Wales in each age group are also shown. The total numbers of deaths in Scotland and Northern Ireland in 1981 were 63 800 and 16 300.

(a) Calculate the crude death rates in England and Wales, Scotland and Northern Ireland in 1981.

(b) Compare the mortality experiences of the three populations using standardized indices.

(c) Comment on your results.

3.5 Table 3E.3 gives age-specific death rates for two regions of a country, together with those for the country as a whole. Would the standardized mortality ratio be a good index to use for comparing the mortality of the two regions? Explain your answer.

3.6 Yerushalmy's index is an alternative index which can be used to compare mortality experiences. This index involves giving the ratios $^Am_x/^Sm_x$ equal weights at every age (where ^Am_x is the age-specific death rate at age x in population A, and ^Sm_x is the corresponding figure for a standard population). If age groups are used, and the width of age group i in years is n_i, then the formula to use for calculating Yerushalmy's

Table 3E.3

Age group	Region		Whole country
	C	D	
0–19	0.010	0.020	0.015
20–39	0.005	0.005	0.005
40–59	0.015	0.010	0.012
60–79	0.035	0.020	0.030
80+	0.100	0.050	0.075

Table 3E.4

Occupation (as described on the death certificate)	Standardized mortality ratio	
	Males	Females
Hairdressers, barbers	263	133
Fishermen	234	–
Travel stewards and attendants, hospital and hotel porters	150	83
Teachers	61	57
Local government officers	49	36
Hairdressing supervisors	49	–

Source: Benjamin and Pollard (1993, pp. 451–452).

index is

$$\text{Yerushalmy's index} = \frac{\sum_i m_i (^A m_i / ^S m_i)}{\sum_i n_i}.$$

(a) Calculate the value of Yerushalmy's index for regions C and D in Exercise 3.5, using the population of the whole country as the standard. State any assumption you are making.

(b) Under what circumstances does Yerushalmy's index have advantages over the directly standardized death rate?

3.7 The data in Table 3E.4 are taken from an investigation of occupational mortality in England and Wales, which used data from the 1981 census to classify the exposed-to-risk and data from deaths registered during the years 1979–80 and 1982–83 to classify the deaths. (The standardized mortality ratios were calculated using 'all occupations' as the standard, and setting the 'all occupations' SMR equal to 100.) Suggest explanations for each of the figures in the table.

Table 3E.5

Age group (years)	Total population (thousands)			Deaths (thousands)		
	Whole country	Graveside	Croakingham	Whole country	Graveside	Croakingham
0–29	10 000	50	40	20	0.15	0.12
30–59	7 000	40	40	30	0.24	0.25
60+	5 000	30	40	200	1.00	3.00

Table 3E.6

Age group (years)	Population in institutions (thousands)			Deaths of people in institutions (thousands)		
	Whole country	Graveside	Croakingham	Whole country	Graveside	Croakingham
0–29	50	0.250	0.250	0.250	0.001	0.001
30–59	100	0.600	0.500	1.200	0.007	0.006
60+	500	3.000	12.000	75.000	0.450	1.800

3.8 Tables 3E.5 and 3E.6 give data about the mortality in two towns, Graveside and Croakingham, in a developed country, in a particular year. The population has been classified by age and by residence in institutions on 30 June in that year. Note that 'institutions' include hospitals, old people's homes, prisons, places where long-term nursing care is offered, etc.

(a) Using standardized death rates and standardized mortality ratios, compare the mortality experiences of the two towns as fully as you can.

(b) Comment on your results.

4

The Life Table

4.1 Introduction

The life table is probably the most widely used method of analysis in demographic work. In the years since the first life table was constructed by John Graunt in the seventeenth century, life tables have been constructed for countless populations, both national and subnational. The aims of this chapter are to describe the calculation of life tables, and to illustrate their usefulness in the analysis of mortality.

A life table is a convenient way of summarizing various aspects of the variation of mortality with age. In essence, it is derived by following a birth cohort of persons through life and tabulating the proportion still alive at various ages. Sections 4.2 and 4.3 of the chapter explain what a life table is, and how the various life table quantities are related to one another.

In practical applications, life tables are based on q-type mortality rates calculated either for groups of people of the same age or for broader age groups. These rates, it will be recalled, make the assumption that mortality does not vary within each of the age groups being used. In practice, of course, mortality varies continuously with age. In Section 4.4, this fact is acknowledged. When mortality is considered as varying continuously with age, a central quantity is the *force of mortality*, which measures the intensity of mortality at a particular exact age x. Section 4.4 may be omitted by those unfamiliar with calculus. Section 4.5 discusses how life tables can be estimated in practice, using census and vital registration data. Section 4.6 presents and discusses a recent national life table produced for the population of England and Wales, and Section 4.7 employs this life table to illustrate some of the ways in which life tables may be used by demographers to do calculations about the chances of dying in specific populations. Finally, Section 4.8 discusses how various aspects of mortality vary with age in real populations.

4.2 The theory of the life table

In Chapter 2 we saw how a q-type mortality rate measures the proportion of those attaining a given birthday within a specific calendar time period who die before they reach their next birthday – that is to say,

$$q_x = \frac{\text{number dying between exact age } x \text{ and exact age } x+1}{\text{number attaining exact age } x}.$$

Now, suppose we consider a cohort of people, all born within the same calendar time period. Let the number of people born in this cohort be l_0, and the number of these who live to experience their xth birthday be l_x. Since everybody ultimately dies, l_x is a curve which takes the value l_0 at age 0, and falls to zero at whatever age represents the maximum attainable human life-span. This maximum age is about 120 years (only two people have ever demonstrably celebrated their 120th birthday).

We can now write

$$q_x = \frac{l_x - l_{x+1}}{l_x}. \tag{4.1}$$

Moreover, if the number of deaths of people among this cohort aged x last birthday when they die is denoted by the symbol d_x (this is rather confusing, because we have already used d to denote the crude death rate – unfortunately, demography has yet to develop a universally accepted algebraic notation), we can write

$$d_x = l_x q_x.$$

Now, making the assumption that deaths are distributed evenly over each year of life, we can define a quantity called the number of *person-years lived* between exact age x years and exact age $x + 1$ years. A person-year is one person living through one year. Two people each living for six months is equivalent to one person-year. This quantity is denoted by the symbol L_x, and (using the assumption of evenly distributed deaths) we can write

$$L_x = \tfrac{1}{2}(l_x + l_{x+1}). \tag{4.2}$$

In other words, the number of person-years lived between exact age x years and exact age $x + 1$ years is equal to the average of the number of people alive at exact age x and the number of people alive at exact age $x + 1$.

Look at this another way. Each person who survives to age $x + 1$ lives one complete year between her xth birthday and her $(x + 1)$th birthday. Assuming an even distribution of deaths between exact ages x and $x + 1$, each person who survives to exact age x but who dies before her $(x + 1)$th birthday lives, on average, half a person-year between these two birthdays. Thus

$$L_x = l_{x+1} + \tfrac{1}{2}d_x = l_{x+1} + \tfrac{1}{2}(l_x - l_{x+1}) = \tfrac{1}{2}(l_x + l_{x+1}).$$

There are a few exceptions to this rule. In the case of the early years of life, the assumption of an even distribution of deaths is unrealistic (this is especially so in the first year of life – see Exercise 2.8). For this reason, the number of person-years lived during the first year of life, L_0, is calculated using the formula

$$L_0 = a_0 l_0 + (1 - a_0)l_1, \tag{4.3}$$

where a_0 is the average age at death of those dying within the first year of life. Typically, values of a_0 between 0.10 and 0.30 are used in practical work, depending on the particular population under investigation. For example, in England and Wales in 1980–82, the value of a_0 was about 0.15 (Office of Population Censuses and Surveys, 1987a).

Equations (4.2) and (4.3) are both, in fact, specific cases of the more general formula

$$L_x = a_x l_x + (1 - a_x)l_{x+1},$$

in which a_x is the average number of person-years lived between exact ages x and $x + 1$ years by those who die within that interval. It turns out that for ages over 2 years, a_x is very close to 0.5, and so equation (4.2) is appropriate for calculating L_x. At age 1 year, a value of a_1 rather less than 0.5 (say, about 0.3) is often used.

Next, consider the total number of person-years lived at ages over exact age x years by the people in the cohort. This is simply equal to the sum of the values L_x at all ages older than exact age x. It is referred to by the symbol T_x, and we can write

$$T_x = \sum_{u=x}^{\omega} L_u, \tag{4.4}$$

where ω is the limiting age, or the oldest age to which anyone survives.

Finally, the average number of years which people have left to live when they celebrate their xth birthday may be calculated by noting that

$$\frac{\text{average number of person-years}}{\text{lived at ages above } x} = \frac{\text{total number of person-years lived at ages above exact age } x}{\text{number of people attaining exact age } x}.$$

Remembering that the numerator of this expression is just T_x, and the denominator is l_x, we have

$$\text{average number of person-years lived at ages above } x = \frac{T_x}{l_x}.$$

Now the average number of years which people have left to live when they celebrate their xth birthday is simply the *life expectation* at age x. We denote it by the symbol e_x. Therefore, we have

$$e_x = \frac{T_x}{l_x}.$$

The life expectation at birth, e_0, is given by the equation

$$e_0 = \frac{T_0}{l_0},$$

where l_0 is the original number of people in the birth cohort. Notice that l_0 can be set arbitrarily. However, it is convenient in practical work to take $l_0 = 1000$, $10\,000$ or $100\,000$, depending on the size of the population which is being analysed.

The quantities q_x, l_x, d_x, L_x, T_x and e_x may be tabulated for a given birth cohort. It is this table which is called a life table.

Occasionally, other quantities are also included in the life table. An important one is the proportion of people who survive from their xth birthday until their $(x+1)$th birthday. This is referred to by the symbol p_x. Clearly,

$$p_x = 1 - q_x,$$

since a person must either die between exact ages x and $x+1$ or survive until his/her $(x+1)$th birthday.

THE LIFE TABLE AND m-TYPE RATES

The *m*-type mortality rate at age x measures the number of deaths of people aged x last birthday divided by the average number of persons alive aged x last birthday. Assuming deaths are evenly distributed within the year of age between exact ages x and $x+1$, the average number of persons alive aged x last birthday is equal to $\frac{1}{2}(l_x + l_{x+1})$. The number of deaths of persons aged x last birthday is just d_x. Therefore, the m-type mortality rate, m_x, is given by the equation

$$m_x = \frac{d_x}{\frac{1}{2}(l_x + l_{x+1})},$$

and, using equation (4.2), this can be written

$$m_x = \frac{d_x}{L_x}.$$

It is seen that $L_x (= \frac{1}{2}(l_x + l_{x+1}))$ is, therefore, not only a measure of the number of person-years lived between exact ages x and $x+1$ years, but also a measure of the number of persons alive at any one time between these two ages. What this means is that in a population subject to a given set of age-specific death rates, given by a set of m_xs or q_xs, in which l_0 births take place each year, the number of people alive at any one time at each age x last birthday is described by the L_x values in the resulting life table. Since $T_x = \sum_{u=x}^{\omega} L_u$, moreover, it is clear that T_0 gives the total number of persons alive in such a population. The population whose age composition at any one time is given by the L_xs is referred to as the *stationary population*. This term is used because it has a constant total size (T_0), and a constant set of mortality rates. In Chapter 13, stationary populations are considered more fully.

4.3   Abridged life tables

Life tables in which the quantities are tabulated by single years of age, as described in the previous section, are very large (there are about 120 rows) and demographers (who do not always need to measure mortality with such precision) often use broader age groups. Commonly, five-year age groups are used (0–4, 5–9, 10–14, etc.). Life tables using broader age groups are called *abridged* life tables to distinguish them from life tables based on single years of age, which we will call full life tables.

Abridged life tables make use of another bit of notation. If we denote the width of an age group in years by the symbol n, the life table quantities q_x, p_x, d_x and L_x are, in the abridged life table, written $_nq_x, _np_x, _nd_x$ and $_nL_x$ respectively, being defined as follows:

- $_nq_x$ is the proportion of those people reaching their xth birthday who die before their $(x+n)$th birthday;
- $_np_x$ is the proportion of those people reaching their xth birthday who survive until their $(x+n)$th birthday;
- $_nd_x$ is the number of deaths occurring between ages x and $x+n$ years;
- $_nL_x$ is the number of person-years lived between exact ages x and $x+n$.

Note that the quantities l_x, T_x and e_x are written in the same way in the abridged life table as in the full life table, since their definition does not in any way depend on the width of the age groups.

The formulae connecting the quantities are exactly the same in the abridged life table as in the full life table, with one exception. This arises with the relationship between l_x and L_x. In the full life table, we made the assumption that deaths were distributed evenly over each year of life to derive equation (4.2). In the abridged life table, we often make the assumption that deaths are distributed evenly within each age group. Clearly, the wider an age group, the less safe this assumption. Thus, the fewer rows we have in the table, the less closely it approximates reality. This is the price of having a smaller table to handle.

Given this assumption, the equivalent of equation (4.2) in the abridged life table is

$$_nL_x = (n/2)(l_n + l_{x+n}),$$ (4.5)

and the equivalent of equation (4.4) is

$$T_x = \sum_{i=x}^{\infty} {}_nL_i,$$

where i takes values x, $x + n$, $x + 2n$, and so on.

There are just two more complications with abridged life tables. The assumption that deaths are distributed evenly over each age group is very poor for the youngest age group (0–4 years). The deaths in this age group are heavily concentrated at the younger end. Indeed, most of the deaths in this age group occur to children aged under 1 year. To allow for this, it is usual for abridged life tables to split the youngest age group into two parts: under 1 year, and 1–4 years. In the age group under 1 year, L_0 is calculated using equation (4.3).

Second, the width of the oldest age group is often unknown, since it is not clear to what age the oldest person survives. Some assumption must be made in order to calculate $_nL_x$ for this age group. The are a number of possible ways to tackle this problem.

One way is to make an assumption about the oldest age to which anyone survives. If this age is denoted by the symbol ω, then this amounts to setting $l_\omega = 0$. The problem with this assumption is that deaths are most unlikely to be evenly distributed over the age range between the lowest age in the oldest age group and age ω.

An alternative approach is to make an assumption about the average number of years a person who reaches the start of the oldest age group has left to live. For example, suppose that the oldest age group consists of persons aged 90 years and over. We make an assumption about e_{90}, and then calculate $_nL_{90}$ using the formula

$$_nL_{90} = l_{90}e_{90}.$$

A third approach makes use of the fact that, logically, $_nq_x$ for the oldest age group must be equal to 1. Provided that $_nm_x$ for this age group is known (and we shall see in Section 4.5 that, in practice, it often is), then assuming that deaths are evenly distributed across this age group, it is possible to set $_nq_x$ equal to 1.0 in the equation which forms the solution of Exercise 2.7, and solve for n. This gives a value of n which ensures internal consistency within the life table given the observed value of $_nm_x$ for the oldest age group, and an assumption that deaths are evenly distributed within that age group. Because deaths in the oldest age group are rarely evenly distributed, the value of n estimated using this method is somewhat artificial. However, it turns out that, provided attention is not being specifically directed towards mortality at the oldest ages, it provides an acceptable practical approximation. This approach is equivalent to using the equation

$$_nL_x = \frac{_nd_x}{_nm_x}$$

for the oldest age group.

4.4 The force of mortality

In Section 4.2, the life table was described by considering successive values of q_x, the proportion of those celebrating their xth birthday who do not live until their $(x + 1)$th birthday. This division of people's lives into years of age is purely for analytical convenience. It implies that the risk of death changes abruptly each birthday. This, of course, is not true for most people. Mortality is really continuously changing with age.

In this section we generalize Section 4.2 by considering the intensity of mortality over an arbitrary, but small, age interval dx.

Consider a birth cohort of l_0 people. At exact age x there are l_x still alive. At some slightly older exact age, $x + dx$, there are l_{x+dx} still alive. The number of deaths between exact ages x and $x + dx$ is $l_x - l_{x+dx}$.

The intensity of the mortality depends on the 'speed' at which these deaths occur with respect to age. The 'speed' at which people are dying depends on the length of the age interval dx. Figure 4.1 illustrates this. The intensity of mortality is greater in situation C than in situation B, since in situation C the $l_x - l_{x+dx}$ deaths occur faster with respect to age (note that the *number* of deaths is the same in situations B and C). We can express this by writing

$$\text{rate at which deaths occur per year of age} = \frac{l_x - l_{x+dx}}{dx}.$$

Finally, the impact of the mortality represented by the deaths $l_x - l_{x+dx}$ depends on the number of people alive to experience that mortality. Look again at Figure 4.1. The deaths $l_x - l_{x+dx}$ will have a greater impact in situation B than in situation A, since in situation B those deaths occur to a smaller number of people l_x. The smaller the number of people left

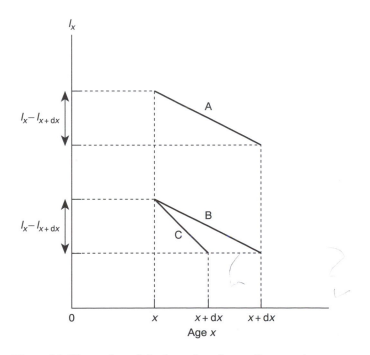

Figure 4.1 Illustration of the intensity of mortality experience

alive at age x, the greater the impact of a given number of deaths between ages x and $x + dx$ (the proportion of people dying will rise).

The intensity of the mortality, therefore, is a combination of the rate at which people die per year of age, and the proportion of people who die. This can be represented by writing

$$\text{intensity of mortality in the age interval } x \text{ to } x + dx = \frac{l_x - l_{x+dx}}{l_x \, dx}.$$

Now, suppose that the length of the interval, dx, becomes very small. We can think of the limiting intensity of mortality as dx becomes very small as being the instantaneous intensity of mortality at age x. In demographic parlance, it has a special name: it is called the *force of mortality* at age x. It is denoted by the symbol μ_x. In symbols, therefore, we have

$$\mu_x = \lim_{dx \Rightarrow 0} \left(\frac{l_x - l_{x+dx}}{l_x \, dx} \right)$$

$$= - \lim_{dx \Rightarrow 0} \left(\frac{l_{x+dx} - l_x}{l_x \, dx} \right)$$

$$= - \frac{1}{l_x} \lim_{dx \Rightarrow 0} \left(\frac{l_{x+dx} - l_x}{dx} \right)$$

$$= - \frac{1}{l_x} \frac{d}{dx} l_x.$$

The force of mortality at age x is a kind of instantaneous probability of dying at that age (strictly speaking, it is not a probability, since it may in theory take a value greater than 1.0 – nevertheless, we shall let this detail pass). It is the demographer's answer to the question 'What is the chance that a person is going to die in the next few minutes?.' It may also be viewed as 'deaths per person alive per year of age'.

Now that we have introduced some calculus, we can re-express a few of the other relationships derived in Section 4.2. Consider l_0 people who are born. Suppose that at every age they experience a force of mortality equal to μ_x. Figure 4.2 shows the value of l_x plotted for all ages between 0 and the oldest age to which anyone survives.

The number of person-years lived between any two subsequent ages, say x and $x + n$, by these l_0 births is equal to the area under this curve between those two ages (see the shaded area in Figure 4.2). It may be written $\int_x^{x+n} l_u \, du$. Thus,

$$_nL_x = \int_x^{x+n} l_u \, du. \tag{4.6}$$

Note that equation (4.6) avoids any assumptions about the distribution of deaths between age x and age $x + n$. The earlier approximation for L_x (see equation (4.2)) required the assumption of an even distribution of deaths within each age group, which amounts to assuming that the probability of surviving to any specific age decreases linearly between ages x and age $x + n$.

The number of person-years lived at ages over age x, T_x, is given by the equation

$$T_x = \int_x^\infty l_u \, du,$$

so the expectation of life at age x, e_x, is given by the equation

$$e_x = \frac{\int_x^\infty l_u \, du}{l_0}.$$

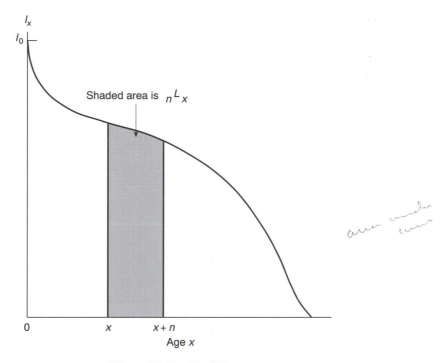

Figure 4.2 Graph of l_x

4.5 The calculation of life tables for specific populations

We have seen that a life table describes the mortality of a group of people born at a particular time, or a birth cohort. This is fine in theory, but it creates problems for calculating life tables for real populations. It would seem that we must wait until every member of a birth cohort has died before we can work out the actual figures in the life table. This might involve waiting over 100 years, which is not very helpful in practice.

One way of getting round this is to split up the life table into single years of age, and work out values of q_x for each single year of age based on the mortality experience at a particular calendar time. This procedure may be illustrated using a Lexis chart, as shown in Figure 4.3. We consider the people who are born during a particular year (represented in Figure 4.3 by the line AB) and follow them through until their first birthday, noting the number who die (the deaths are in the parallelogram ABKD). This will give us a value for q_0.

We then consider the people who celebrate their first birthday during *the same year* (represented in Figure 4.3 by the line CD), and follow them through until their second birthday, noting the number who die (the deaths are in the parallelogram CDJF). This will give us a value for q_1. We repeat this procedure for all ages, obtaining a complete set of values for q_x, based in each case on the people who celebrated their xth birthday during *the same year*.

Once we know all the q_x values, we can work out the values of l_x, d_x, L_x, T_x and e_x. Putting them together then gives us a life table based on the experience of the population during a specific period of calendar time. Such a life table is known as a *period life table*, because it is based on the experience of the population during a particular time period.

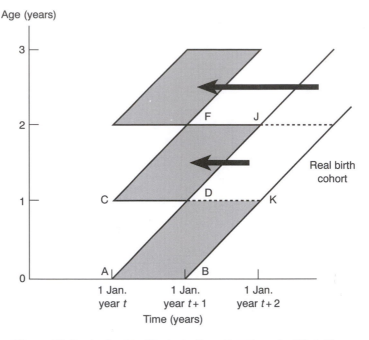

Figure 4.3 Lexis chart to illustrate the estimation of a life table

Clearly, a period life table will not represent the experience of any actual people unless the rates of mortality q_x are constant from one calendar year to the next over a period of over 100 years. In practice, this is nowhere near true, since mortality in the United Kingdom and elsewhere has been declining almost constantly during the twentieth century, and is still falling. What the period life table does illustrate is the mortality experience of a population during a particular period. In other words, it represents the experience of a *hypothetical cohort* of people who, at each age x last birthday, experience the mortality rates prevailing in a population among people of that age x last birthday during that period. Such hypothetical cohorts of people are frequently used in demography.

This procedure seems fine, but there is a further problem. In many countries (including the United Kingdom), no records are kept of the number of people who celebrate their xth birthday during a particular period. However, data are usually much more readily available from which we can calculate the m-type mortality rates for each age x during a particular year. We normally have estimates of the mid-year population aged x last birthday, and the number of deaths during the year of people aged x last birthday when they died, for all ages x. This means that we can calculate the values of m_x in a given year for all ages x, and then use the approximation

$$q_x = \frac{2m_x}{2 + m_x} \tag{4.7}$$

described in Chapter 2 to estimate the q-type rates.

This approximation involves the assumption that deaths are uniformly distributed across each year of age. For the youngest age groups, this is a bad assumption. It can be shown, however (see Exercise 4.8), that if the average number of person-years lived between exact ages x and $x + 1$ by persons who die between those two ages is a_x, then equation (4.7) can be generalized to read

$$q_x = \frac{1 \; m_x}{1 + (1 - a_x)m_x}.$$ (4.8)

It is easy to show that, if $a_x = 0.5$ (as is the case if deaths are evenly distributed), then equation (4.8) reduces to equation (4.7).

We must also assume that mortality does not change over calendar time, but the period of calendar time over which we need to assume mortality to be constant is only two years at most, so that this assumption is not too bad (to see that the period of calendar time over which we assume mortality to be constant is only two years at most, you should look again at Figure 4.3).

4.6   English Life Table 14

A good example of a national life table is English Life Table 14. Life tables are calculated for England and Wales every 10 years by the Government Actuary. Because mortality varies by sex, the life tables are calculated separately for males and females.

The exposed-to-risk for the m-type mortality rate is based on census data, and censuses take place every 10 years. English Life Table 14 is based on the 1981 census. An extract from this life table is shown in Table 4.1. Note that in English Life Table 14, l_0 is taken to equal 100 000. This does not reflect the actual number of births per year in England; it is merely a matter of convenience.

In practice, a slight variation from the procedure outlined above was adopted in the calculation of the m-type rates for English Life Table 14. This is because the English Life Tables are designed to represent the general mortality experience of the population in a particular period. It may be that the census years are years of abnormal mortality (because of epidemics of influenza, and so on). To reduce the impact of exceptional events on the life table estimates, the data used are based on the three years surrounding each census year. So we have the following formula for English Life Table 14:

$$m\text{-type rate} = \frac{\text{deaths in the years } 1980-82 \text{ to people aged } x \text{ last birthday when they died}}{\text{average population aged } x \text{ last birthday during the years } 1980-82}.$$

The denominator was estimated from the census data using a rather complex procedure, which is described in detail in Office of Population Censuses and Surveys (1987a).

4.7   Using the life table in practical work

Table 4.2 is an extract from English Life Table 14 for males. It is an abridged version of the full life table. This table can be used to illustrate some applications of the life table.

To begin with, we can work out the probability that a man aged 20 years will die before his 50th birthday. The algebraic expression that is needed to find this probability is

$$\text{probability of a man aged 20 years dying before his 50th birthday} = \frac{l_{20} - l_{50}}{l_{20}}.$$

Table 4.1 Values of q_x, l_x and e_x from English Life Table 14 for males ($l_0 = 100\,000$)

Age x	q_x	l_x	e_x
0	0.01271	100 000	71.043
1	0.00085	98 729	70.956
2	0.00051	98 645	70.016
3	0.00038	98 594	69.051
4	0.00035	98 557	68.077
5	0.00032	98 522	67.101
6	0.00030	98 490	66.123
7	0.00027	98 461	65.142
8	0.00025	98 434	64.160
9	0.00024	98 409	63.176
10	0.00024	98 385	62.191
11	0.00024	98 362	61.206
12	0.00026	98 338	60.221
13	0.00029	98 312	59.237
14	0.00034	98 283	58.254
15	0.00041	98 250	57.274
16	0.00053	98 210	56.297
17	0.00102	98 158	55.326
18	0.00111	98 057	54.382
19	0.00102	97 948	53.442
20	0.00093	97 849	52.496
21	0.00087	97 757	51.545
22	0.00083	97 672	50.589
23	0.00081	97 591	49.631
24	0.00081	97 511	48.671
25	0.00081	97 432	47.710
26	0.00082	97 353	46.749
27	0.00083	97 273	45.787
28	0.00084	97 192	44.824
29	0.00086	97 110	43.862
30	0.00088	97 027	42.899
31	0.00091	96 941	41.936
32	0.00094	96 853	40.974
33	0.00099	96 762	40.012
34	0.00105	96 666	39.051
35	0.00113	96 564	38.092
36	0.00123	96 455	37.134
37	0.00134	96 337	36.179
38	0.00148	96 208	35.227
39	0.00165	96 065	34.279
40	0.00184	95 907	33.335
41	0.00206	95 731	32.395
42	0.00231	95 534	31.461
43	0.00260	95 313	30.532
44	0.00293	95 066	29.611
45	0.00332	94 787	28.696
46	0.00376	94 472	27.790
47	0.00425	94 117	26.893

Table 4.1 Continued

Age x	q_x	l_x	e_x
48	0.00481	93 717	26.006
49	0.00545	93 266	25.129
50	0.00615	92 758	24.264
51	0.00694	92 187	23.411
52	0.00781	91 548	22.571
53	0.00877	90 833	21.744
54	0.00982	90 037	20.932
55	0.01098	89 152	20.135
56	0.01224	88 173	19.353
57	0.01361	87 094	18.586
58	0.01509	85 909	17.836
59	0.01670	84 612	17.101
60	0.01843	83 199	16.383
61	0.02028	81 666	15.681
62	0.02229	80 010	14.995
63	0.02448	78 226	14.326
64	0.02687	76 312	13.672
65	0.02949	74 261	13.036
66	0.03238	72 071	12.417
67	0.03555	69 738	11.815
68	0.03903	67 259	11.232
69	0.04285	64 634	10.668
70	0.04703	61 864	10.123
71	0.05160	58 955	9.597
72	0.05658	55 913	9.092
73	0.06198	52 749	8.607
74	0.06783	49 480	8.143
75	0.07416	46 123	7.699
76	0.08096	42 703	7.275
77	0.08827	39 246	6.872
78	0.09610	35 781	6.489
79	0.10445	32 343	6.126
80	0.11334	28 965	5.782
81	0.12278	25 682	5.458
82	0.13278	22 528	5.152
83	0.14333	19 537	4.865
84	0.15440	16 737	4.596
85	0.16591	14 153	4.345
86	0.17776	11 805	4.112
87	0.18986	9 706	3.895
88	0.20215	7 863	3.693
89	0.21453	6 274	3.506
90	0.22693	4 928	3.331
91	0.23929	3 810	3.167
92	0.25153	2 898	3.012
93	0.26374	2 169	2.863
94	0.27632	1 597	2.718
95	0.28971	1 156	2.574
96	0.30430	821	2.431

Table 4.1 Continued

Age x	q_x	l_x	e_x
97	0.32044	571	2.288
98	0.33844	388	2.145
99	0.35853	257	2.004
100	0.38087	165	1.865
101	0.40551	102	1.729
102	0.43241	61	1.597
103	0.46140	34	1.471
104	0.49214	19	1.350
105	0.52414	9	1.236
106	0.55667	4	1.129
107	0.58874	2	1.029
108	0.61896	1	0.935

Source: Office of Population Censuses and Surveys (1987a, p. 8).

From the data in Table 4.2, this can be calculated as

$$\frac{\text{probability of a man aged 20 years}}{\text{dying before his 50th birthday}} = \frac{97\,849 - 92\,758}{97\,849} = 0.052.$$

Thus, according to English Life Table 14, a man aged 20 has just over a 5% chance of dying before his 50th birthday.

Recall from Section 4.3 that, in an abridged life table, the assumption that deaths are evenly distributed across each age group is not always very good. How justifiable this assumption is may be examined by calculating the average age at death of those who die within a specific age group. If the assumption is a good one, then the average age at death of those who die between ages x and $x + n$ should be close to $x + \frac{1}{2}n$.

We can write down an algebraic expression, in terms of l_x and T_x, for the average age at death of those who die between 1 and 5 years of age. This is a little bit complicated to do,

Table 4.2 Extract from English Life Table 14 for males

Exact age x (years)	Number of men still alive per 100 000 births (l_x)	Number of person-years lived over age x by the original 100 000 born (T_x)
0	100 000	7 104 298
1	98 729	7 005 379
5	98 522	6 610 963
10	98 385	6 118 716
20	97 849	5 136 675
30	97 027	4 162 347
40	95 907	3 197 022
50	92 758	2 250 658
60	83 199	1 363 058
70	61 684	626 235
80	28 965	167 477
90	4 928	16 415
100	165	307

Source: Office of Population Censuses and Surveys (1987a, p. 8).

and is best approached in stages. First, note that everyone who dies between exact ages 1 and 5 years has lived at least one year. Therefore, we can write

$$\begin{array}{l}\text{average age at death of those}\\ \text{who die between exact}\\ \text{ages 1 and 5 years}\end{array} = 1 + \dfrac{\begin{array}{c}\text{person-years lived between exact ages}\\ \text{1 and 5 years by those who die}\\ \text{between exact ages 1 and 5 years}\end{array}}{\begin{array}{c}\text{number of persons who die between}\\ \text{exact ages 1 and 5 years}\end{array}}$$

Consider now just the numerator of the quotient on the right-hand side of this equation. The total number of person-years lived between exact ages 1 and 5 years is equal to $(T_1 - T_5)$. However, this includes the person-years lived between these two ages by those who will survive to ages over 5 years. We need to subtract these to get the right numerator. To put it another way,

$$\begin{array}{c}\text{person-years lived between}\\ \text{exact ages 1 and 5 years}\\ \text{by those who die between}\\ \text{1 and 5 years}\end{array} = \begin{array}{c}\text{total number of person-years}\\ \text{lived between exact}\\ \text{ages 1 and 5 years}\end{array} - \begin{array}{c}\text{person-years lived}\\ \text{between 1 and 5}\\ \text{years by those}\\ \text{who die} > 5 \text{ years}\end{array}$$

The last quantity in this equation is equal to $4l_5$, because we know that l_5 people die at ages over 5 years, and each of them must live exactly four person-years between exact ages 1 and 5 years.

Thus we have

$$\begin{array}{l}\text{person-years lived between exact}\\ \text{ages 1 and 5 years by those who}\\ \text{die between 1 and 5 years}\end{array} = T_1 - T_5 - 4l_5.$$

The number of persons who die between exact ages 1 and 5 years is simply $(l_1 - l_5)$.

Putting all this together, we have, therefore,

$$\begin{array}{l}\text{average age at death of those who}\\ \text{die between 1 and 5 years of age}\end{array} = 1 + \dfrac{T_1 - T_5 - 4l_5}{l_1 - l_5}$$

Using the data in Table 4.2, this is

$$\begin{array}{l}\text{average age at death of those who}\\ \text{die between 1 and 5 years of age}\end{array} = 1 + \dfrac{7\,005\,379 - 6\,610\,963 - 4(98\,522)}{98\,729 - 98\,522}$$

$$= 1 + \frac{328}{207}$$

$$= 2.58.$$

Thus, according to English Life Table 14, the average age at death of a boy dying between exact ages 1 and 5 years is 2.58 years. The assumption of an even distribution of deaths implies that this average age is 3.00 years. So we can see that the assumption is rather unsatisfactory in this case.

4.8 The general shape of life table quantities

It is a feature of human mortality that the age pattern of mortality exhibits regularities. Although the level of mortality varies widely, from life expectations below 40 years in

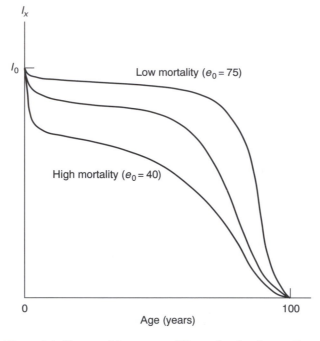

Figure 4.4 Shapes of l_x curve at different levels of mortality

some central African populations to life expectations of almost 80 years in many European countries, the age pattern displays some common features.

Consider the proportion surviving to particular ages, l_x. This is shown in Figure 4.4 for several different populations, with different levels of mortality. The general pattern is that l_x initially falls rapidly as age increases. This is because of high rates of infant and child mortality. At ages above about 5 years, however, the rate of decline reduces markedly. This reflects the low chance of death at ages between about 5 and 40 years. At older ages, the rate of decline gradually accelerates, reflecting the higher mortality at progressively older ages.

We can also demonstrate regularities in the values of e_x, the life expectation at age x. Figure 4.5 shows how e_x varies with age x in populations with different levels of mortality.

One important feature of the variation of life expectation with age is that, at most levels of mortality, the value of e_x in childhood (at ages from about 1 to 5 years) is greater than the life expectation at birth, e_0. At first sight, it might seem implausible that a child attaining his/her first birthday has longer left to live than he/she did at birth.

However, a little thought will reveal why this is. The value of e_0 is a kind of average of the individual life expectations of all the people who are born. It is, therefore, an average of all their ages at death. The value of e_1 is a similar average of the remaining expected lifetimes of the people who survive until their first birthday. But because, in most populations, the rate of mortality at age 0 last birthday, q_0, is very high relative to the average of the q_xs across the whole age range, the collection of lifetimes on which e_0 is based contains a lot of very short lives. These drag the value of e_0 downwards. The collection of lifetimes on which e_1 is based contains far fewer very short lifetimes, and so we can find that $e_1 > e_0$.

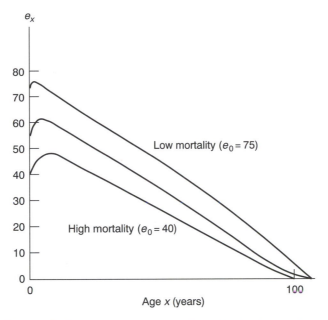

Figure 4.5 Shapes of e_x curve at different levels of mortality

Further reading

More formal expositions of the theory of the life table are contained in Benjamin and Pollard (1993, pp. 20–28) and Keyfitz (1977). Both these expositions make extensive use of calculus. A good simple introduction to life tables is contained in Cox (1975a).

Exercises

4.1 The q-type age-specific death rates in Table 4E.1 relate to the male population of a developing country. (They are actually taken from a model life table – see Chapter 14.) Use them to calculate the life expectation at birth for males in that country, assuming $l_0 = 10\,000$. Assume that no man survives until his 100th birthday – that is, $l_{100} = 0$.

4.2 Table 4E.2 gives estimates of the life expectation, e_x, at various ages x, for females in Nicaragua during 1990–95 and the United States in 1989. Use them to estimate q_0, $_4q_1$, $_5q_5$ and $_5q_{10}$ for these two countries.

4.3 Show that the crude birth rate in a stationary population corresponding to a life table is equal to $1/e_0$, where e_0 is the life expectation at birth.

4.4 In a life table, e_0 is equal to e_1, and

$$l_x = \left(\frac{l_1}{l_0}\right)^x l_0, \qquad \text{for } 0 \le x < 1.$$

Show that p_0 is equal to $\exp(-1/e_1)$.

4.5 Derive an algebraic expression relating q_x to the force of mortality μ_x.

4.6 Table 4E.3 gives m-type death rates per thousand of the population for males in England and Wales in 1992. Use these data to calculate a life table, taking

Table 4E.1

Age x	Width of age group, n	q-type age-specific death rate, $_nq_x$
0	1	0.1402
1	4	0.0714
5	5	0.0207
10	5	0.0150
15	5	0.0220
20	5	0.0313
25	5	0.0343
30	5	0.0393
35	5	0.0466
40	5	0.0578
45	5	0.0715
50	5	0.0947
55	5	0.1250
60	5	0.1751
65	5	0.2412
70	5	0.3362
75	5	0.4549
80	5	0.6126
85	5	0.7707
90	5	0.8984
95	5	1.0000

Source: Coale and Demeny (1983, p. 48).

$l_0 = 100\,000$. Assume that a person who reaches the age of 90 years has a remaining life expectation of 5 years.

4.7   The abridged life table data in Table 4E.4 refer to the female population of England and Wales in 1980–82. They are based on an assumed 100 000 births, and have been abridged from a full life table which used information by single years of age. Using these data, calculate:

(a)   the probability that a woman aged exactly 20 years will survive until her 40th birthday;

(b)   the infant mortality rate, q_0;

(c)   the life expectation at birth and at exact age 1 year;

Table 4E.2

Age x	Life expectation, e_x	
	Nicaragua, 1990–95	United States, 1989
0	67.7	78.6
1	70.0	78.3
5	67.7	74.4
10	63.2	69.5
15	58.5	64.6

Source: Wilkie *et al.* (1996, p. 154).

Table 4E.3

Age last birthday	Death rate	Age last birthday	Death rate	Age last birthday	Death rate
0	7.32	30	0.91	60	13.58
1	0.53	31	0.98	61	15.14
2	0.34	32	0.96	62	17.10
3	0.26	33	1.02	63	19.21
4	0.25	34	1.11	64	21.50
5	0.21	35	1.21	65	23.44
6	0.18	36	1.26	66	26.38
7	0.20	37	1.31	67	29.74
8	0.15	38	1.65	68	32.60
9	0.16	39	1.64	69	36.25
10	0.17	40	1.68	70	39.28
11	0.16	41	1.88	71	43.90
12	0.22	42	1.90	72	45.53
13	0.19	43	2.15	73	53.74
14	0.26	44	2.38	74	59.07
15	0.33	45	2.41	75	60.98
16	0.42	46	2.91	76	68.40
17	0.64	47	3.33	77	76.25
18	0.78	48	3.57	78	82.75
19	0.80	49	3.96	79	88.29
20	0.76	50	4.50	80	99.95
21	0.84	51	5.04	81	106.18
22	0.85	52	5.70	82	118.13
23	0.78	53	6.35	83	127.96
24	0.85	54	6.72	84	137.29
25	0.75	55	7.85	85	149.03
26	0.82	56	8.53	86	163.65
27	0.83	57	9.53	87	181.04
28	0.88	58	11.04	88	194.02
29	0.87	59	12.13	89	212.59
				90 +	266.00

Source: Office of Population Censuses and Surveys (1994, p. 6).

Table 4E.4

Exact age x	Survivors to age x, l_x	Woman-years lived above exact age x, T_x
0	100 000	7 700 187
1	99 016	7 601 014
10	98 746	6 711 410
20	98 497	5 725 004
30	98 105	4 741 877
40	97 346	3 764 073

Source: Office of Population Censuses and Surveys (1987a, pp. 8, 10).

(d)   the probability that a girl who survives until her first birthday will die between her 10th and 20th birthdays;

(e)   the expected age at death of those who die between their 20th and 30th birthdays;

(f)   the expected age at death of those who die when they are aged under 1 year.

What do the results of (e) and (f) tell you about the validity of assuming that deaths are evenly distributed across each age group?

4.8   If the average age at death of those who die between exact ages x and $x + 1$ is $x + a_x$, show that

$$q_x = \frac{m_x}{1 + (1 - a_x)m_x}.$$

5

Multiple-Decrement Life Tables

5.1 Introduction

In Chapter 1, we introduced the idea of representing demographic processes as transitions between states. Using this notion, the simple life table may be depicted as the transition between two states, 'alive' and 'dead' (see Figure 1.2a). Some features of this representation are worthy of comment:

1   Only one transition is possible: from 'alive' to 'dead'.
2   'Dead' is an absorbing state. People do not come back to life. Thus the transition between the states 'alive' and 'dead' is only made in one direction.
3   Everybody ultimately makes the transition. There are no immortal people.

In other words, the simple life table (of the sort we have analysed so far) is an analysis of a single, universal and unidirectional transition.

In this chapter life table analysis is extended to look at situations where more than one transition is possible, and situations where not everybody makes a transition. Life tables where more than one transition is possible are called *multiple-decrement life tables*. Section 5.2 of this chapter discusses the basic ideas underlying multiple-decrement life tables. Section 5.3 presents some relevant theory. Two examples of multiple-decrement life tables are described in Section 5.4. Section 5.5 introduces the important distinction between dependent and independent death rates, and Section 5.6 derives formulae to relate the two. The use of these formulae is illustrated using a prospective investigation of mortality in which some people withdraw from the investigation before they are observed to die. In Section 5.7 it is shown that data of this kind of investigation are common, and the ability of life table analysis to handle them renders it a very powerful tool. Finally, Section 5.8 briefly considers the estimation of multiple-decrement life tables from data on m-type mortality rates. Such data are also very common.

5.2 The idea of the multiple-decrement life table

In the ordinary life table, the movement from the state 'alive' to the state 'dead' is often called a *decrement*. This word is used because the number of people alive decreases by one for every death. In the ordinary life table, death is the only decrement, and so such a life table is sometimes called a *single-decrement life table*.

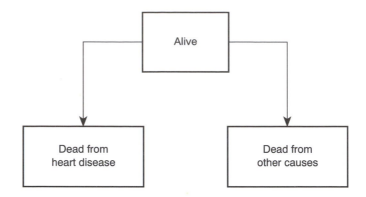

Figure 5.1 Multiple-state representation of a double-decrement life table

Sometimes, however, there are several possible states into which someone can move, and we are interested in analysing the transitions into each of them. Figure 5.1 shows a situation in which deaths from heart disease have been separated from deaths from other causes. People may die either of heart disease or of some other cause. If they die of heart disease, they move from the state 'alive' to the state 'dead from heart disease'. If they die from some other cause, they move from the state 'alive' to the state 'dead from other causes'. There are thus two decrements: deaths from heart disease, and deaths from other causes.

An important feature of this situation is that a person cannot die both from heart disease and from some other cause. That is, although everyone who is presently alive has two possible transitions available to him/her, each person will actually make one, and only one, of these. A person will either die of heart disease or die from some other cause. In such a situation, death from heart disease and death from other causes are said to be *competing risks*. They cannot both occur to any one person.

Situations like those shown in Figure 5.1 can be analysed using life tables. Because there are two decrements operating, the life tables used to analyse them are called *double-decrement life tables*. Of course, there might be more than two decrements. For example, we might have 'deaths from heart disease', 'deaths from cancer' and 'deaths from other causes'. In general, therefore, we refer to *multiple-decrement life tables* whenever there is more than one decrement.

5.3   The algebra of the multiple-decrement life table

The ideas of the multiple-decrement life table are very straightforward, but are complicated by a rather awkward algebraic notation.

For convenience, we shall assume that we have two decrements, which we shall refer to as α and β. Suppose that they are two different causes of death. No one can die of both causes; everyone must die of one of them.

In the ordinary single-decrement life table, a fundamental quantity is the proportion of people attaining some exact age x years who die before they reach the exact age $x + 1$ years. In the ordinary life table, this proportion is denoted by the symbol q_x. In the multiple-decrement situation, there are two causes of death, α and β. A person cannot die from both. Therefore, we can define two quantities:

- the proportion of people attaining some exact age x years who die before they reach their $(x + 1)$th birthday, and whose death is attributed to cause α (let us call this the q-type death rate from cause α);
- the proportion of people attaining some exact age x years who die before they reach their $(x + 1)$th birthday, and whose death is attributed to cause β (let us call this the q-type death rate from cause β).

Now (and this is a very important point) these two quantities are dependent on each other. That is, if the q-type death rate from cause α increases, then, other things being equal, the q-type death rate from cause β will decrease. Why? Because a person cannot die from both causes. The people who die from cause α cannot die from cause β. In the example shown in Figure 5.1, if the death rate from heart disease increases, the death rate from other causes will decrease.

Because of this dependency between the death rates from cause α and cause β, we refer to the death rates in the multiple-decrement life table as *dependent rates*. They are also given special symbols, $(aq)_x^\alpha$ and $(aq)_x^\beta$:

- $(aq)_x^\alpha$ is the proportion of people attaining some exact age x years who die before they reach their $(x + 1)$th birthday, and whose death is attributed to cause α;
- $(aq)_x^\beta$ is the proportion of people attaining some exact age x years who die before they reach their $(x + 1)$th birthday, and whose death is attributed to cause β.

Like the ordinary life table, the multiple-decrement life table is constructed by assuming that we start with some number of births. In the ordinary life table, this number was denoted by the symbol l_0. In the multiple-decrement life table, it is given the symbol $(al)_0$.

The numbers of deaths between exact ages x and $x + 1$ from each of the two causes, α and β, which are denoted by the symbols $(ad)_x^\alpha$ and $(ad)_x^\beta$ respectively, are given by the equations

$$(ad)_x^\alpha = (al)_x (aq)_x^\alpha \tag{5.1}$$

and

$$(ad)_x^\beta = (al)_x (aq)_x^\beta. \tag{5.2}$$

(These equations follow directly from the definitions of $(aq)_x^\alpha$ and $(aq)_x^\beta$ given above.)

The total number of deaths between exact ages x and $x + 1$ from *both causes combined*, which is denoted by the symbol $(ad)_x$, is given by the equation

$$(ad)_x = (ad)_x^\alpha + (ad)_x^\beta. \tag{5.3}$$

Now, if $(aq)_x$ is the q-type death rate at age x last birthday for both causes combined, we have

$$(aq)_x = \frac{(ad)_x}{(al)_x},$$

so that

$$(ad)_x = (al)_x (aq)_x. \tag{5.4}$$

Substituting from equations (5.1), (5.2) and (5.4) into equation (5.3) produces

$$(al)_x (aq)_x = (al)_x (aq)_x^\alpha + (al)_x (aq)_x^\beta.$$

Dividing this equation by $(al)_x$ produces

$$(aq)_x = (aq)_x^\alpha + (aq)_x^\beta. \tag{5.5}$$

Equation (5.5) says that the q-type death rate at age x last birthday from both causes combined is equal to the sum of the dependent q-type death rates at age x last birthday from the individual causes of death.

ABRIDGED MULTIPLE-DECREMENT LIFE TABLES

Just as with the ordinary single-decrement life table, we can have abridged multiple-decrement life tables. Suppose that the width of an age group is n years. We refer to the proportion of people attaining some exact age x years who die before they reach their $(x + n)$th birthday, and whose death is attributed to cause α, denoting this by the symbol $_n(aq)_x^\alpha$; and the proportion of people attaining some exact age x years who die before they reach their $(x + n)$th birthday, and whose death is attributed to cause β, denoting this by the symbol $_n(ad)_x^\beta$. Similarly, the numbers of deaths from causes α and β respectively between exact ages x and $x + n$ are denoted by the symbols $_n(ad)_x^\alpha$ and $_n(ad)_x^\beta$. Equation (5.3) becomes

$$_n(ad)_x = {}_n(ad)_x^\alpha + {}_n(ad)_x^\beta,$$

where $_n(ad)_x$ is the total number of deaths between exact ages x and $x + n$, and equation (5.5) becomes

$$_n(aq)_x = {}_n(aq)_x^\alpha + {}_n(aq)_x^\beta,$$

where $_n(aq)_x$ is the q-type death rate between exact ages x and $x + n$ for both causes combined.

5.4   Some examples

EXAMPLE 1

The abridged multiple-decrement life table in Table 5.1 gives some data for females in England and Wales in the year 1995. It is a period life table, since it refers to deaths during a particular period of calendar time. It shows the q-type death rates in various age groups from cerebrovascular disease and all other causes of death. It is clear from the life table that cerebrovascular disease was a very rare cause of death among young women, but much more common among older women during this period.

EXAMPLE 2

Once we view multiple-decrement life tables using a multiple-state framework, it becomes clear that they may be used to analyse any situation in which people all start in a specific state, and must move at some age to one of a number of other absorbing states. We are not restricted to cases where all the relevant transitions involve death.

One of the most common of these situations arises in actuarial work in connection with investigations of the mortality of people who have taken out life insurance policies. Consider a group of such people, all of whom attain a certain age within a particular period of time – in other words, a birth cohort. In the investigation, these people will be followed

Table 5.1 Multiple-decrement life table of deaths of females from cerebrovascular disease and other causes in England and Wales, 1995

Age x	Width of age group n	Survivors to age x $(al)_x$	q-type death rate Combined $_n(aq)_x$	Cerebrovascular $_n(aq)_x^s$	Other causes $_n(aq)_x^c$
0	1	100 000	0.00528	0.00002	0.00527
1	4	99 472	0.00100	0.00001	0.00098
5	5	99 372	0.00059	0.00001	0.00058
10	5	99 314	0.00067	0.00002	0.00066
15	5	99 247	0.00130	0.00002	0.00128
20	5	99 118	0.00147	0.00002	0.00145
25	5	98 972	0.00188	0.00008	0.00180
30	5	98 786	0.00266	0.00014	0.00252
35	5	98 523	0.00394	0.00021	0.00373
40	5	98 135	0.00661	0.00043	0.00618
45	5	97 486	0.01049	0.00056	0.00993
50	5	96 464	0.01725	0.00103	0.01622
55	5	94 800	0.02717	0.00157	0.02560
60	5	92 224	0.04488	0.00286	0.04202
65	5	88 085	0.07663	0.00577	0.07087
70	5	81 335	0.12603	0.01258	0.11345
75	5	71 084	0.19427	0.02471	0.16956
80	5	57 275	0.31128	0.04821	0.26307
85	5	39 446	0.46210	0.07601	0.38609
90		21 218	1.00000	0.14824	0.85176

Source: Estimated from data in Office for National Statistics (1997b, pp. 1, 92–93, 126–127).

through until they die. At each age, x, the number who die before their $(x + 1)$th birthday, divided by the number who reach their xth birthday, is our estimate of q_x.

A common problem with such investigations is that some people will be lost to the study for reasons other than death. It may not then be possible to follow these people until they die. Reasons why people might be lost to the study include failure to maintain their insurance premiums (because they can no longer afford them, or because they no longer wish to have life insurance) and emigration. Such individuals are said to *withdraw* from the study.

What can be done about this? Well, we can assume that people who withdraw from the study (because they stop paying their premiums before they die, or because they emigrate) are alive at the point at which we lose track of them. This means that the situation may be analysed as if death and withdrawal were competing risks, for, from the point of view of the investigation, a person can only do one of two things: die while in the investigation, or withdraw from the study before death (an individual cannot do both of these). Figure 5.2 depicts the situation in multiple-state terms.

Of course the people who withdraw from the study will subsequently die. This will involve their making the transition marked with a dotted arrow in Figure 5.2. However, we shall not be able to observe this transition, and so we shall never know at what age these people die. Therefore, we cannot use information about this transition to calculate q-type death rates. All we have to go on is information about the transitions marked with solid

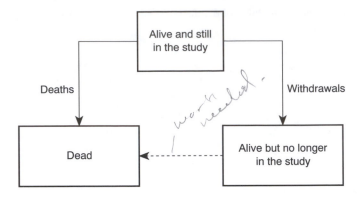

Figure 5.2 Multiple-state representation of an investigation of mortality in which some people withdraw from the study

lines. These transitions form a situation which we can analyse using a double-decrement life table.

5.5   Dependent and independent death rates

We noted in Section 5.3 that the q-type death rates from each decrement in a multiple-decrement situation are dependent on the q-type death rates from the other decrements.

However, what is frequently of interest is not these dependent death rates themselves, but what are called the *independent* death rates from each decrement. The independent death rate from a specific decrement is the death rate which we would observe if the other, competing, decrements did not exist.

Let us reconsider the examples in the previous section. The first of these concerned English and Welsh mortality in 1995 (Table 5.1). Deaths from cerebrovascular disease were distinguished from those due to other causes. We might ask what the death rate from other causes would be if deaths from cerebrovascular disease were eliminated. In the second example, which concerned a mortality investigation where some people withdraw prior to death, we might ask what the q-type death rate would be if there were no withdrawals.

The independent q-type rate from one decrement is not normally the same as the dependent q-type rate from the same decrement. To see why this is, consider the cerebrovascular disease example. If we eliminate cerebrovascular disease as a cause of death, the people who previously died of cerebrovascular disease will die of some other cause, increasing the q-type death rates from other causes.

Therefore, the independent q-type rates from a particular decrement are greater than the dependent q-type rates from the same decrement. (The only – trivial – exceptions to this arise if both are zero, or if the rates for a competing risk are zero.)

5.6   The relationship between dependent and independent rates of decrement

In order to understand the relationship between dependent and independent rates of decrement, suppose that we have two decrements, α and β. Denote the independent rate of decrement at age x last birthday from decrement α by the symbol q_x^α, and the independent rate of decrement at age x last birthday from decrement β by the symbol q_x^β. These independent rates of decrement can themselves be used to construct two life tables, one based on the q_x^αs and the other on the q_x^βs. These life tables are known as the *associated single-decrement life tables* of the multiple-decrement life table with decrements α and β.

If a certain number of people attaining their xth birthday, say l_x, are subject to the independent rates of decrement q_x^α and q_x^β, then the numbers of them who will die between their xth and $(x+1)$th birthdays from causes α and β respectively are denoted by the symbols d_x^α and d_x^β. Clearly,

$$d_x^\alpha = q_x^\alpha l_x \tag{5.6}$$

and

$$d_x^\beta = q_x^\beta l_x.$$

We are now in a position to consider the relationship between the dependent and the independent rates of decrement.

Consider a group of people who attain their xth birthday within a specific period. Suppose that there are $(al)_x$ of these people. Suppose, further, that between the exact ages x and $x+1$ these people are subject to two causes of death, α and β. We will make two assumptions:

- that the deaths from decrement α are distributed evenly across the year of age from exact age x to exact age $x+1$;
- that the deaths from decrement β all take place when people are aged exactly $x+\frac{1}{2}$ years.

This situation is depicted in Figure 5.3.

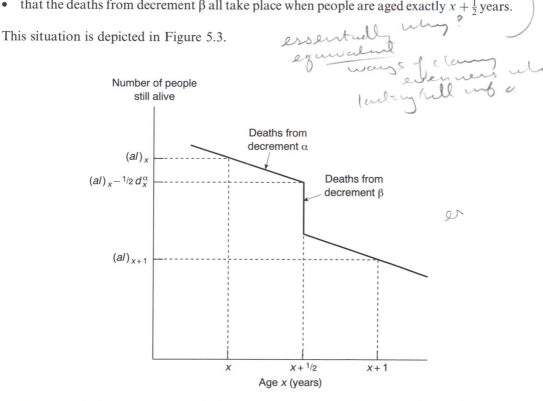

Figure 5.3 Diagram illustrating the derivation of the relationship between dependent and independent rates of decrement

We are going to find an expression for the number of deaths we will observe from cause β *when both causes are operating*. That is – using the notation introduced in Section 5.3 – we are going to find an expression for $(ad)_x^\beta$.

Between exact ages x and $x + \frac{1}{2}$, only cause α is operating (because there are no deaths from cause β until exact age $x + \frac{1}{2}$). Therefore, the rate of death between exact ages x and $x + \frac{1}{2}$ from cause α is just $\frac{1}{2}q_x^\alpha$. The number of deaths between exact ages x and $x + \frac{1}{2}$ is $\frac{1}{2}d_x^\alpha$ (because we are only considering half a year, and deaths are evenly distributed across each year of age).

So, at exact age $x + \frac{1}{2}$, when the deaths from cause β take place, there are $(al)_x - \frac{1}{2}d_x^\alpha$ people left alive. These people are subject to an independent rate of death from cause β of q_x^β. Thus we have:

$$\text{number of deaths from cause } \beta \atop \text{when both causes are operating} = [(al)_x - \tfrac{1}{2}d_x^\alpha)]q_x^\beta,$$

or

$$(ad)_x^\beta = [(al)_x - \tfrac{1}{2}d_x^\alpha)]q_x^\beta. \tag{5.7}$$

Now, from equation (5.6), we know that

$$d_x^\alpha = (al)_x q_x^\alpha. \tag{5.8}$$

(Note that notation for the original number of people has been changed from l_x to $(al)_x$ because we are now dealing with a multiple-decrement situation.)

Moreover, from equation (5.2),

$$(ad)_x^\beta = (al)_x (aq)_x^\beta. \tag{5.9}$$

Equation (5.9) says that the number of deaths observed from cause β when both causes are operating is equal to the original number of people $(al)_x$ multiplied by the dependent rate of decrement from cause β, $(aq)_x^\beta$.

Substituting from equations (5.8) and (5.9) into equation (5.7) produces

$$(al)_x (aq)_x^\beta = [(al)_x - \tfrac{1}{2}(al)_x q_x^\alpha] q_x^\beta, \tag{5.10}$$

and dividing equation (5.10) by $(al)_x$ gives

$$(aq)_x^\beta = q_x^\beta (1 - \tfrac{1}{2}q_x^\alpha). \tag{5.11}$$

Equation (5.11) expresses the relationship between the independent and dependent rates of decrement for cause β.

We can use an analogous argument to find the relationship between the dependent and independent rates of decrement for cause α, by simply interchanging causes α and β in the derivation. This produces the equation

$$(aq)_x^\alpha = q_x^\alpha (1 - \tfrac{1}{2}q_x^\beta). \tag{5.12}$$

It should be noted that these derivations are not rigorous. We have used the device of assuming that the deaths from one cause of decrement are evenly distributed across the year of age, and that the deaths from the other cause are concentrated at a point half-way through the year of age. A more rigorous derivation involving calculus assumes that deaths from both decrements are evenly distributed across the year of age (see Benjamin and Pollard, 1993, pp. 129–138). Clearly, if this is the case, then on the average the deaths will take place half-way through the year of age.

Equations (5.11) and (5.12) enable us to work out the dependent rates of decrement once we know the independent rates. How about the situation (which is more usual in practice) where we know the dependent rates, and wish to work out the independent rates?

The obvious way to work out expressions for the independent rates in terms of the dependent rates is to solve the simultaneous equations (5.11) and (5.12) for q_x^α and q_x^β. Unfortunately, the solution produces quadratic equations which must be solved using the quadratic equation formula. However, it can be shown (the algebra is very tedious, and will not be reproduced here) that

$$q_x^\alpha \cong \frac{(aq)_x^\alpha}{1 - \frac{1}{2}(aq)_x^\beta} \qquad (5.13)$$

and

$$q_x^\beta \cong \frac{(aq)_x^\beta}{1 - \frac{1}{2}(aq)_x^\alpha}. \qquad (5.14)$$

These approximations are better the smaller the rates of decrement.

A PRACTICAL APPLICATION

An application should help fix an understanding of the formulae derived above. The example we shall use is of a life insurance company's investigation of the mortality of people aged between 40 and 45 years. The data are given in Table 5.2. We can see that the investigation was hindered by the fact that a number of people were lost to the study – these people are called withdrawals in actuarial jargon.

The two decrements are death, which we can call decrement α, and withdrawal, which we can call decrement β. The life insurance company wishes to work out the q-type death rates at these ages for its policyholders. Because of the withdrawals, the data in Table 5.2 only permit the calculation of the dependent death rates. The company really wants to know the independent death rates of those who do not withdraw, because it wishes to use the death rates calculated from this investigation in order to calculate future premiums, which need to apply to situations in which the number of withdrawals may differ from those in Table 5.2.

The first step is to work out the dependent q-type rates of decrement (Table 5.3). These are just the deaths and withdrawals at each age x last birthday divided by the number attaining exact age x (see equations (5.1) and (5.2)). As an example, the dependent rate of decrement for deaths at age 40 last birthday is equal to $10/1000 = 0.01$. The independent

Table 5.2 An actuarial investigation into mortality

Age last birthday in years x	Number attaining exact age x years $(al)_x$	Number dying between exact ages x and $x+1$ $(ad)_x^\alpha$	Number of withdrawals between exact ages x and $x+1$ $(ad)_x^\beta$
40	1000	10	20
41	970	20	20
42	930	18	22
43	890	20	40
44	830	22	20
45	788		

Table 5.3 Dependent rates of decrement for the actuarial investigation of mortality

Age last birthday in years x	Dependent q-type rates of decrement	
	Deaths $(aq)_x^\alpha$	Withdrawals $(aq)_x^\beta$
40	0.0100	0.0200
41	0.0206	0.0206
42	0.0194	0.0237
43	0.0225	0.0449
44	0.0265	0.0241
45		

death rates may then be found using equation (5.13), and the independent rates of withdrawal found using equation (5.14) (Table 5.4). As expected, the independent rates are a little bit higher than the dependent rates in each case.

5.7 Censoring

The mortality investigation considered in Section 5.6 is an example of a prospective study of mortality (see Chapter 1 for a definition of what a prospective study is). It is very common in such studies for some study members to be lost to follow-up prior to their death. The lives of these study members are said to be *censored* at the point when they are lost to follow-up. In Section 5.6 we have shown how the use of a double-decrement life table enables the estimation of mortality rates which are independent of the censoring of the lives of a proportion of the people in the study.

Of course, usually the investigator has little or no interest in estimating censoring rates (either dependent or independent). In such cases, a formula for calculating the independent death rates directly from the data may be derived (assuming death is decrement α and censoring is decrement β). Take equation (5.13):

$$q_x^\alpha \cong \frac{(aq)_x^\alpha}{1 - \frac{1}{2}(aq)_x^\beta}.$$

Table 5.4 Independent rates of decrement for the actuarial investigation of mortality

Age last birthday in years x	Independent q-type rates of decrement	
	Deaths q_x^α	Withdrawals q_x^β
40	0.0101	0.0201
41	0.0208	0.0208
42	0.0196	0.0239
43	0.0230	0.0454
44	0.0268	0.0244
45		

5.5 Direct estimation of independent death rates in the actuarial investigation of mortality

t birthday x	Number attaining exact age x years $(al)_x$	Deaths $(ad)^\alpha_x$	Withdrawals $(ad)^\beta_x$	Independent death rates q^α_x
	1000	10	20	$10/(1000 - 10) = 0.0101$
	970	20	20	$20/(970 - 10) = 0.0208$
	930	18	22	$18/(930 - 11) = 0.0196$
	890	20	40	$20/(890 - 20) = 0.0230$
	830	22	20	$22/(830 - 10) = 0.0268$
45	788			

Since $(aq)^\alpha_x = (ad)^\alpha_x/(al)_x$ (from equation (5.1)) and $(aq)^\beta_x = (ad)^\beta_x/(al)_x$, we can rewrite this equation as

$$q^\alpha_x \cong \frac{(ad)^\alpha_x/(al)_x}{1 - \frac{1}{2}[(ad)^\beta_x/(al)_x]},$$

which may be simplified to

$$q^\alpha_x \cong \frac{(ad)^\alpha_x}{(al)_x - \frac{1}{2}(ad)^\beta_x}. \tag{5.15}$$

Equation (5.15) says that the independent q-type death rate at age x last birthday may be calculated by dividing the observed number of deaths at age x last birthday by the number still alive at exact age x minus half the observed number of persons censored between exact ages x and $x + 1$ years. The application of equation (5.15) to the data of Table 5.2 is shown in Table 5.5.

The subject of censoring is considered further in the next chapter, and will recur again in later chapters. The ability of life table analysis to deal with censored observations is of vital importance for demographic analysis.

5.8 Estimating multiple-decrement life tables from data in the form of *m*-type rates

In practical situations, it is common to have period data in the form of *m*-type rates. In Chapter 4 it was shown how such data could be used to estimate period life tables with a single decrement. When there is more than one decrement, these data typically take the form of dependent *m*-type rates – they are dependent rates because they come from observation of the real world, in which all the decrements are operating. In order to use these rates to estimate multiple-decrement life tables, we need formulae relating *m*-type and *q*-type rates in the multiple-decrement situation. Two useful formulae in this context are

$$(aq)^\alpha_x = \frac{(am)^\alpha_x}{1 + \frac{1}{2}[(am)^\alpha_x + (am)^\beta_x]}$$

and

$$(aq)^\beta_x = \frac{(am)^\beta_x}{1 + \frac{1}{2}[(am)^\alpha_x + (am)^\beta_x]}.$$

Exercises

5.1 (a) Using the data in Table 5.1, calculate the life expectation at birth for females in England and Wales in 1995. You may assume that $e_{90} = 5.5$ years.

(b) Estimate what the life expectation at birth would be if cerebrovascular disease were eliminated as a cause of death, and all other causes retained the same independent rates of decrement.

5.2 A life insurance company carried out an investigation of the mortality of its male policyholders. The company followed a sample of 1000 such men from their 60th birthday until either their 70th birthday, or their death, or their withdrawal from the scheme. Table 5E.1 shows the results of the investigation.

(a) Calculate the age-specific q-type independent death rates by single years of age from 60 to 69 years.

(b) A recent national life table shows that the remaining life expectation of males aged exactly 70 years, e_{70}, is 12 years. Assuming that the mortality of the policy-holders when they are aged 70 years and over will be the same as that in the national life table, estimate their remaining life expectation at exact age 60 years, e_{60}.

(c) According to the results of the investigation above, what is the average age at death of male policyholders dying between exact ages 60 and 70 years?

5.3 A population of small furry animals lives on a remote island. Every year, on 30 June, it is surveyed by naturalists who take a census. By means of a tagging system the naturalists can identify the age of each animal: thus they can work out how many animals celebrated their xth birthday in the previous year ($x = 0, \ldots, 4$). Their results for the year ended 30 June 1995 are shown in Table 5E.2.

The animals have lived undisturbed by humans for many years (apart from the annual censuses). Therefore, their mortality at each age has been constant for a long period. No animal has yet been observed to survive until a fifth birthday.

All this changed very early on the morning of 1 July 1995. A ship carrying toxic chemicals ran aground on the island. The chemicals polluted the animals' only supply of fresh water. Laboratory studies have shown that ingesting this toxic chemical

Table 5E.1

Exact age x	Number still under observation	Number of deaths between exact ages x and $x+1$	Number of withdrawals between exact ages x and $x+1$
60	1000	18	30
61	952	19	20
62	913	18	10
63	885	22	30
64	833	24	20
65	789	20	50
66	719	25	30
67	664	25	10
68	629	25	80
69	524	21	50
70	453		

Table 5E.2

Age x	Number of animals celebrating their xth birthday in the year ended 30 June 1995 (l_x)
0	1618
1	1000
2	730
3	515
4	160

causes small furry animals like the ones on this island to suffer an *independent* q-type mortality rate of 0.100 at all ages (that is, the independent $q_x = 0.100$, $x = 0, \ldots, 4$).

(a) Calculate the life expectation at birth of the animals prior to the pollution incident.

(b) Assuming that the same number of animals are born in the year ended 30 June 1996 as were born in the year ended 30 June 1995, calculate the number of animals born in the year ended 30 June 1996 who will survive to exact age 1 years.

(c) Estimate the change in the life expectation at birth of the animals caused by the pollution incident.

5.4 Carefully describe how you would evaluate the effect on the expectation of life at birth in a given population of eliminating a major cause of death such as cancer. You should clearly explain each step of the analysis you would do and state any assumptions or formulae you would use.

5.5 Which of the following would be most likely to have the largest impact on the expectation of life at birth in a developing country:
(a) eliminating maternal deaths (that is, deaths associated with childbirth);
(b) eliminating infant deaths from infectious diseases;
(c) eliminating deaths from cancer?
Explain your reasoning.

6

Survival Analysis

6.1 Introduction

In this chapter we consider another way of studying the life table. This approach makes use of what is called *survival analysis*. Survival analysis has become increasingly widely used in the disciplines of social statistics, medical statistics, demography and economics. It is a very powerful method of studying life history data.

The material in this chapter is technically rather more demanding than anything in earlier chapters. The early sections, therefore, concentrate very much on some basic ideas. Section 6.2 introduces the model of mortality which survival analysis embodies. (This section and Section 6.3 owe much to the ideas of Angus Macdonald of Heriot-Watt University.) Sections 6.3–6.5 introduce three functions which describe this distribution – the survivor function, the probability density function and the hazard function – and Section 6.6 shows how these three are related to one another. We show how these functions are related to quantities already described in Chapter 4. Fundamental to survival analysis is the idea of censoring: Section 6.7 extends the discussion of censoring which was begun in Section 5.7. In Section 6.8 the application of survival analysis to estimating mortality from real data is described. Section 6.9 shows how the resulting estimates of mortality can be combined to produce a life table. Finally, Section 6.10 describes briefly some of the advantages which survival analysis has over the conventional approach to the life table which was the subject of Chapters 4 and 5.

An understanding of survival analysis requires a knowledge of basic statistical theory. Readers who have not studied much statistical theory will find the following sections quite a challenge. Nevertheless, the amount of prior knowledge assumed has been kept to a minimum. Readers without any prior acquaintance with statistical theory are advised to consult a basic text in that subject (see the suggestions for further reading at the end of this chapter). Later in this chapter, we shall need to use the maximum likelihood method of estimating statistical parameters. Because many readers may be unfamiliar with this technique, its use in the context of survival analysis is described essentially from first principles.

6.2 A model of mortality

Recall that, in calculating the life table, we imagined a birth cohort of people – a group of people born within a particular period – and followed them through until they died. We then calculated q-type death rates at each age x, q_x, where

$$q_x = \frac{\text{number dying between exact age } x \text{ and exact age } x+1}{\text{number attaining exact age } x}.$$

Now, in order to calculate all these q_xs, we did not need to know each person's exact age at death, we simply needed to know how many were still alive at each age x.

Suppose, however, that we did know each person's exact age at death (this will often be the case, for example, if we have prospective survey data). Let us suppose that we follow a sample of n births and note the age at death of each member of the sample. This sample of n births is equivalent to the group of l_0 lives we considered when looking at the traditional life table. We use the symbol n to remind us that we are considering a *sample*. Let the age at death of person i $(i = 1, \ldots, n)$ be x_i.

Now the individual ages at death, the x_is, can be seen as individual values taken from the distribution of a random variable X, which is *length of life*. The approach to the life table we describe in this chapter involves constructing statistical models of the distribution of the random variable X. This variable, being length of life, can take any value between 0 and (about) 120 years. We are going to specify this distribution by using three functions of the length of life, X: the survivor function, the probability density function and the hazard function.

6.3 The survivor function

Without having any data at all, it is possible to say at least two things about the random variable X: first, X is always positive (people do not die before they are born); and second, X cannot be greater than (about) 120 years.

The variable X also has a *distribution function*, which we can call $F(x)$. The distribution function is the probability that the length of a person's life is less than (or equal to) a specific value x. Thus $F(70)$ is the probability that a person does not survive beyond his/her 70th birthday. Clearly, $F(0) = 0$ (everybody survives for at least some time) and $F(120) \cong 1$ (nobody's lifetime is greater than about 120 years). In symbols, we can write

$$F(x) = \Pr[X \le x].$$

We can also define another function, called the *survivor function*, or $S(x)$. This function describes the probability that a person will still be alive at exact age x. We can write

$$S(x) = \Pr[X > x].$$

Since a person must either be alive at exact age x or have died by that age, we can say that

$$S(x) = 1 - F(x).$$

Clearly, $S(0) = 1$ (everybody survives for at least some time), and $S(120) \cong 0$ (nobody survives longer than about 120 years).

RELATIONSHIP OF THE SURVIVOR FUNCTION TO THE TRADITIONAL LIFE TABLE FUNCTIONS

In Chapter 4, we denoted the number of people out of an original l_0 births who survive to a particular exact age x by the symbol l_x. The proportion of the l_0 births who survive to exact age x is therefore just l_x/l_0. This proportion can also be viewed as the probability that some

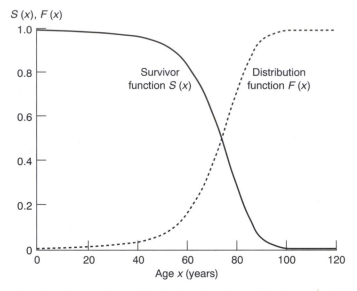

Figure 6.1 The survivor function and the distribution function of the length of life: the survivor function shown is that of English males, 1980–82. Source: Office of Population Censuses and Surveys (1987a, p. 8)

individual will survive to exact age x. For example, if, out of 1000 births, 700 are still alive at exact age 40 years, then the probability that any of these births will survive to exact age 40 years is just 700/1000 or 0.7. Thus we can write that

$$S(x) = \frac{l_x}{l_0}. \tag{6.1}$$

Equation (6.1) shows that survival analysis is very simply related to the traditional approach to analysing life tables. (The function $S(x)$ can also be written another way. Using traditional life table notation, we can define a quantity called $_x p_0$, which is the probability that someone will survive from birth to exact age x. In this notation $_x p_0$ is also equal to l_x/l_0. Thus $S(x)$ is the same as $_x p_0$.) Examples of the survivor function and the distribution function of the variable length of life are plotted in Figure 6.1.

6.4   The probability density function

We are often interested in knowing the probability that the lifetime of a person lies between two specific values of X, say x and $x + dx$ – in other words, the probability that a person will die between exact age x and exact age $x + dx$.

There are two ways of looking at this probability. First, we can look at it from the point of view of a new-born baby. We ask the question: 'What is the probability that a new-born baby will die between exact ages x and $x + dx$?'. To answer this question, note that the probability that the baby will survive to exact age x is $S(x)$, and the probability that the baby will survive to exact age $x + dx$ is $S(x + dx)$. Therefore, the probability that the baby will die between these two ages is just the difference between these two probabilities, or

$$\begin{array}{c} \text{probability that a new-born baby will die} \\ \text{between exact ages } x \text{ and } x + dx \end{array} = S(x) - S(x + dx).$$

Now the probability that a new-born baby will die between exact ages x and $x + dx$ depends on the length of the age interval, dx. Other things being equal, the probability will be greater if dx is greater. Thus, it is helpful to think of the rate of change of the survivor function with age. This rate of change is

$$\frac{S(x) - S(x + dx)}{dx}.$$

We are now in a position to define the *probability density function* of X, $f(x)$. The probability density function is, in effect, the instantaneous rate of change of the survivor function. That is,

$$f(x) = \lim_{dx \Rightarrow 0} \left[\frac{S(x) - S(x + dx)}{dx} \right]$$

$$= - \lim_{dx \Rightarrow 0} \left[\frac{S(x + dx) - S(x)}{dx} \right],$$

so

$$f(x) = \frac{d}{dx} S(x). \tag{6.2}$$

It may be viewed as the probability that a person will die in a very small age interval between exact ages x and $x + dx$. Alternatively, it might be thought of as lives ending per year of age.

6.5 The hazard function

We have just looked at the probability of dying between exact ages x and $x + dx$ from the point of view of a new-born baby. However, of equal interest is this probability *conditional* on survival to age x. We might ask: 'What is the probability that someone will die in the age interval between exact age x and exact age $x + dx$, given that he/she is still alive at exact age x?'.

This may be written as

$$\Pr \left[\begin{array}{c|c} \text{dies between} & \text{alive at} \\ \text{ages } x \text{ and } x + dx & \text{age } x \end{array} \right] = \Pr[x < X \leq (x + dx)| X > x]$$

$$= \frac{\Pr[x < X \leq x + dx]}{\Pr[X > x]}$$

$$= \frac{S(x) - S(x + dx)}{S(x)}.$$

Again, this probability will depend on the length of the age interval, dx. So it is helpful to consider the rate of change of this probability with age, or

$$\frac{S(x) - S(x + dx)}{S(x) \, dx}.$$

The *hazard function* is simply the instantaneous rate of change of this conditional probability of death. It is denoted by the symbol $h(x)$. In other words, the hazard function at age x is a measure of the *intensity* of the mortality in the very small interval dx between exact ages x and $x + dx$. It is defined by the equation

$$h(x) = \lim_{dx \Rightarrow 0} \left[\frac{S(x) - S(x+dx)}{S(x)\,dx} \right]$$

$$= -\lim_{dx \Rightarrow 0} \left[\frac{S(x+dx) - S(x)}{S(x)\,dx} \right]$$

$$= -\frac{1}{S(x)} \lim_{dx \Rightarrow 0} \left[\frac{S(x+dx) - S(x)}{dx} \right]. \tag{6.3}$$

Thus

$$h(x) = -\frac{1}{S(x)} \frac{d}{dx} S(x). \tag{6.4}$$

THE HAZARD FUNCTION AND THE FORCE OF MORTALITY

In Section 4.4, we defined a quantity known as the force of mortality. This is a measure of the intensity of the mortality at any exact age x. In the traditional life table notation, the force of mortality at age x, μ_x, is defined by the equation

$$\mu_x = \lim_{dx \Rightarrow 0} \left[\frac{l_x - l_{x+dx}}{l_x\,dx} \right]. \tag{6.5}$$

Now, using equation (6.1), we can rewrite equation (6.5) using our new notation for the survivor function, $S(x)$:

$$\mu_x = \lim_{dx \Rightarrow 0} \left[\frac{l_0 S(x) - l_0 S(x+dx)}{l_0 S(x)\,dx} \right]$$

$$= \lim_{dx \Rightarrow 0} \left[\frac{S(x) - S(x+dx)}{S(x)\,dx} \right]. \tag{6.6}$$

Comparing equations (6.3) and (6.6) reveals that the hazard function is exactly the same as the force of mortality.

6.6   The relationships between the three functions

Equation (6.4) shows that the hazard function is related to the survivor function by the equation

$$h(x) = -\frac{1}{S(x)} \frac{d}{dx} S(x). \tag{6.7}$$

Equation (6.2) describes the relationship between the probability density function and the survivor function

$$f(x) = -\frac{d}{dx} S(x). \tag{6.8}$$

It is not difficult to see that substituting from equation (6.8) into equation (6.7) gives

$$h(x) = \frac{1}{S(x)} f(x). \tag{6.9}$$

Equations (6.7), (6.8) and (6.9) show that the survivor function, the probability density function and the hazard function all contain the same information about the mortality experience of the birth cohort. If any one of these is known, the other two can be worked out. This must be so because, from equation (6.8) above,

$$f(x) = -\frac{d}{dx} S(x).$$

Integrating this equation produces

$$S(x) = 1 - \int_0^x f(u)\,du.$$

Further, equation (6.7),

$$h(x) = -\frac{1}{S(x)} \frac{d}{dx} S(x),$$

may be rewritten as

$$h(x) = -\frac{d}{dx}[\ln S(x)]. \tag{6.10}$$

By integrating equation (6.10), we obtain

$$\ln S(x) = -\int_0^x h(u)\,du,$$

so that

$$S(x) = \exp\left[-\int_0^x h(u)\,du\right].$$

6.7 Censoring

The final principle we need to look at before we can apply survival analysis to the analysis of mortality is the idea of censoring. We shall illustrate this using two examples of mortality investigations.

EXAMPLE 1

Consider Figure 6.2. This Lexis chart depicts an investigation of mortality in which we follow a set of people from birth until death. Unfortunately, some people withdraw from the investigation while they are still alive. In the previous chapter, this situation was analysed using a double-decrement life table. A feature of survival analysis is that withdrawals may be accommodated within the same general framework as deaths.

What do we know about the people who withdraw? Well, assuming their age at withdrawal has been recorded, we know that they were still alive at that age. Thus, we know that their length of life was greater than their age at withdrawal. Because they withdraw, we do not observe their exact age at death.

People who withdraw from the investigation while they are still alive are said to be *censored*. The word 'censored' is used because part of their life is hidden from our view (the part following their withdrawal from the investigation) – we do not observe it.

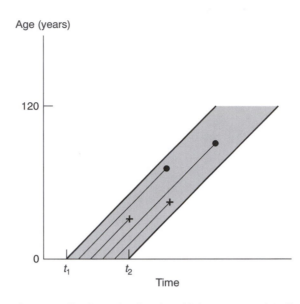

Figure 6.2 Illustration of a mortality investigation in which some people's lives are censored: the cohort being studied consists of all persons born between times t_1 and t_2 (dots denote deaths, crosses denote times of censoring)

EXAMPLE 2

In the previous example, we assumed that we followed people through until either they died or they withdrew from the investigation while still alive. However, there are other situations in which censoring happens. Consider the investigation depicted in Figure 6.3. Here we follow a group of people through from their 40th birthday for the next 10 years of their life only. During this period, one of three things may happen to these people:

1   they die;
2   they withdraw from the investigation while they are still alive;
3   they survive until their 50th birthday, at which age they are still in the investigation.

We can treat a person who survives until his or her 50th birthday as censored, because all we know about that person is that he/she survived until at least age 50 years. We do not continue to observe people who are older than 50 years. Thus, in this investigation, people in categories **2** and **3** in the list above are censored.

Thus in any investigation of mortality, there may be individuals whose exact age at death is unknown, but of whom it is known that they survived until at least some particular age. These individuals' lives are said to be censored.

6.8   The estimation of mortality using survival analysis

In this section, we show how mortality may be estimated by means of survival analysis using a specific example. The example is very similar in form to Example 2 in Section 6.7 (see Figure 6.3), save that we suppose that a sample of lives has been followed only from exact age 40 years to exact age 41 years. As was pointed out in that example, during this

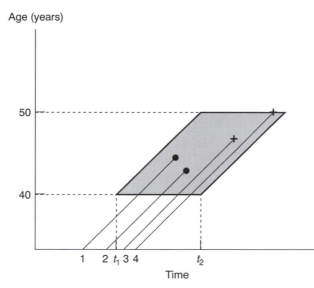

Figure 6.3 Another mortality investigation in which some people's lives are censored. The cohort being studied consists of all persons who celebrated their 40th birthday between times t_1 and t_2. Dots denote deaths, crosses denote times of censoring. Lives 1 and 2 are terminated by death at the times shown with the dots. Lives 3 and 4 are censored, life 3 by withdrawal from the study, and life 4 by reaching exact age 50 years

period one of three things might happen to the individuals whose lives are included in our sample: death; loss to the study while still alive; or survival until exact age 41 years.

For each person who dies before his/her 41st birthday, we have information on that person's exact age at death. Let us also suppose that, for each person who is lost to the study while still alive, we know his/her exact age at withdrawal. We are going to use this information to estimate the force of mortality or the hazard function between exact ages 40 and 41 years.

Before beginning, it will be helpful to redefine the random variable X to be equal to age minus 40 years. In other words, we measure length of life in years since the start of the investigation.

First, we must make an assumption about the variation in the force of mortality in this age interval. This involves making an assumption about the hazard function, $h(x)$. The simplest form which the hazard function can take is where the force of mortality is constant. Thus, we shall assume that the force of mortality between exact ages 40 and 41 years is constant, and denote its value by λ. In symbols, we write that

$$h(x) \; (\equiv \mu_x) = \lambda,$$

where $0 < x \le 1$ (because x measures years since the 40th birthday). Our task is to estimate λ using the data we have available to us. To do this we use the method of *maximum likelihood*.

MAXIMUM LIKELIHOOD ESTIMATION OF THE HAZARD FUNCTION

Consider one of the people in our investigation. As we have shown, one of three things might happen to him/her:

1   Death before his/her 41st birthday. If this happens, we know his/her exact age at death. Suppose that the age at death of the ith person is x_i years. What is the probability that the ith person will die at exactly x_i years of age? It is equal to the probability density $f(x_i)$.

2   Withdrawal from the study for other reasons before his/her 41st birthday. If this happens, then all we know about the person is that he/she was still alive when he/she withdrew. So, if the exact age of person i was x_i when he/she withdrew, then the probability of his/her being still alive at this age is just $S(x_i)$, the value of the survivor function at exact age x_i.

3   Survival until his/her 41st birthday. The probability of this happening is $S(1)$.

Now we can combine **2** and **3** by noting that if the person does not die, then all we know is that he/she survives until some age x_i years, where

$$x_i = \begin{cases} \text{age at withdrawal} & \text{if he/she withdraws} \\ 1 & \text{if he/she survives until his/her 41st birthday} \end{cases}. \quad (6.11)$$

So the probability of either **2** or **3** happening is just $S(x_i)$, where x_i is defined by equation (6.11).

Now suppose we study a sample of n people, and we suppose that they are all subject to the same underlying force of mortality, λ. What we want to do is to find the most *likely* estimate of λ, given the data that we have.

If we only had data on a single person, i, who died at exact age x_i, the most likely value of λ would be the value which maximized $f(x_i)$, the probability of dying at exact age x_i. Why? Because this value of λ is the one which comes closest to the data that we have.

If we only had data on a single person, i, who was not observed to die (that is, that person's life was censored), the most likely value of λ would be that value which maximized $S(x_i)$, where x_i is defined by equation (6.11). In other words, if we only had data for one person, i, who did not die, then our best estimate of the force of mortality, λ, would be the value which maximized the probability that the person would still be alive at x_i – for this is the value which comes closest to the data that we have. (Clearly, this best estimate is zero, from the point of view of the single person i.)

Of course, data on just one person are not much use for estimating forces of mortality. Generally, in a mortality investigation, we observe a sample of people, say n, and we want to find the most likely value of λ given the data on all the people.

Well, for the people who die, the most likely value of λ will maximize the quantity

$$[f(x_1) \quad f(x_2) \quad f(x_3) \quad \cdots \quad f(x_{n_1})],$$

where there are n_1 people who die between their xth and $(x+1)$th birthdays. In other words, we want to maximize the product of the $f(x_i)$s for all the people who die.

Note that the quantity

$$[f(x_1) \quad f(x_2) \quad f(x_3) \quad \cdots \quad f(x_{n_1})]$$

is just the combined probability that person 1 dies at age x_1, and person 2 dies at age x_2, and person 3 dies at age x_3, and so on. Thus it is the combined probability of the data that we observe actually arising. We are therefore trying to find the value of λ which maximizes the chance that the data we actually observe will be observed. This is the meaning of the term 'maximum likelihood'.

For the people who do not die, the most likely value of λ will maximize the quantity

$$[S(x_1) \quad S(x_2) \quad S(x_3) \quad \cdots \quad S(x_{n_2})],$$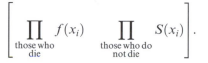

where there are n_2 people who do not die between their xth and $(x+1)$th birthdays ($n_2 = n - n_1$). In other words, we want to maximize the product of the $S(x_i)$s for all the people who do not die.

We can put all this together by saying that, as far as the whole sample of n people is concerned, we want to maximize the quantity

$$\left[\prod_{\substack{\text{those who} \\ \text{die}}} f(x_i) \quad \prod_{\substack{\text{those who do} \\ \text{not die}}} S(x_i) \right].$$

This quantity is rather ugly to write down. We can make the symbols rather more elegant by defining a variable δ_i, such that if person i dies, δ_i is equal to 1, and if person i does not die, δ_i is equal to zero.

Consider now the quantity $[f(x_i)]^{\delta_i}[S(x_i)]^{1-\delta_i}$. If person i dies, then δ_i is equal to 1, so $1 - \delta_i$ is equal to 0. So,

$$[f(x_i)]^{\delta_i}[S(x_i)]^{1-\delta_i} = f(x_i).$$

If person i does not die, then δ_i is equal to 0, so $1 - \delta_i$ is equal to 1. So,

$$[f(x_i)]^{\delta_i}[S(x_i)]^{1-\delta_i} = S(x_i).$$

Thus the product of the function $[f(x_i)]^{\delta_i}[S(x_i)]^{1-\delta_i}$ over all the n individuals in the sample is equal to the quantity

$$\left[\prod_{\substack{\text{those who} \\ \text{die}}} f(x_i) \quad \prod_{\substack{\text{those who do} \\ \text{not die}}} S(x_i) \right].$$

This product over all the individuals in the sample is called the *likelihood*, L. In symbols,

$$L = \prod_{i=1}^{n} [f(x_i)]^{\delta_i}[S(x_i)]^{1-\delta_i}. \tag{6.12}$$

Our problem, then, is to find the value of the hazard, λ, which maximizes the likelihood, L. How can we do it? We proceed in three stages:

1 We express L as a function of λ.
2 We differentiate L with respect to λ and set the derivative equal to zero.
3 We find the second-order derivative of L with respect to λ and check that we do indeed have a maximum.

Stage 1

We know from Section 6.6 above that

$$S(x) = \exp\left[-\int_0^x h(u)\, du \right]. \tag{6.13}$$

Now, using our assumption that the hazard is constant, and equal to λ, equation (6.13) becomes

$$S(x) = \exp\left[-\int_0^x \lambda \, du\right].$$

Performing the integration gives us

$$S(x) = e^{-\lambda x}. \tag{6.14}$$

We also know from Section 6.6 that

$$h(x) = \frac{1}{S(x)} f(x).$$

Rearranging this equation, we obtain

$$f(x) = h(x)S(x). \tag{6.15}$$

Thus, if $h(x)$ is equal to λ, and $S(x)$ is equal to $e^{-\lambda x}$, then, substituting these values into equation (6.15), we obtain

$$f(x) = \lambda e^{-\lambda x}. \tag{6.16}$$

We now use equations (6.14) and (6.16) to substitute into equation (6.12) expressions for $f(x_i)$ and $S(x_i)$ in terms of λ. This produces the following equation for the likelihood:

$$L = \prod_{i=1}^{n} [\lambda e^{-\lambda x_i}]^{\delta_i} [e^{-\lambda x_i}]^{1-\delta_i}. \tag{6.17}$$

Stage 2

We now have to differentiate equation (6.17) with respect to λ. This does not look as if it is going to be easy. It is perfectly possible – but not easy.

However, the value of λ which maximizes the likelihood will be the same value as that which maximizes the logarithm of the likelihood. By the look of equation (6.17), it is going to be much easier to differentiate the logarithm of the likelihood than it is to differentiate the likelihood itself.

Taking natural logarithms of equation (6.17) produces

$$\ln L = \sum_{i=1}^{n} [\delta_i(\ln \lambda) - \delta_i \lambda x_i] + \sum_{i=1}^{n} [(1 - \delta_i)(-\lambda x_i)]$$

$$= \sum_{i=1}^{n} \delta_i(\ln \lambda) - \sum_{i=1}^{n} \delta_i \lambda x_i - \sum_{i=1}^{n} \lambda x_i + \sum_{i=1}^{n} \delta_i \lambda x_i,$$

which simplifies to

$$\ln L = \sum_{i=1}^{n} \delta_i(\ln \lambda) - \sum_{i=1}^{n} \lambda x_i. \tag{6.18}$$

We can now differentiate equation (6.18) with respect to λ to obtain

$$\frac{d(\ln L)}{d\lambda} = \frac{\sum_{i=1}^{n} \delta_i}{\lambda} - \sum_{i=1}^{n} x_i.$$

This is equal to zero if

$$\frac{\sum_{i=1}^{n} \delta_i}{\lambda} = \sum_{i=1}^{n} x_i,$$

or if

$$\lambda = \frac{\sum_{i=1}^{n} \delta_i}{\sum_{i=1}^{n} x_i}. \tag{6.19}$$

Stage 3

The second-order derivative of equation (6.18) with respect to λ is given by the equation

$$\frac{\mathrm{d}^2 (\ln L)}{\mathrm{d}\lambda^2} = -\frac{\sum_{i=1}^{n} \delta_i}{\lambda^2}.$$

Since both $\sum_{i=1}^{n} \delta_1$ and λ^2 are positive, $\mathrm{d}^2(\ln L)/\mathrm{d}\lambda^2$ must be negative, and we have a maximum. (It is, in principle, possible that $\sum_{i=1}^{n} \delta_i$ is zero. This will only happen, however, if there are no deaths at all. In such a case, the maximum likelihood estimate of λ is zero.)

THE INTERPRETATION OF THE MAXIMUM LIKELIHOOD ESTIMATE OF THE HAZARD

Look again at equation (6.19). The numerator of the right-hand side of this equation is simply the total number of deaths during the age interval. The denominator is the total length of time during this age interval for which the people in the sample are exposed to the risk of dying. Thus, we have

$$\begin{array}{c} \text{maximum likelihood estimate} \\ \text{of the hazard} \end{array} = \frac{\text{total number of deaths}}{\begin{array}{c}\text{total length of time exposed to} \\ \text{the risk of dying}\end{array}},$$

which is intuitively sensible. It also accords with the principle of correspondence.

Note that, in calculating the total length of time for which the people in the sample are exposed to the risk of dying, we have included the censored cases: the people who are not observed to die. Moreover, we have used the exact length of time for which each person is exposed to the risk of dying, making no assumption about deaths being evenly distributed across the year of age. Thus, we have used all the information we have available.

6.9 Using survival analysis to estimate a life table

So far, we have only discussed the estimation of the hazard, or the force of mortality, for a single year of age (between exact ages 40 and 41 years). However, we can use the same principle to estimate the hazard, or the force of mortality, for all ages. We just apply the method described in Section 6.8 to each single year of age.

By applying the method to each single year of age between birth and (about) 120 years, we can obtain estimates of the hazard of death during every single year of age. These can be used to estimate the survivor function at each exact age x, by repeatedly applying equation (6.14). To illustrate this, suppose we have estimated the values of the hazard at every age last birthday. Denote the hazard at age 0 last birthday by the symbol λ_0, the hazard at age 1 last birthday by λ_1, and so on. Then, using equation (6.14), we can calculate the value of the survivor function at exact age 1 year, $S(1)$:

$$S(1) = e^{-\lambda_0 \cdot 1} = e^{-\lambda_0}.$$

The value of the survivor function at exact age 2 years is then given by the equation

$$S(2) = S(1) e^{-\lambda_1} = e^{-\lambda_0} e^{-\lambda_1},$$

and the value of the survivor function at exact age x is given by the equation

$$S(x) = e^{-\lambda_0} e^{-\lambda_1} \ldots e^{-\lambda_{x-1}}.$$

This gives us our life table, since the values of $S(x)$ may be related to the l_x values of the life table using equation (6.1). Note that survival analysis may be applied either to mortality data relating to a cohort of people, or to data derived by studying groups of people who celebrate their xth birthdays during the same time period (where x ranges from 0 up to the last birthday celebrated by persons with the longest lives), and following each such group for a period of 1 year. In the latter case, the resulting life table refers to a particular period.

6.10   Advantages of survival analysis

To conclude, let us list some advantages which survival analysis has over the traditional approach to the life table.

First, censored cases (such as withdrawals) may be incorporated into the analysis without bothering about double decrements, dependent and independent rates, and so on.

Second, we use all the information available to us, specifically that about the exact ages at death or censoring.

Third, the maximum likelihood method is generalizable. Suppose, for example, that we did not like the assumption that the hazard (or force of mortality) is constant over each single year of age. This is not a problem, for one or both of two things can be done.

1   Use shorter age intervals, and assume the hazard to be constant within each of the shorter age intervals.
2   Use a different assumption about the hazard (or the force of mortality) within each age interval. For example, we could assume that the hazard varies within each age interval according to some function such as

$$h(x) = \alpha + \beta x.$$

We can then calculate $S(x_i)$ and $f(x_i)$ from this expression, and substitute these values into equation (6.12). Partially differentiating the resulting equation with respect to α and β, and setting the partial derivatives equal to zero, produces maximum likelihood estimates of α and β. Of course, the algebra becomes more complicated, but the method is the same in principle.

This shows that the method of maximum likelihood is very powerful and flexible indeed.

Further reading

This chapter provides only a basic introduction to the principles of survival analysis, and its use to estimate mortality. Survival analysis is the subject of a number of excellent textbooks. Elandt-Johnson and Johnson (1980) and Lee (1980) are fairly elementary introductions. Kalbfleisch and Prentice (1980) is a more advanced treatment. Kleinbaum (1996) is designed as a self-learning text, with an emphasis on the analysis of real data.

All of these books, however, require some prior knowledge of statistical theory. Of the many textbooks providing introductions to statistical theory, DeGroot (1986) and Larson (1982) are recommended. Mood *et al.* (1974) provide a more comprehensive but somewhat terser treatment.

Exercises

6.1 A life insurance company carried out a mortality investigation. It followed a sample of n policyholders, observing them from their 40th birthday until either they died, or they withdrew from the investigation while still alive, or they celebrated their 45th birthday (whichever of these events occurred first). Table 6E.1 gives some data about 20 individuals in the investigation.

(a) Illustrate the investigation using a Lexis chart and the data for individuals 1, 3 and 7.

(b) Assuming that the force of mortality does not vary with age between exact ages 40 and 45 years, find the maximum likelihood estimate of this constant force of mortality, given the data on the 20 individuals in the table.

6.2 Look again at the data on the small furry animals described in Exercise 5.3 (Table 5E.2). Assuming that the force of mortality is constant within each year of age (but that it varies between years of age), calculate this force of mortality at ages 0, 1, 2 and 3 years last birthday.

Table 6E.1

Person number	Last age at which person was observed (years)	Fate of person
1	41.0	Died
2	42.0	Died
3	45.0	Survived
4	45.0	Survived
5	40.5	Withdrew
6	41.0	Withdrew
7	44.0	Withdrew
8	45.0	Survived
9	45.0	Survived
10	44.5	Died
11	44.0	Died
12	43.0	Died
13	42.0	Withdrew
14	45.0	Survived
15	42.5	Died
16	40.5	Died
17	41.0	Withdrew
18	42.0	Withdrew
19	43.0	Withdrew
20	43.0	Withdrew

6.3   Suppose that a group of people are subject to a hazard of death which changes linearly with age x:

$$h(x) = \alpha + \beta x.$$

Derive expressions for the survivor function, $S(x)$, and the probability density function, $f(x)$, in terms of α and β.

6.4   The Gompertz mortality 'law', which is known to describe human mortality at ages above 30 years quite well, supposes that the hazard of death (or the force of mortality) is given by the equation

$$h(x) = \beta \gamma^x.$$

Derive two simultaneous equations which may be solved to obtain maximum likelihood estimates of β and γ.

6.5   Assuming that the Gompertz model holds (see Exercise 6.4), show that, if the chance of surviving from exact age 60 years to exact age 70 years is double that of surviving from exact age 60 years to exact age 80 years, then

$$\beta = \ln 2 \left(\frac{\ln \gamma}{\gamma^{10}(\gamma^{10} - 1)} \right).$$

7

The Analysis of Marriage

7.1 Introduction

In this chapter we turn to analyse marriage. Although marriage is not one of the basic components of demographic change listed in Chapter 1, there are good reasons why demographers consider the analysis of marriage patterns important. The principal of these is that marital status is one of the most important sources of population heterogeneity: people may be classified as single (that is, never married), married, divorced or widowed, and the behaviour of these marital statuses with respect to the basic components of population change is very different. This is most obviously seen in relation to fertility. The fertility of currently married women is usually much greater than the fertility of women in other marital statuses. Indeed, in some populations, fertility is principally determined by the proportion of the female population that is married – this was notably the case in England in the seventeenth and eighteenth centuries (Wrigley and Schofield, 1989). Marital status also affects mortality: typically the mortality of married persons is lower than that of unmarried persons, after controlling for age compositional differences (Office of Population Censuses and Surveys, 1987a).

Demographers use the term *nuptiality* to refer to the propensity of a population to get married. Populations in which nuptiality is high tend to have a large proportion of people getting married, and doing so at a young age (for example, many contemporary populations in south Asia). In populations where nuptiality is low, marriage typically takes place later in life, and a substantial proportion of the population never marries (a good example would be the English population of the seventeenth century).

The analysis of marriage uses many of the same techniques which we have described in the previous few chapters in the context of mortality analysis. However, as will be seen, it raises some new issues. These new issues are, in many cases, also relevant to the analysis of fertility.

In Section 7.2, the marriage process is compared with the mortality process, and its additional complexity illustrated. Section 7.3 discusses the calculation of marriage rates, and shows how the distinction between two types of rate – those in which the denominator is reduced by the event in the numerator, and those in which it is not – is important. Section 7.4 looks at two approaches to analysing nuptiality: that based on analysing the experience of a particular cohort, and that based on analysing the experience of a population during a particular time period. Section 7.5 considers the importance of taking mortality into account when analysing the marriage process.

It is common to use measures of the average age at marriage as single-figure indices of the level of nuptiality in a population. Section 7.6 looks at some of these. Their estimation is hampered in many populations by the absence of registration data on ages at marriage. However, this problem can be overcome by using an indirect method of estimating nuptiality, outlined in Section 7.7. This method leads to an estimate of the mean age at marriage, known as the singulate mean age at marriage. Sections 7.8 and 7.9 consider the analysis of divorce and widowhood, and the question of cohabitation and separation as distinct marital statuses.

7.2   The marriage process

When analysing mortality, it was found helpful to consider the mortality process in terms of states and transitions between states. For example, the life table can be seen as a way of analysing the transition between two states: 'alive' and 'dead'. In Chapter 5, this idea was extended to consider more than one destination state by using multiple-decrement life tables. Although the stress on the use of multiple-state diagrams in the context of the mortality process may have seemed rather excessive, since the mortality process is simple, it was emphasized because of its general usefulness in demography. This usefulness becomes more apparent when the more complex marriage process is considered.

Figure 7.1 shows the marriage process in terms of the possible transitions between five states: 'never married', 'married', 'divorced', 'widowed' and 'dead'. These five states are those which are legally recognized in most societies, and, as a result, those for which official statistics are typically collected.

The first point to notice about the marriage process is that every state may be left in more than one way. This suggests that multiple decrements will be involved in the analysis. The second point to notice is that it is possible to move between certain states in both directions (for example, between the states 'married' and 'divorced').

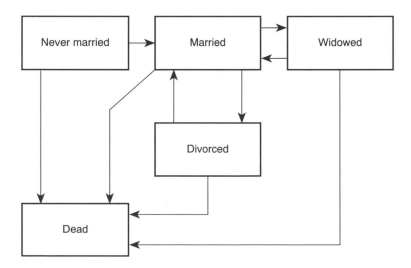

Figure 7.1 The marriage process, showing possible transitions between states

7.3   Marriage rates

In this section we describe the various kinds of marriage rate which may be used to measure the intensity of the transitions shown in Figure 7.1. Because the number of transitions is large, the number of potential rates is large. The description which follows concentrates on *marriage rates*, that is, rates which measure the transition into the 'married' state. The general principles outlined in this section, however, may be applied to rates measuring other transitions (for example, that from the 'married' to the 'divorced' state).

The discussion in this section draws considerably on the excellent summary of marriage rates and the issues surrounding their calculation in Lutz (1993).

THE CRUDE MARRIAGE RATE

We can define a *crude marriage rate* in much the same way as the crude death rate. The crude marriage rate is given by the formula

$$\text{crude marriage rate} = \frac{\text{number of marriages in a given year}}{\text{total mid-year population}}.$$

As with the crude death rate, the crude marriage rate may be multiplied by 1000 to give a rate per thousand of the population.

There are at least four potential limitations of this rate:

1   The total mid-year population includes some persons who are too young to marry.
2   The crude marriage rate will be affected by the age structure of the population in the denominator, since the risk of getting married is not the same at all ages.
3   The crude marriage rate is actually measuring the combined intensity of three transitions in Figure 7.1, since it does not distinguish between first marriages, in which the persons marrying are previously 'never married', and remarriages, involving 'widowed' or 'divorced' persons.
4   The denominator also includes some persons who are already married, and who are therefore not at risk of marrying again (unless polygamy is practised).

THE GENERAL MARRIAGE RATE

The first of these shortcomings can be overcome by replacing the denominator by the total mid-year population of legally marriageable age. For convenience, this may be taken as age 15 years and over (although in some countries an older age may be used, depending on the minimum legal age at marriage). This gives a rate called the *general marriage rate*:

$$\text{general marriage rate} = \frac{\text{number of marriages in a given year}}{\text{total mid-year population aged 15 years or more}}.$$

The general marriage rate is often calculated separately for males and females. If this is done, then the numerator becomes the number of males (or females) marrying in a given year, and the denominator becomes the total mid-year population of males (or females) aged 15 years or more.

AGE-SPECIFIC MARRIAGE RATES

Provided data are available on the ages at marriage of those marrying, age-specific marriage rates may be calculated. There is an additional complication, however, since marriage involves two persons who may not be the same age. In view of this, age-specific marriage rates are defined in terms of persons marrying, rather than marriages, using the formula

$$\text{age-specific marriage rate at age } x \text{ last birthday} = \frac{M_x}{P_x},$$

where M_x is the number of persons marrying at age x last birthday, and P_x is the mid-year population aged x last birthday. If age groups are used, of width n years, then an analogous formula is applied:

$$\text{age-specific marriage rate between exact ages } x \text{ and } x + n \text{ years} = \frac{_nM_x}{_nP_x},$$

where $_nM_x$ and $_nP_x$ are the number of persons marrying, and the mid-year population, between exact ages x and $x + n$ years.

Because the variation of the risk of marriage with age is different for men and women (men typically marry at an older age than women), age-specific marriage rates are usually calculated separately for males and females.

FIRST MARRIAGE AND REMARRIAGE RATES

All the rates so far described refer to all marriages. As has been pointed out, they therefore combine more than one transition. It is often helpful to be able to distinguish first marriages from remarriages. *First marriage rates* measure only the intensity of the transition from the 'never married' state to the 'married' state, whereas *remarriage rates* measure the combined intensities of the transitions from the 'divorced' state to the 'married' state and from the 'widowed' state to the 'married' state. First marriage rates may be defined in the same way as the rates applying to all marriages, save that only first marriages are included in the numerator. However, because many marriages involve one partner marrying for the first time and one partner who has been married before, they are almost always calculated separately for males and females. Thus

$$\text{crude first marriage rate} = \frac{\text{number of males (or females) marrying for the first time in a given year}}{\text{total male (or female) mid-year population}},$$

$$\text{general first marriage rate} = \frac{\text{number of males (or females) marrying for the first time in a given year}}{\text{total male (or female) mid-year population aged 15 years or more}},$$

and

$$\text{age-specific first marriage rate at age } x \text{ last birthday} = \frac{FM_x}{P_x}, \tag{7.1}$$

where FM_x is the number of people marrying for the first time in the given year when they were aged x last birthday.

Let us think a bit more about the age-specific first marriage rate. The first point to note about it is that it is an *m*-type rate. The denominator is an estimate of the average number of persons exposed to the risk of being married for the first time in the given year. However, in an important sense it is not a very good estimate. This is because once a person has married for the first time, he/she can no longer be at risk of again marrying for the first time. The implication of this is that the P_x in the denominator of equation (7.1) becomes a worse and worse estimate of the true population at risk as x increases. At young ages, very few people are already married, and so P_x is quite a good estimate; at older ages, the true population exposed to the risk of marrying for the first time is only a small fraction of the total population.

This problem can be taken into account by modifying the denominator of equation (7.1) so that it only includes 'never married' persons. This modified age-specific marriage rate is defined by the formula

$$\text{age-specific first marriage rate at age } x \text{ last birthday} = \frac{\text{number of persons marrying for the first time at age } x \text{ last birthday}}{\text{mid-year population of never married persons aged } x \text{ last birthday}}.$$

In the context of Figure 7.1, what this rate does is to restrict the exposed-to-risk in the denominator to those persons in the state which forms the origin state for the transition being measured.

The importance of making this restriction becomes magnified when using remarriage rates. This is because at all ages, the number of persons at risk of marrying for a second or subsequent time is a very small proportion of the total population. It therefore makes little sense to use the total population in the denominator for remarriage rates. Instead, we can define an age-specific remarriage rate as

$$\text{age-specific remarriage rate at age } x \text{ last birthday} = \frac{\text{number of persons marrying for the second or subsequent time at age } x \text{ last birthday}}{\text{mid-year population of widowed and divorced persons aged } x \text{ last birthday}}.$$

There are, therefore, two types of marriage rate. The first type uses the whole population in the denominator, while the second type only uses that proportion of the population exposed to the risk of experiencing the particular event in question. The distinction between the two types of rate depends on whether the events in the numerator reduce the population in the denominator. If they do, as in the rates of the second type, we can refer to the rates as *decremental rates*, because the denominator is decreased or decremented by the events in the numerator. If they do not, as in the rates of the first type, we can call the rates *non-decremental* rates.

Notice that, in the context of measuring mortality, the distinction between these two types of rate was not relevant, because a death by definition removes a person from the population. All mortality rates are, therefore, decremental rates.

m-*TYPE AND* q-*TYPE FIRST MARRIAGE RATES*

So far, all the rates described have been *m*-type rates. The question arises as to whether we can construct *q*-type marriage rates analogous to the *q*-type mortality rates discussed in Chapter 2. The answer is that we can, and in practice demographers often do this. The

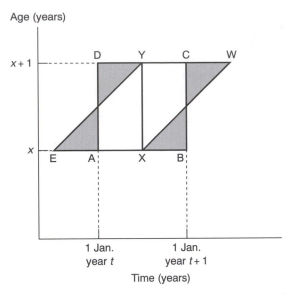

Figure 7.2 Lexis chart illustrating the relationship between *m*-type and *q*-type first marriage rates. The assumptions that the marriage rate between exact ages *x* and *x* + 1 years does not vary with calendar time, and that marriages are evenly distributed within this year of age, mean that the number of marriages represented by each of the shaded areas is the same

relationship between *m*-type and *q*-type rates, however, is different for decremental and non-decremental rates.

In decremental rates, we have already seen in Section 2.6 that, making certain assumptions, a fairly simple relationship between *m*-type and *q*-type rates can be derived in the context of mortality analysis:

$$q_x = \frac{2m_x}{2 + m_x}.$$

By replacing 'death' by 'first marriage', exactly the same relationship can be derived to relate *m*-type and *q*-type first marriage rates. The assumptions necessary to establish this relationship are that first marriages are evenly distributed across each year of age, and that first marriage rates do not vary with calendar time (at least over a limited period surrounding the year to which the *m*-type rates apply).

The case of non-decremental rates can be illustrated using the Lexis chart in Figure 7.2. The *m*-type rate is calculated using the first marriages in the square ABCD in the numerator, and the mid-year population XY in the denominator. A *q*-type rate referring approximately to the same period may be calculated using the first marriages in the parallelogram EXWY in the numerator, and the number of persons attaining exact age *x* years during the year represented by the line EX in the denominator. Making the same two assumptions as before, it is clear that the number of first marriages represented by the parallelogram EXWY is the same as that represented by the square ABCD. Moreover, because the rates are non-decremental, the fact that a person marries for the first time does not remove him/her from the population at risk. As a result, the population XY consists of the same number of persons as does the population EX. This means that the denominators of the *m*-type and the *q*-type rates are also the same. Therefore, for non-decremental rates, the *m*-type and *q*-type rates are the same.

OTHER DECREMENTAL RATES

Decremental rates can be calculated for other transitions shown in Figure 7.1. For example, an *m*-type age-specific rate measuring the remarriage of divorced persons can be calculated using the number of divorced persons remarrying aged *x* last birthday in the numerator, and the mid-year population of divorced persons aged *x* last birthday in the denominator.

7.4 Period and cohort analysis of marriage

Consider again the non-decremental age-specific first marriage rate:

$$\text{age-specific first marriage rate at age } x \text{ last birthday} = \frac{FM_x}{P_x}.$$

Since the *m*-type and *q*-type non-decremental rates are the same, this rate is a measure of the proportion of persons who attain their *x*th birthday and marry for the first time before their $(x + 1)$th birthday. Imagine that we take a birth cohort, follow its members through their lives, and work out the age-specific non-decremental first marriage rate at each age. Given this set of rates, it is possible to add them together over all ages. This is possible because the denominators will be the same throughout. The result is called the *total first marriage rate* (TFMR) for the birth cohort, and its formula is

$$\text{TFMR} = \sum_x \frac{FM_x}{P_x}.$$

Since a person may only marry for the first time once in his/her life, it is easy to see that the TFMR measures the proportion of the birth cohort who ever marry for the first time. By extension, if we restrict the summation to any particular age range, the resulting 'partial TFMR' measures the proportion of persons in the birth cohort who marry for the first time within that age range. For example, the quantity $\sum_{x=0}^{30} FM_x/P_x$ measures the proportion of the birth cohort who marry for the first time prior to their 30th birthday.

Summed cohort first marriage rates like this are very useful in demography as summary measures of the nuptiality experience of particular birth cohorts. A problem with them is that we have to wait many years after the birth of the cohort in order to calculate them. They are, therefore, not much use for assessing the *current* propensity of the population to marry for the first time.

Since the *m*-type and *q*-type rates are the same, however, it is also possible to add together the *m*-type rates calculated using first marriages in a particular year and the mid-year population of that year. The result gives an indication of the proportion who would ever marry for the first time in a hypothetical, or synthetic, cohort of persons who, at each age *x* last birthday, experienced the relevant age-specific first marriage rates applying in the year in question. It is, therefore, a very useful summary measure of the nuptiality of the population in a given period. Clearly, the set of rates in question may not apply to any real people: hence the use of the phrase 'hypothetical cohort', or 'synthetic cohort'.

The TFMR calculated on a period basis, using a synthetic cohort, has the advantage that it is up to date and is a good indicator of current nuptiality. However, it often happens that the result exceeds 1.0. Clearly, this is illogical, in that it implies that the members of the synthetic cohort have, on average, more than one first marriage! It occurs because the period TFMR is sensitive to changes in the timing of marriage. This may be seen in an extreme form by considering two situations.

First, suppose that no marriages take place in a particular year. This will result in a period TFMR of 0. It may well be, however, that the marriages which did not take place in that year will merely be postponed until a later year. In that case, the TFMRs of the *cohorts* who were alive during the year in question may be unchanged by the moratorium on marriages in that year.

Now suppose that the government were to decree that all never married persons were to get married during a particular year (other than a few persons who could not find spouses because of an imbalance in the sexes). In such a case the TFMR for that year would be enormous, but the rate for the next few years would be close to zero.

The critical point is that non-decremental rates do not take into account the previous experience of the population. The denominators are based on the total population, regardless of the proportion of that population really at risk of marrying for the first time. Therefore, the cost of having, in the period TFMR, a very useful measure of current nuptiality is the possibility that the result will be heavily sensitive to short-term changes in the timing of marriage. This point will be taken up again in Chapter 8 in relation to fertility rates.

DECREMENTAL RATES

It is not possible to add decremental first marriage rates together in the same way as non-decremental rates, on either a period or a cohort basis, because the denominators of these rates at different ages are not the same. However, we have seen that decremental first marriage rates are, in principle, just the same as mortality rates. In order to get an overall impression of mortality at all ages, demographers use the life table. Given the formal analogy between decremental first marriage rates and mortality rates, it should clearly be possible to use life table analysis in the same way to study first marriage rates. The resulting tables are called *nuptiality tables* (an example is shown in Table 7.1).

The method of construction of a nuptiality table relating to first marriages follows exactly that described in Chapter 4 for the construction of a life table, save that first marriages are used instead of deaths. All the life table functions have exactly analogous counterparts in the nuptiality table. For example, the life expectation at birth in the life table has as its counterpart the 'life expectation in the never married state'. Note that, since no one will marry for the first time at an age below the minimum legal age at marriage, the counterpart of the l_x function of the life table does not depart from its initial value, l_0, until that age.

One difference between a nuptiality table and a life table is that, whereas everyone ultimately dies, not everyone ultimately marries for the first time. This creates a problem in the calculation of the 'life expectation in the never married state'. Consider the nuptiality table's counterpart to the T_x function of the ordinary life table. In the life table, T_x measures the number of person-years lived between exact ages x and the oldest age to which people survive (about 120 years). In the nuptiality table, the counterpart to T_x measures the number of person-years lived in the never married state at ages greater than exact age x. Because some people never marry, this quantity is, essentially, infinite.

The solution to this problem is to restrict the calculation of the life expectation in the never married state to persons who ultimately marry. Because very few first marriages occur at ages over 50 years, we also usually assume that the number of person-years lived in the never married state at ages over 50 years is zero (that is, using life table notation, by assuming that $L_x = 0$ for all ages $x \geq 50$). From Table 7.1, it can be seen that about 79% of males and 83% of females marry by age 50 years.

Table 7.1 Gross nuptiality table for England and Wales, 1987 ($l_0 = 1000$)

Exact age x	Number not yet having married	
	Males	Females
16	1000	1000
17	1000	997
18	999	987
19	992	959
20	978	911
21	950	845
22	904	765
23	846	686
24	778	609
25	709	542
26	639	482
27	577	432
28	522	389
29	473	353
30	429	321
35	299	231
40	246	195
45	223	178
50	212	169

Source: Office of Population Censuses and Surveys (1989, p. 32).

7.5 Death and marriage combined

The nuptiality table in Table 7.1, and the age-specific first marriage rates from which it is derived, have been calculated without reference to mortality. Consider Figure 7.1 again. The first marriage rates are measuring the transition from the 'never married' state to the 'married' state. However, it is also possible to move from the 'never married' state to the 'dead' state. In truth, therefore, we have a double-decrement situation, with two decrements: first marriage prior to death, and death prior to first marriage. (Note that these two decrements are mutually exclusive and exhaustive: a person must make one, and only one of them.)

NET NUPTIALITY TABLES

This double-decrement situation can be analysed using the methods described in Chapter 5. The resulting double-decrement nuptiality table is called a *net nuptiality table*. A nuptiality table which ignores mortality (as in Table 7.1) is referred to as a *gross nuptiality table*.

A net nuptiality table may be constructed from data on age-specific *m*-type first marriage rates and *m*-type death rates for single people. These form dependent *m*-type rates of decrement. These may then be converted into dependent *q*-type rates of decrement using a formula such as that given in Section 5.8. From these dependent rates, the multiple-decrement life table may be constructed. Independent rates of decrement can be calculated using equations (5.13) and (5.14). Table 7.2 shows an example of a net nuptiality table

Table 7.2 Net nuptiality table for England and Wales, 1987 ($l_0 = 1000$)

Exact age x	Males			Females		
	Deaths of never married persons before next age	Persons marrying for the first time before next age	Survivors in the never married state at age x	Deaths of never married persons before next age	Persons marrying for the first time before next age	Survivors in the never married state at age x
0	15	0	1000	12	0	1000
16	1	0	985	0	3	988
17	1	1	984	0	9	985
18	0	7	982	1	29	976
19	1	14	975	0	47	946
20	1	27	960	0	65	899
21	1	45	932	0	79	834
22	1	57	886	1	78	755
23	0	66	829	0	75	676
24	1	68	763	0	67	601
25	1	68	694	0	59	534
26	0	61	625	1	49	475
27	1	54	564	0	42	425
28	0	48	509	0	36	383
29	1	42	461	0	31	347
30	3	126	418	2	89	316
35	4	51	289	1	35	225
40	4	21	234	2	16	189
45	7	11	209	4	9	171
50			191			158

Source: Adapted from Office of Population Censuses and Surveys (1989, p. 32).

corresponding to the gross nuptiality table shown in Table 7.1. One of the features of this table is that mortality is very low throughout most of the age range in which first marriages take place. The result is that the independent and dependent marriage rates are very similar to one another.

MORTALITY AND NON-DECREMENTAL RATES

The introduction of death into the scenario also creates complications for the non-decremental rates. In fact, it turns out that, if we take into account mortality, then the m-type and q-type rates are no longer the same. It is intuitively clear that this must be true, because death is by definition a decremental rate, and so the combined 'death and first marriage' rates cannot be non-decremental. This point is illustrated further in Chapter 8. Suffice it to say for the moment that demographers get around this problem by assuming that mortality is zero when using non-decremental marriage rates. Only by doing this can the useful additive property of the non-decremental rates be preserved.

7.6   The average age at marriage

When nuptiality in many populations is to be compared, it is helpful to have single-figure indices. In previous sections, a number of such indices have been described, including the crude marriage rate, the general marriage rate and the total first marriage rate. Although these indices all have their uses as estimates of nuptiality, none of them gives any indication of the distribution of ages at marriage.

One of the principal reasons for analysing nuptiality is its implications for understanding fertility, and, in this context, as will be seen in Chapter 10, it is the distribution of ages at marriage which is of importance.

There are a number of possible single-figure indices which give an indication of the distribution of ages at marriage. One obvious candidate is the *mean age at first marriage*. Because not everyone ultimately marries, the mean age at marriage can only be calculated for those people who marry. It may be either calculated directly, using information on age at marriage provided in vital registration data, or estimated from a nuptiality table by taking an age at which everybody is still single (for example, age 16 years in Tables 7.1 and 7.2) and adding that age to the remaining life expectation in the single state at that age – that is, using the formula

$$\text{mean age at first marriage} = 16 + e_{16},$$

where e_{16} is estimated from the nuptiality table.

A shortcoming of the mean age at first marriage is that it is prone to distortion by, for example, a relatively small number of persons who marry for the first time at very old ages. Put another way, the distribution of ages at first marriage is often (though not always) positively skewed. Under such circumstances, the *median age at first marriage* provides a convenient alternative. The median age at first marriage is simply the age by which half the population have ever married. It may be calculated either using only those who ultimately marry, or using the whole population, since those who never marry can be treated as if they marry for the first time at some very advanced age. The median age at first marriage can be read from a nuptiality table as that age at which the counterpart of the l_x column of the life table is equal to 50% of whatever value of l_0 is used in the construction of the nuptiality table. Thus, using the data in Table 7.1, for example, the median age at marriage for males in England and Wales in 1987 was about 28.5 years.

7.7   The analysis of marriage using current status data

So far, this chapter has focused on estimating marriage rates using the conventional type of demographic data: in the numerator the events of interest, and in the denominator some measure of the population exposed to the risk of experiencing the events of interest. Such data are widely available in developed countries. In these countries, censuses ask questions about current marital status which enable the denominators to be calculated, and vital registration systems collect the necessary details to enable the numerators to be calculated.

In countries where vital registration schemes are absent or poorly developed, however, it is not possible to obtain accurate estimates of the numerators of the rates described so far. However, it is usually the case that data on the population classified by marital status at some point in time are available, either from censuses or from surveys. Table 7.3 displays some data from the west African country of Mali, obtained from a nationally representative sample survey in 1995–96.

Now consider the single transition between the two states 'never married' and 'married'. These data reveal the state of play in this transition, according to age, at a single point in time. The column headed 'never married' in Table 7.3 shows the percentage of males in each age group who have yet to make the transition.

Table 7.3 Males in Mali classified by marital status, 1995–96

Age group	Percentage of men		
	Never married	Married	Widowed or divorced
10–14	95.4	4.1	0.7
15–19	71.1	26.2	2.7
20–24	32.1	64.1	3.7
25–29	17.1	81.3	1.6
30–34	1.7	96.8	1.4
35–39	1.7	97.5	0.7
40–44	0.4	97.9	1.7
45–49	0.7	96.0	3.2
50–54	0.0	98.5	1.4

Source: Coulibaly *et al.* (1996, p. 88).

Such data are called *current status data*. They may be obtained with respect to any process which involves a single transition being made by different persons at various ages (see Figure 7.3).

Clearly, the persons in the population from which the data have been collected are subject to a set of (unobserved) age-specific first marriage rates. Each age group for which the current status data have been collected may be considered as a birth cohort, and the observed proportions never married in each age group reflect the operation of that set of age-specific first marriage rates up to the ages at which that cohort is observed at the time of the census or survey. If we are prepared to make the assumption that the cohort age-specific first marriage rates are not changing over time, then it is possible to use the current status information to estimate the relevant set of age-specific first marriage rates.

This must be so because, assuming constant age-specific rates across cohorts, the curve depicted in Figure 7.3 can be interpreted as a survivor function of a random variable, length of time spent in the never married state. At any exact age, x, it is measuring the proportion of persons whose first marriages take place at ages older than x. Interpreting the curve in Figure 7.3 in this way, we showed in Chapter 6 that we can estimate other characteristics of the distribution of ages at first marriage, in particular the probability density function, $f(x)$, and the hazard function, $h(x)$.

In this context, the probability density function, $f(x)$, represents the chance that a person will marry for the first time within a small age interval between exact ages x and $x + dx$. The hazard function denotes the chance that a person will marry for the first time within the corresponding small interval given that he/she has not married prior to reaching exact age x. Notice that $f(x)$ has an analogy in the non-decremental rates discussed in Section 7.3 above, and that $h(x)$ has an analogy in the decremental rates.

Now, in practice, real current status data tend not to be as well behaved as the survivor function plotted in Figure 7.3. Various difficulties are typically present:

1 We tend to have data in age groups, rather than in infinitely small subdivisions of age.
2 The age-specific rates usually are changing both across cohorts and across time periods.

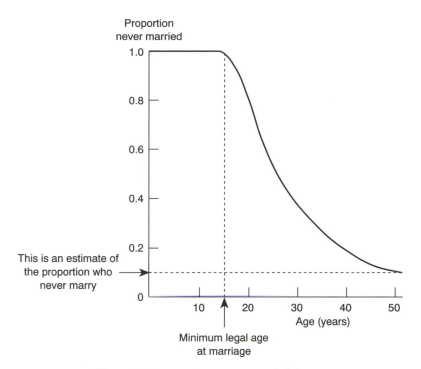

Figure 7.3 Proportions never married, by age

3 There is some sampling error in survey data.

All of these make the analysis of current status data quite complicated in practice.

THE SINGULATE MEAN AGE AT MARRIAGE

Although the problems posed by current status data mean that using them to estimate a set of age-specific rates is quite difficult, they may be used to estimate a number of simpler quantities quite easily. In particular, they may be used to estimate the mean age at first marriage. The method which is used to estimate this quantity from current status data was first proposed by Hajnal (1953), and is known as the *singulate mean age at marriage* (SMAM).

In order to calculate the SMAM we require data on the proportions never married by age (five-year age groups are normally used) over ages ranging from the youngest age at which marriage takes place to an age beyond which very few first marriages take place. Normally this range is from 15 to 50 years, but other limits may be used (in the Malian data in Table 7.3 a range of 10–50 years is more appropriate).

The SMAM may be defined, in words, as the average number of years lived in the single state by those who marry prior to age 50 years. Therefore, it is clear that

$$\text{SMAM} = \frac{\text{number of person-years lived in single state by those who marry before age 50}}{\text{number of persons marrying before age 50}}.$$

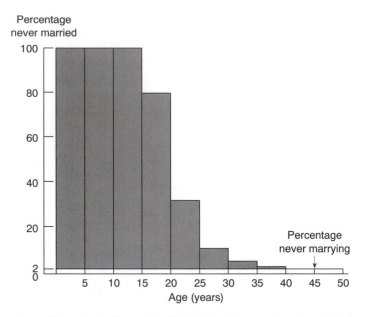

Figure 7.4 Illustration of the calculation of the SMAM for Kenyan females, 1989. Source: National Council for Population and Development (1989 p. 9).

The calculation of the SMAM may be illustrated by the diagram in Figure 7.4. This shows the proportion of single females by five-year age groups in Kenya in 1989.

Figure 7.4 has age on the horizontal axis, and the percentage of women on the vertical axis. If we imagine that we are dealing with a population of exactly 100 women, then areas in Figure 7.4 are proportional to numbers of woman-years. The shaded area represents the numerator of the SMAM, and the height of the line denoting the percentage (which, given a population of 100, is the same as the absolute number) of women never marrying by age 50 years, subtracted from 100, is the denominator of the SMAM.

We therefore simply need to calculate the area of the shaded section of Figure 7.4. If we denote the number ever marrying by age 50 by the symbol E, then it is clear that this area is given by the following sum:

$$\begin{matrix} \text{number of woman-years lived} \\ \text{in the single state by those} \\ \text{who marry before age 50 years} \end{matrix} = 1500 + \sum_{i=15-19}^{i=45-49} 5 \times \begin{matrix} \text{number never} \\ \text{married in} \\ \text{age group } i \end{matrix} - 50(100 - E).$$

This quantity, divided by E, gives the SMAM.

The only remaining question which arises is how to calculate E, the percentage ever married by age 50. Usually, E is taken to be the average of the percentages ever married in the age groups 45–49 years and 50–54 years. In some cases, however (notably in Demographic and Health Survey data), women aged 50 years and over were not interviewed. In these cases, the percentage ever married by exact age 50 years has to be estimated from the percentages never married at ages 40–44 and 45–49. In the case of Kenyan females in 1989, only 2% of women in the age groups 40–44 years and 45–49 years were never married, so it seems reasonable to conclude that 98% of women had married for the first time by their 50th birthday.

Therefore, for Kenyan females in 1989 we have

$$\begin{array}{l}\text{number of woman-years lived in}\\ \text{the single state by those who}\\ \text{marry before age 50 years}\end{array} = (15 \times 100)$$

$$+ 5(80 + 32 + 11 + 5 + 3 + 2 + 2) - 100$$

$$= 1500 + 675 - 100$$

$$= 2075$$

and

$$\text{SMAM} = \frac{2075}{98} = 21.2.$$

With certain applications, one or two complications can arise. Because of sampling errors, or because of changes in the age-specific marriage rates across cohorts, it sometimes happens that the percentages never married increase with age. This clearly violates the assumption we have made that the current status data reflect the experience of a hypothetical cohort of persons. An example is seen in Table 7.3, where the percentage of males never married in the age group 45–49 years is greater than that for the age group 40–44 years. A simple solution to the problem is to amalgamate the adjacent age groups which violate the assumption. Thus, using the data in Table 7.3, we would combine the age groups 40–44 years and 45–49 years, work out the percentage never married in the pooled age groups, and use this in the calculations. We can then consider the two age groups as both having the same percentage never married, and proceed to calculate the SMAM as before.

Sometimes it happens that data on marital status are missing for some persons. We do not know whether or not they have ever married. One possible approach in this case is to calculate the SMAM twice, once assuming that all the persons for whom data are missing have married, and once assuming that none of them has married.

It should be noted that using the 15th birthday as the minimum age at which first marriages take place and the 50th birthday as the maximum age at which first marriages take place is simply a matter of convention. Other ages could be used, and the details of the calculation modified accordingly. Similarly, five-year age groups need not be used in the calculations. Age groups of any width can be used. If wider age groups are used, the SMAM will tend to be less accurate as a guide to the true mean age at marriage. Using a larger number of narrower age groups will give a more accurate approximation, but there is a greater chance that adjacent age groups will have to be pooled in order to ensure that the proportions never married do not increase with age. Five-year age groups have been found to provide a good compromise.

Finally, the assumption that age-specific marriage rates are constant across the cohorts whose experience is reported in the current status data must be emphasized. Although the SMAM is reasonably robust to slow change, if marriage patterns are changing rapidly, then it is not such a useful measure. An extension to the SMAM which may be used in such circumstances is described in United Nations (1983, pp. 225–229).

7.8 The analysis of other transitions in the marriage process

So far, we have been considering only the analysis of transitions out of the 'never married' state to the states 'married' or 'dead'. However, interest frequently centres on the other transitions in Figure 7.1. For example, demographers are often interested in divorce rates or widowhood rates.

When analysing first marriage rates, or, more accurately, the transition out of the 'never married' state, it is clear that initially everyone is at risk of making the transitions in question. Put another way, everyone is born in the 'never married' state. Of course, as transitions take place, the population at risk decreases. Nevertheless, this can be handled readily in the analysis, as we have seen.

When analysing divorce or widowhood rates, it is not true that every member of the population is, at some stage, at risk of experiencing the events of interest. For example, people who never marry will never be at risk of experiencing a divorce. When analysing transitions of this kind, the appropriate method to use and the appropriate rates to calculate are determined by the precise question in which the analyst is interested. Broadly speaking, these questions can be classified into two groups.

Sometimes interest is focused on the incidence of the event in the population as a whole. In such a case, an appropriate measure would be a rate calculated with the number of events in the numerator, and the whole population in the denominator (Lutz, 1993). For example, an age-specific divorce rate could be calculated, for a given year, as

age-specific divorce rate

$$= \frac{\text{total number of divorces to persons aged } x \text{ last birthday}}{\text{mid-year population aged } x \text{ last birthday}}.$$

These rates, which are non-decremental, may then be summed to give a total divorce rate.

Often, however, interest focuses not on the event itself, but on the length of time a person spends in the original state. For example, one might be interested in the stability of marriages. In this case, the population at risk consists only of married people, and the events of interest are those events which cause people to leave the married state. There are three such events: divorce, widowhood and death (Figure 7.1). In this case, an appropriate method of analysis is to create what are called marital dissolution tables.

MARITAL DISSOLUTION TABLES

Marital dissolution tables are simply multiple-decrement life tables in which there are three decrements: divorce, widowhood and death. However, instead of being constructed from age-specific rates, they are constructed from rates *specific to the duration of marriage*. In other words, rather than using age itself to formulate the specific rates, we use the 'age of the marriage'. The data required for constructing these tables are rather detailed. For example, we require the number of divorces, deaths of spouses and deaths themselves classified by how long the person experiencing the event has been married. Details of their construction are given in Preston (1975).

7.9   Cohabitation and separation

In Figure 7.1, the marriage process is depicted using only those marital statuses which are legally recognized in most countries. However, in some countries other marital statuses exist and may even be legally recognized. In recent years it has become much more common than formerly for couples in developed countries to live together, often on a long-term basis, without being legally married. Such couples are said to be *cohabiting*.

The mirror-image of cohabitation is the situation when people who are legally married are, in fact, living apart. Such married couples are said to be *separated*. Apart from its sociological importance, the extent to which legal marriage is not a good estimate of the prevalence of sexual unions has implications for the analysis of fertility (see Chapter 10 for more discussion of this).

Including the additional states of cohabitation and separation in the marriage process makes it considerably more complex (see Exercise 7.7). The difficulty of incorporating cohabitation and separation into the analysis of the marriage process is compounded in most societies by a lack of accurate data on the prevalence of these states. Normally, censuses do not ask questions about these statuses, and special surveys are necessary to elicit the required information. Nevertheless, in principle, the methods of analysis described in this chapter may be applied.

Further reading

The singulate mean age at marriage was first proposed in Hajnal (1953) and is also described in detail in United Nations (1983, pp. 225–229). Readers interested in the analysis of current status data should consult Diamond and McDonald (1992).

Exercises

7.1 The data in Table 7E.1 relate to certain Latin American countries. Use them to calculate crude marriage rates.

7.2 Table 7E.2 shows the number of never married males and females of various ages in England and Wales in 1991, and the number of persons marrying in that year in each

Table 7E.1

Country	Total population (thousands)	Number of marriages
Mexico, 1986	81 200	579 895
Paraguay, 1987	2 270	17 741
Uruguay, 1987	3 058	22 728

Table 7E.2

Age group	Never married population		Number of persons marrying for the first time	
	Males	Females	Males	Females
16–19	1 326 324	1 255 983	4 630	17 704
20–24	1 628 987	1 425 269	74 378	103 689
25–29	1 060 650	777 869	91 675	72 523
30–34	506 547	333 376	34 560	21 000
35–39	275 695	168 780	10 252	5 785
40–44	209 217	117 324	3 998	2 075
45–49	141 150	78 021	1 520	911

Sources: Office of Population Censuses and Surveys (1993a, p. 64; 1993b, pp. 10, 12).

Table 7E.3

Age group	Percentage never married	
	Males	Females
15–19	99.6	98.6
20–24	89.1	77.7
25–29	58.6	42.9
30–34	33.2	21.7
35–39	20.8	12.5
40–44	13.7	8.4
45–49	10.9	6.5
50–54	10.0	7.0

Source: Office of Population Censuses and Surveys (1992, p. 51).

age group. Use them to calculate age-specific first marriage rates. Are the rates you have calculated decremental or non-decremental?

7.3 Use the data in Table 7.1 to calculate the remaining expectation of life in the never married state at each age x for persons who ultimately marry.

7.4 Use the net nuptiality table in Table 7.2 to calculate:
   (a) the dependent q-type first marriage and 'death in the single state' rates in each age group;
   (b) the independent q-type first marriage rates in each age group (that is, the first marriage rates assuming no mortality).

7.5 Calculate the SMAM for males in Mali in 1995–96, using the data in Table 7.3.

7.6 Table 7E.3 gives percentages never married for males and females in the English county of East Sussex in 1991. Use them to calculate the SMAM for males and females.

7.7 Draw a multiple-state representation of the marriage process similar to Figure 7.1, but incorporating the additional states 'cohabiting (but not legally married)' and 'legally married, but separated'.

8

The Measurement of Fertility

8.1 Introduction

This and the next three chapters will be concerned with the analysis of fertility, or the propensity of the women in a population to bear children. Many of the ideas already developed in the context of the analysis of mortality and nuptiality can be applied to the analysis of fertility. In two respects at least, however, the analysis of fertility presents more difficulties than the analysis of mortality.

1 Whereas people only die once, women may have several children. Thus fertility involves the analysis of a repeated event.
2 Whereas death is normally determined by physiological and medical factors which are only influenced to a limited degree by human action (in other words, people do not usually choose to die), fertility in most populations is subject to individual choice (although the precise extent to which individual choice governs fertility varies widely).

This chapter describes some of the ways in which demographers measure fertility. In Sections 8.2 and 8.3, some elementary fertility measures are described. Section 8.4 discusses standardization in the context of fertility rates. In Section 8.5 a widely used measure of fertility called the total fertility rate is defined.

As with the measurement of mortality, fertility may be analysed from a period perspective (births in a given time period) or from a cohort perspective (births to a group of women born within a particular time period). Sections 8.6–8.8 explore the advantages and disadvantages of these two perspectives.

8.2 Some simple single-figure indices of fertility

As with the measurement of mortality, it is convenient to have single-figure indices of fertility to facilitate comparisons between different populations.

THE CRUDE BIRTH RATE

The simplest single-figure index is the *crude birth rate*. This is equal to the number of births occurring in a given population in a year, divided by the total mid-year population. In symbols, denoting the crude birth rate in year t by b_t, the total number of births in year

t by B_t, and the total population on 30 June in year t by P_t, we have

$$b_t = \frac{\text{total number of births in year } t}{\text{total mid-year population}} = \frac{B_t}{P_t}.$$

For simplicity, the subscripts t are usually omitted because, unless otherwise stated, we shall always be considering single calendar years. Thus, we can write

$$b = \frac{B}{P}.$$

Moreover, since birth is a relatively rare event in most populations, the crude birth rate is often small. For this reason, we often express it as the number of births per thousand of the population, or

$$b = \frac{B}{P} \times 1000.$$

Note that in working out the numerator of the crude birth rate, and all other fertility rates described in this and subsequent chapters, twins count as two births, triplets as three births, and so on.

THE GENERAL FERTILITY RATE

The crude birth rate would not seem to be a very good measure of fertility. It violates the principle of correspondence in two ways.

1   It is an *m*-type rate, and all *m*-type rates which use mid-year populations to approximate the population at risk violate the principle of correspondence because of deaths during the period of observation. A birth occurring, say, on 30 January to a woman who died on 2 March in the same year would appear in the numerator, but the woman would not appear in the denominator. In practice, however, because mortality at childbearing age is rare in most populations, this is not a major issue.
2   The denominator includes a great many people who cannot bear children: for example, men, little girls and women aged over 50 years. This is clearly a serious problem.

The second of these problems is overcome using a rate called the *general fertility rate* (GFR). The GFR for year t, GFR_t, is defined as

$$GFR_t = \frac{\text{total number of births in year } t}{\text{mid-year population of women of childbearing age}}.$$

What do we mean by 'of childbearing age'? In practice, very few women bear children at ages under 15 years and over 50 years, so these are usually taken to mark the limits of the childbearing age range. Thus the GFR is defined as

$$GFR = \frac{\text{total number of births in a year}}{\text{mid-year population of women aged 15–49 years last birthday}}.$$

Here we have omitted the subscript t, since it is usually clear that the GFR refers to a particular year. Like the crude birth rate, the GFR is sometimes multiplied by 1000 to give births per thousand women aged 15–49 years last birthday. It is worth noting that the age range 15–49 years is not universally used. In some populations, very few women have children after their 45th birthday. In such populations, the GFR may be calculated using the age range 15–44 years last birthday in the denominator.

8.3 Age-specific fertility rates

The chance of a woman bearing children varies with age. It is highest when she is in her twenties and early thirties, but it declines increasingly rapidly at ages older than 35 years, reaching zero at about age 50 years. A more complete picture of the fertility of a population can therefore be gained by looking at *age-specific fertility rates*.

The age-specific fertility rate for women aged x last birthday, f_x, is defined as

$$f_x = \frac{\text{births in year } t \text{ to women aged } x \text{ last birthday at the time of birth}}{\text{mid-year population of women aged } x \text{ last birthday}}.$$

Denoting the numerator of this expression by the symbol B_x, and the mid-year population of women aged x last birthday by the symbol P_x^f (the superscript f denotes that we are only considering women), we have

$$f_x = \frac{B_x}{P_x^f}. \tag{8.1}$$

Occasionally, we multiply f_x by 1000 to give births per thousand women aged x last birthday.

The rate f_x is an *m*-type rate. As such, it suffers from the problem that a woman who gives birth to a child on, say, 2 February, when she is aged x years, but whose birthday is on, say, 2 March, will be $x + 1$ years old on 30 June, when the mid-year population is worked out. So her birth will be included in the numerator of the age-specific fertility rate at age x last birthday, but she will be in the denominator of the age-specific fertility rate at age $x + 1$ last birthday. Only if we know the exact dates of each birth included in the numerator, and the exact date of birth of each woman included in the denominator, can we obtain *m*-type fertility rates in which numerator and denominator correspond exactly.

We saw in Chapter 2 that *q*-type rates to some extent overcome this difficulty (at least at the 'woman' level). Can we, then, calculate *q*-type age-specific fertility rates?

In principle, we can. We could calculate a fertility rate by considering a group of women who all had their xth birthday within a particular time period, following them through until their $(x + 1)$th birthday, and noting down how many children they had between exact ages x and $x + 1$. The *q*-type rate would then be calculated as

$$q\text{-type fertility rate} = \frac{\text{number of births to women aged between}}{\text{exact ages } x \text{ and } x + 1 \text{ years}}{\text{number of women attaining exact age } x \text{ years}}.$$

How is this *q*-type rate related to the *m*-type rate defined in equation (8.1)? In Chapter 7 we showed that, for non-decremental processes, the q_xs and m_xs at any age are equal. Having a child is a repeatable event for many women; so women do not leave the population at risk of giving birth once they have had a child, and thus fertility rates are non-decremental. However, there is a necessary assumption here, which is that female mortality during the childbearing age range is zero.

To see why this assumption is necessary, consider Figure 8.1. The numerator and denominator of the *m*-type age-specific fertility rate for women aged x last birthday

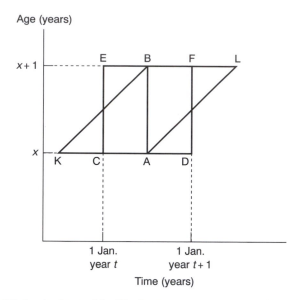

Figure 8.1 Lexis chart of fertility between exact ages x and $x + 1$ years

in year t are represented by the births in the square CDFE and the vertical line AB, respectively. Thus the vertical line AB represents the population P_x^f.

The denominator of the q-type fertility rate for women aged x last birthday which most closely overlaps with year t may be represented by the horizontal line KA. The births representing the numerator of this q-type fertility rate are those occurring in the parallelogram KALB.

Now, a complication arises when we try to relate the m-type and the q-type fertility rate for women aged x last birthday. This complication is caused by mortality. Between exact ages x and $x + 1$, some of the women represented by the line KA in Figure 8.1 die. The deaths of women aged x last birthday in year t may also be represented by the square CDFE, and the deaths between exact ages x and $x + 1$ of the women represented by the line KA may be represented by the parallelogram KALB. Areas on the Lexis chart may thus be used to represent both births and deaths, although the actual number of births represented by a given area will not normally be the same as the number of deaths represented by the same area.

We make the conventional assumptions that deaths are evenly distributed across each year of age, and that mortality does not vary with calendar time. If we do this, then, remembering that the population represented by the vertical line AB in Figure 8.1 is P_x^f, the population represented by the horizontal line KA must be equal to $P_x^f + \frac{1}{2}\theta_x^f$, where θ_x^f is the number of deaths between exact ages x and $x + 1$ occurring to the women represented by the horizontal line KA.

If we make the further assumptions that births are evenly distributed across each year of age, and that fertility does not vary with calendar time, then the number of births represented by the parallelogram KALB is equal to the number of births represented by the square CDEF, which, as we have seen, is B_x.

Thus, we have

$$q\text{-type fertility rate at age } x \text{ last birthday} = \frac{B_x}{P_x^f + \frac{1}{2}\theta_x^f}. \tag{8.2}$$

From equation (8.1), we have

$$B_x = f_x P_x^f.$$ (8.3)

Substituting from equation (8.3) into equation (8.2) produces

$$q\text{-type fertility rate at age } x \text{ last birthday} = \frac{f_x P_x^f}{P_x^f + \frac{1}{2}\theta_x^f}.$$ (8.4)

Equation (8.4) tells us that the relationship between the m-type and q-type fertility rates depends on mortality.

If we assume that there is no mortality, though, θ_x^f in equation (8.2) will be zero. If θ_x^f is zero, then the right-hand side of equation (8.2) becomes simply B_x/P_x^f, and the q-type rate and the m-type rate are equal.

The assumption that there is no mortality is, in fact, only an assumption that there is no mortality among women in the childbearing age groups. In practice, this is not too bad an assumption, as we know that mortality among women between the ages of 15 and 50 years is generally low.

ABRIDGED AGE-SPECIFIC FERTILITY RATES

Age-specific fertility rates can also be calculated for age groups. The most commonly used age groups are five-year age groups. For example, the age-specific fertility rate between exact ages 15 and 20 years is equal to

$$\frac{\text{births in a year to women aged } 15-19 \text{ years last birthday at the time of the birth}}{\text{mid-year population of women aged } 15-19 \text{ years last birthday}}.$$

8.4   Standardization applied to fertility rates

Both the crude birth rate and the general fertility rate are affected by variations in the age structure. This is because fertility varies with age. We can overcome this problem by using standardization to calculate *standardized fertility rates* and *standardized fertility ratios*. The procedure for calculating these is exactly the same as the one we used for calculating standardized death rates and standardized mortality ratios, save that we use just the female population instead of the total population, and the age-specific fertility rates (the f_xs) instead of the age-specific death rates (the m_xs).

However, the only variations in the age structure which affect the crude birth rate and the general fertility rate are the variations within the childbearing age range. These are usually less in magnitude than the variations, over the whole age range between 0 and (about) 120 years, which affect single-figure indices of mortality. Thus standardization is less commonly used with respect to fertility rates than it is with respect to mortality rates.

One situation in which standardization has been widely used in the context of fertility rates arises in the historical demography of nineteenth-century Europe. In most nineteenth-century European populations, data on births by age of the mother are not available, so age-specific fertility rates cannot be calculated. Regular censuses, however, enable us to know the age structure of the female population, and the total number of births is normally available. These are precisely the circumstances in which a standardized fertility ratio can be calculated, provided a standard population, the age-specific fertility rates of which are known, and known to be reliable, can be chosen.

Coale (1967) had the idea of choosing as a standard population the population with the highest reliably recorded fertility rates ever. This population is that of married Hutterite women between the two world wars (the Hutterites are a North American religious sect who do not practise birth control within marriage).

Using the Hutterite age-specific fertility rates, standardized fertility ratios were calculated for all the provinces of Europe during the nineteenth century (see Coale and Watkins, 1986). The results measure the fertility of each European province as a proportion of the fertility of married Hutterite women, after controlling for differences in the age structure of the childbearing female population between the European provinces. Coale (1967) denoted this standardized fertility ratio by the symbol I_f.

8.5   The total fertility rate

Recall that the age-specific fertility rate, f_x, is defined as

$$f_x = \frac{\text{births in year } t \text{ to women aged } x \text{ last birthday at the time of birth}}{\text{mid-year population of women aged } x \text{ last birthday}}.$$

Assuming that no women die between exact ages x and $x + 1$ years, we also know that

$$f_x = \frac{\text{number of births to women aged between exact ages } x \text{ and } x + 1}{\text{number of women attaining exact age } x}.$$

Thus the age-specific fertility rate measures the number of children that the average woman has between her xth and $(x + 1)$th birthdays.

Now, consider an average woman living through her entire span of childbearing years, from her 15th to her 50th birthdays. The number of children she has between her 15th and 16th birthdays is equal to f_{15}, the number of children she has between her 16th and 17th birthdays is equal to f_{16}, and so on.

Thus, summing the f_xs over the ages between 15 and 49 last birthday, we obtain the total number of children a woman would have in her life (provided that she survives until age 50 years). This sum is known as the *total fertility rate* (TFR). The principle by which it is calculated is the same as that used in Chapter 7 to calculate the total first marriage rate. Notice that it is the non-decremental nature of age-specific fertility rates that makes their summation over a range of ages possible.

In symbols, then, we have

$$\text{TFR} = \sum_{x=15}^{49} f_x. \tag{8.5}$$

The TFR is the most widely used single-figure index of fertility. Some demographers think that it is poorly named; they claim that it is not a rate at all. While this is strictly true, it may be argued that these demographers are rather pedantic, since the TFR does, in a sense, measure 'events' divided by 'length of time exposed to risk'. Specifically, we can write

$$\text{TFR} = \frac{\text{births}}{\text{childbearing lives}},$$

where one childbearing life is one woman living through the age range between exact ages 15 and 50 years.

If we use abridged age-specific fertility rates, with five-year age groups, then equation (8.5) has to be modified slightly to read

$$\text{TFR} = 5 \sum_{i=15\text{–}19}^{45\text{–}49} f_i.$$

The multiplication by 5 is required because each woman spends five years in each five-year age group.

Total fertility rates by country in the 1990s range from over 6.0 in several sub-Saharan African countries to under 1.5 in some countries in southern Europe.

8.6 Period and cohort analysis of fertility

In Section 8.3, we saw that an age-specific fertility rate for women aged x last birthday, f_x, may be defined in two ways, either as an m-type rate or as a q-type rate. If we assume that there is no mortality among women in the childbearing age range, these two rates are the same.

We also saw that the sum of the age-specific fertility rates over the whole of the childbearing age range, or the total fertility rate, is a measure of the number of children a woman would have in her life, assuming that she lived through the whole of the childbearing age range. (The childbearing age range is usually taken to be from exact age 15 years to exact age 50 years.)

Since the m-type and q-type age-specific rates are equal, we can calculate the total fertility rate in two ways. These are illustrated in Figure 8.2.

One way is to sum (over all values of x from 15 to 49) the age-specific fertility rates which apply in a given calendar period (shown by the vertical band in Figure 8.2). This gives us an

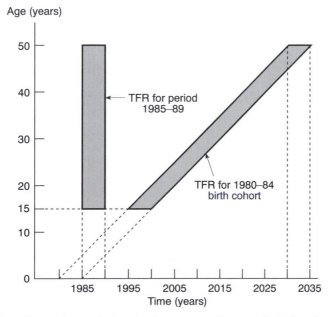

Figure 8.2 Lexis chart illustrating period and cohort approaches to calculating the total fertility rate

estimate of the number of children a woman would end up with if, at each age x last birthday, she experienced the age-specific fertility rate of women aged x last birthday in that given period. Thus, the TFR for the period from 1 January 1985 to 31 December 1989 is calculated by adding up the age-specific fertility rates for that period over all the ages from 15 last birthday to 49 last birthday. Exercise 8.5 asks for summations of this kind.

The other way is to sum (over all values of x from 15 to 49) the age-specific fertility rates experienced by a given birth cohort of women. A birth cohort is represented in Figure 8.2 by a diagonal band running from bottom left to top right. The resulting sum is the average number of children that the women in the given birth cohort end up with, assuming they survive to exact age 50 years. Thus the TFR for the women born during the period 1 January 1980 to 31 December 1984 is calculated by adding up the age-specific fertility rates for that birth cohort over all the ages from 15 last birthday to 49 last birthday.

The first approach, which sums up the age-specific fertility rates for a particular period, is known as the *period approach*. The second approach, which sums up the age-specific fertility rates for a particular birth cohort, is known as the *cohort approach*.

The period approach produces a TFR which does not apply to any real women. It applies only to a hypothetical birth cohort of women who, at each age x last birthday, experience the age-specific fertility rate of women aged x last birthday in a particular period (usually, but not necessarily, a particular calendar year). It is a hypothetical or synthetic measure. Indeed, in France, it is known as the 'synthetic index of fertility'. The cohort approach, on the other hand, produces a TFR which does apply to a real group of women.

8.7   Advantages and disadvantages of the period approach

The period approach to the analysis of fertility is probably the more commonly used of the two. It has three major advantages.

1   The data are easily available. Annual numbers of births classified by the age of the mother are routinely produced by governmental statistical agencies, and we can obtain the number of women at each age x last birthday from census data, or estimates produced from census data by the same governmental agencies.
2   It is up to date. The period approach gives us an estimate of the current fertility prevailing in a population. We can base the period estimate of the TFR on the most recent year for which data are available.
3   It is useful for forecasting purposes. To forecast the number of births that will occur in a particular year, period measures of fertility are needed, since the number of births which will take place in a particular year depends on the fertility rates in that year. Similarly, the number of people of a given age alive in some future year depends on the number of people born in a previous year.

There are, however, disadvantages to the period approach. The main disadvantage arises from the fact that fertility varies substantially from year to year in most populations (especially in developed countries). The period TFR will only represent the fertility of any real women if fertility is constant over time for many years. This means that we can sometimes obtain quite misleading estimates of the long-run fertility of a given population if we only consider a single year's experience. This problem was mentioned in Section 7.4 in the context of the total first marriage rate.

In Japan in the early and mid-1960s the period TFR was about 2.0. In 1966, however, it fell to around 1.6 for a single year, before rising to about 2.2 in 1967 and remaining at 2.2 in 1968 (Bureau of Statistics, 1972, p. 32). Why did the sudden fall in 1966 happen? It is tempting to think that there is a printing error in the tables, but this is not so. In Japan, 1966 was the Year of the Fiery Horse. It was believed that girls born in that year would suffer from ill fortune. To avoid this risk, many Japanese decided not to have children in 1966.

The fact that the Japanese period TFRs in 1967 and 1968 were rather higher than the average value for years prior to 1966 may be because parents who did not want to have children in 1966 raised their fertility in subsequent years to compensate. In other words, in 1966 births were not cancelled, but just postponed until a more favourable year. If we had used only data for the year 1966 to estimate fertility in Japan, our estimate would not have reflected the general level of fertility during the 1960s.

This example is a graphic illustration of the fact that the period TFR is subject to distortions caused by changes in the distribution of births within the childbearing age range. This distortion is illustrated further in Exercise 8.7.

8.8   Advantages and disadvantages of the cohort approach

The advantages of the cohort approach are, first, that cohort TFRs reflect the experience of real people; and second, that cohort TFRs are not distorted by transient period effects, like that in Japan in 1966. (There is actually another advantage, which is that they are more relevant to the theory of population growth: see Chapter 12.)

Practical problems with the cohort approach mainly arise from the data requirements. We need data stretching over many years in order to be able to calculate cohort TFRs. In fact, we may have to wait for 35 years from the date at which women in a birth cohort begin to have children until we can be sure that they will have no more (35 years is the length of the childbearing age range). Thus it is impossible to obtain up-to-date cohort TFRs.

Table 8.1 gives the age-specific fertility rates for five-year age groups for a number of birth cohorts in England and Wales.

The rates marked 'na' in Table 8.1 are not yet known, because they refer to births which have not yet happened. The women born in 1955, for example, did not celebrate their 40th

Table 8.1: Age-specific fertility rates for certain birth cohorts in England and Wales

Year of birth	Age-specific fertility rates (ASFRs) by age last birthday (years)					
	15–19	20–24	25–29	30–34	35–39	40–44
1935	0.040	0.174	0.168	0.076	0.022	0.004
1940	0.056	0.196	0.148	0.054	0.016	0.002
1945	0.070	0.176	0.124	0.050	0.016	0.002
1950	0.072	0.140	0.118	0.058	0.022	0.002
1955	0.062	0.122	0.120	0.070	0.026	na
1960	0.048	0.114	0.122	0.076	na	na
1965	0.040	0.102	0.116	na	na	na

Source: Office for National Statistics (1997c, pp. 52–53). Note that hardly any children are born in England and Wales to women aged over 45 years. Thus the ASFRs for the age group 45–49 years can be assumed to be close to zero.

birthdays until 1995. Therefore, we do not yet know how many children they will have between their 40th and 45th birthdays. Women born in 1960 did not reach their 35th birthday until 1995. Thus their fertility at ages 35–39 years last birthday is not yet known. The most recent birth cohort for which we can calculate a cohort TFR is that born in 1950. In other words, the data for recent birth cohorts are censored. We do not know what will happen to these cohorts in the future.

Of course, we can still calculate the age-specific fertility rates at younger ages for more recent birth cohorts, and compare them with previous birth cohorts (Table 8.1). Nevertheless, we do not know what will be the completed fertility of censored birth cohorts. Clearly, the greater the amount of censoring, the greater our uncertainty about the likely cohort TFR of a birth cohort. Thus, for the 1955 birth cohort in Table 8.1, we can be fairly sure that the cohort TFR will be close to the sum of the age-specific fertility rates over the age range 15–39 years (because fertility rates for women over 40 years are very low). For the 1965 birth cohort, however, it is far from clear what the final cohort TFR will turn out to be.

Further reading

The measurement of fertility is discussed in Shryock and Siegel (1975, pp. 462–486). Brass (1989) and Ní Bhrolcháin (1992) provide discussions of the relative merits of the period and cohort approaches to measuring fertility.

Exercises

8.1   Table 8E.1 shows the population of Scotland on 30 June in various years, together with the number of births taking place in those years. Use the data in the table to calculate crude birth rates and general fertility rates for the years in question.

8.2   If the crude birth rate in a country remains constant over a number of years, but the general fertility rate increases steadily, what does this tell you about the country's population?

8.3   The data in Table 8E.2 relate to fertility in England and Wales in 1976 and 1993.
   (a)   Calculate the general fertility rate for 1976 and 1993.
   (b)   Calculate age-specific fertility rates for the two years.
   (c)   Using the 1976 population as the standard, calculate a standardized fertility rate for 1993.

Table 8E.1

Year	Total population (thousands)	Female population aged 15–44 years (thousands)	Total births in year
1971	5236	1011	86 700
1981	5180	1094	69 100
1991	5107	1122	67 000
1993	5121	1103	63 300
1994	5132	1102	61 700
1995	5136	1101	60 100

Source: *Population Trends* 87 (1997), pp. 43, 48 and 51.

Table 8E.2

Age group	1976		1993	
	Number of births (thousands)	Mid-year female population (thousands)	Number of births (thousands)	Mid-year female population (thousands)
15–19	57.9	1809	45.1	1455
20–24	182.2	1672	152.0	1831
25–29	220.7	1855	236.0	2070
30–34	90.8	1593	171.1	1967
35–39	26.1	1374	58.8	1729
40–44	6.5	1300	10.5	1750

Source: *Population Trends* 87 (1997), p. 52.

8.4 Table 8E.3 gives information on the number of births to women in various age groups and the age-specific fertility rates (ASFRs) for Egypt and Tunisia in the late 1980s.
   (a)   Calculate the general fertility rates for Egypt and Tunisia.
   (b)   Calculate a standardized fertility rate for Tunisia, using the female population of Egypt as the standard population.
   (c)   Calculate a standardized fertility ratio for Tunisia, using the female population of Egypt as the standard population.
   (d)   Comment briefly on your results.

8.5 Using the data for Egypt and Tunisia presented in Exercise 8.4 (Table 8E.3), calculate total fertility rates for Egypt and Tunisia.

8.6 The data in Table 8E.4 relate to the African country of Malawi. They come from a large-sample survey of the population of this country which took place in 1992. You are also told that the total number of urban women in the survey is 1334, and that the total number of rural women in the survey is 10 518.
   (a)   Calculate the general fertility rates for rural and urban areas.
   (b)   Calculate total fertility rates for urban and rural areas.
   (c)   Calculate standardized fertility rates for urban areas, using the rural areas as the standard.
   (d)   What do your results tell you about fertility in Malawi?

Table 8E.3

Age group	Egypt, 1988		Tunisia, 1989	
	Number of births (thousands)	ASFR	Number of births (thousands)	ASFR
15–19	43.6	0.021	6.3	0.017
20–24	402.8	0.194	43.6	0.131
25–29	578.9	0.317	55.7	0.195
30–34	403.4	0.269	41.1	0.176
35–39	242.4	0.191	21.6	0.113
40–44	77.7	0.073	5.7	0.041
45–49	25.1	0.026	1.1	0.009

Source: Adapted from United Nations (1993).

Table 8E.4

Age group	Percentage of all women in age group		Age-specific fertility rates (per woman)	
	Urban areas	Rural areas	Urban areas	Rural areas
15–19	9.7	9.4	0.135	0.165
20–24	10.1	7.8	0.268	0.291
25–29	9.0	6.3	0.242	0.273
30–34	6.3	5.3	0.210	0.261
35–39	4.7	4.4	0.149	0.202
40–44	3.0	4.4	0.086	0.123
45–49	1.9	3.1	0.012	0.062

Source: Republic of Malawi (1994, pp. 7, 20).

Table 8E.5

Age group	Calendar time (years)						
	1960–64	1965–69	1970–74	1975–79	1980–84	1985–89	1990–94
15–19	0.040	0.049	0.045	0.030	0.030	0.031	0.033
20–24	0.175	0.165	0.130	0.108	0.095	0.094	0.088
25–29	0.183	0.164	0.135	0.125	0.127	0.125	0.120
30–34	0.105	0.090	0.067	0.065	0.075	0.084	0.087
35–39	0.049	0.040	0.025	0.020	0.024	0.030	0.033
40–44	0.015	0.010	0.007	0.005	0.005	0.005	0.006
45–49	0.000	0.000	0.000	0.000	0.000	0.000	0.000

Sources: Derived from data in Office of Population Censuses and Surveys (1987b, p. 54); *Population Trends* 60 (1990), p. 50; and *Population Trends* 83 (1996), p. 64.

8.7 Table 8E.5 gives age-specific fertility rates for England and Wales.
   (a) Calculate period total fertility rates for each of the five-year periods between 1960–64 and 1990–94.
   (b) Use these data to estimate cohort total fertility rates for women born during the following periods: 1943–47, 1948–52 and 1953–57. State any assumptions you are making.
   (c) Calculate the proportion of all births which take place to women aged under 30 years for each of the five-year periods between 1960–64 and 1990–94.
   (d) What do your results tell you about the interpretation of period and cohort total fertility rates?

8.8 Using the results of Exercise 8.7, describe the main trends in the fertility experience of England and Wales since 1960.

9
Parity Progression

9.1 Introduction

In the previous chapter, the measurement of fertility using age-specific fertility rates and total fertility rates (TFRs) was described. Age-specific fertility rates are widely used in demography (indeed, they are still the most widely used measures of fertility). They reflect the important fact that fertility varies with age. In Section 8.1, however, it was noted that one thing which distinguishes fertility from mortality is that fertility is much more the subject of individual choice than is mortality (notwithstanding suicides, and people who deliberately engage in dangerous pursuits).

In developed countries, and increasingly in developing countries, the most important aspect of this choice is the number of children that people decide to have. Some people decide to have no children at all; others decide to have a particular number and then stop childbearing. (A few people cannot have children for physiological reasons, but these people are a very small proportion of the total population in the majority of countries.)

In the light of this, it would seem desirable to analyse fertility in a way which takes account of the number of children a woman has already had. This approach to the analysis of fertility is the subject of this chapter. The term *parity* is used by demographers to denote the number of children a woman has already had, and the term *birth order* is used to refer to the children women have in the order in which they appear (first births, second births, and so on). Section 9.2 looks at fertility rates specific to birth order. Section 9.3 describes measures known as parity progression ratios, which are becoming increasingly widely used in fertility analysis. Parity progression ratios may be calculated on a period or a cohort basis. Their calculation on a period basis has recently been receiving attention in the demographic literature, and forms the subject of Section 9.4. Section 9.5 makes a number of general observations about the two approaches to the measurement of fertility: age-based measures and parity-based measures.

9.2 Order-specific birth rates

As noted in Section 9.1, *birth order* is used by demographers to distinguish children who are the first born to a particular mother from those who are the second born, the third born, and so on. It is possible to calculate *order-specific fertility rates* by considering in the numerator only births of a specific order. Order-specific fertility rates may be calculated for all women of childbearing age (using the same denominator as the general fertility rate), or for women in specific age groups (making age-order-specific fertility rates). The formula for all women of childbearing age is

$$GFR_j = \frac{\text{total number of births of order } j \text{ in a year}}{\text{mid-year population of women aged 15–49 years last birthday}},$$

where GFR_j is the general fertility rate for births of order j. For age-order-specific fertility rates, the formula is

$$f_{x,j} = \frac{B_{x,j}}{P_x^f},$$

where $f_{x,j}$ is the age-order-specific fertility rate for births of order j to women aged x years last birthday in a given year, $B_{x,j}$ is the number of births in that year of order j to women aged x years last birthday, and P_x^f is the mid-year population of women aged x years last birthday. The population P_x^f may be defined on either a period or a cohort basis.

One potential difficulty with age-order-specific fertility rates is that the number of births in some of the age-order groups is typically very small. For example, fourth births to women aged 20 years last birthday are very rare! Similarly, first births to women aged 40 years and over are rare. Indeed, unless one is dealing with a very large population, the numbers of births in very many of the age-order categories are likely to be small if single years of age are used. It is therefore very common to use five-year age groups in the calculation of age-order-specific fertility rates.

TOTAL FERTILITY RATES BY BIRTH ORDER

The age-order-specific fertility rates may be summed over the childbearing age range to give total fertility rates by birth order. The calculation may be carried out on either a period basis or a cohort basis. If TFR_j is the total fertility rate for birth order j,

$$TFR_j = \sum_{x=15}^{49} f_{x,j},$$

or, if five-year age groups are used,

$$TFR_j = 5 \sum_{i=15-19}^{45-49} f_{i,j}.$$

When TFR_j is calculated for a birth cohort, it denotes the total number of births of order j that the women in the cohort will have. When TFR_j is calculated for births in a given time period, it denotes the total number of births of order j that a hypothetical cohort of women would have if, at each age, they had the age-order-specific fertility rates experienced during the time period in question. Note that the overall TFR is simply the sum of the order-specific TFRs:

$$\text{overall total fertility rate} = \sum_j TFR_j.$$

A logical difficulty sometimes arises with period TFR_js. It is possible (and quite often happens) that the period TFR_j, especially for low birth orders, is greater than 1.0. This suggests that the average woman in a hypothetical cohort would have more than one birth of that order. This is impossible: a woman cannot have more than one first birth, one second birth, and so on. The problem arises because period TFR_js are obtained by summing non-decremental age-order-specific rates. The denominators of these rates are not the correct exposed-to-risks as they include all women in the relevant age groups, whereas only a subset of the women are actually at risk of having a birth of a particular order. We return to this issue in Section 9.5. Thus, in the same way as the total first marriage rate was sensitive to changes in the timing of first marriages (see Chapter 7), total fertility rates by birth order are sensitive to changes in the timing of births.

9.3 Parity progression ratios

As we have seen, the total fertility rate for a given birth order, when calculated for a birth cohort, is a measure of the average number of children of that birth order which each woman in that cohort will have. This order-specific cohort TFR will range between 0 and 1. Although it is perfectly legitimate to view it as an 'average', it is rather artificial, since babies come in whole units, and each individual woman either has, or does not have, a birth of that order. Because of this, it is often more helpful to interpret the order-specific TFR as indicating the proportion of women in the cohort who have a birth of that order. Thus the TFR_j for a birth cohort is just the proportion of women in that birth cohort who have a jth birth.

It is easy to see, therefore, that if the original number of women in the birth cohort is n_0, then the number of women who have a child of birth order j, n_j, is given by the formula

$$n_j = \text{TFR}_j \times n_0.$$

Further, because each woman only has one child of each birth order, the number of women in the cohort who have a jth child is equal to the number of jth children born to the women in the cohort.

A particularly interesting and useful set of measures of fertility can be obtained by calculating the proportion of women who have already had a certain number of children and go on to have another child. The number of children a woman has already had is referred to by demographers as her *parity*, and the proportion of women of a given parity who go on to have another child is known as a *parity progression ratio* (PPR).

Formally, we can say that, for any birth cohort of women,

$$\text{PPR from } j \text{ births to } j+1 \text{ births} = \frac{\text{number of women who have a } (j+1)\text{th child}}{\text{number of women who have a } j\text{th child}}.$$

Denoting the PPR from j births to $j + 1$ births as a_j, we have

$$a_j = \frac{n_{j+1}}{n_j}. \tag{9.1}$$

As mentioned above, the subscript j denotes the order of the birth. Thus n_1 is the number of children who have a first child. In other words, n_j is the number of women who have j *or more* children. Thus, in order to calculate PPRs, we need to know the number of women who have j or more children, for all values of j, for a particular birth cohort.

Parity progression ratios can help us understand the distribution of completed family sizes among a cohort of women. By the distribution of completed family sizes we mean the proportion of women in the cohort who end up with exactly no children, exactly one child, exactly two children, and so on, at the end of the childbearing years. This is important, because the same TFR can be produced by very different distributions of completed family sizes. For example, a cohort of ten women, of which three have nine children each, and seven have no children at all, will have the same cohort TFR as another cohort of ten women, nine of whom have three children and one of whom has no children.

Finally, we can derive a relationship between the PPRs for a given birth cohort and the cohort total fertility rate.

The cohort TFR is equal to the total number of children they have, divided by the total number of women in the birth cohort. In symbols

$$\text{cohort TFR} = \frac{\text{total number of births}}{\text{total number of women in birth cohort}}.$$

But remembering that the number of births of order j to the cohort of women is the same as the number of women who have a jth child, that is n_j, then the total number of births is equal to $n_1 + n_2 + n_3 + \cdots$ (summed until we reach the highest parity attained by the women in that birth cohort).

Thus

$$\text{cohort TFR} = \frac{n_1 + n_2 + n_3 + \cdots}{\text{total number of women in birth cohort}}.$$

The denominator, however, is equal to n_0, the number of women who have zero or more children (every woman must, by definition, have at least zero children).

So

$$\text{cohort TFR} = \frac{n_1 + n_2 + n_3 + \cdots}{n_0},$$

or

$$\text{cohort TFR} = \frac{n_1}{n_0} + \frac{n_2}{n_0} + \frac{n_3}{n_0} + \cdots. \tag{9.2}$$

It is easily seen from this that

$$\text{cohort TFR} = \frac{n_1}{n_0} + \left[\frac{n_1}{n_0}\frac{n_2}{n_1}\right] + \left[\frac{n_1}{n_0}\frac{n_2}{n_1}\frac{n_3}{n_2}\right] + \cdots. \tag{9.3}$$

Substituting from equation (9.1) into equation (9.3) produces

$$\text{cohort TFR} = a_0 + a_0 a_1 + a_0 a_1 a_2 + \cdots, \tag{9.4}$$

which demonstrates that we can calculate the cohort TFR once we know the parity progression ratios.

9.4   Period parity progression ratios

Parity progression ratios have been described in Section 9.3 in relation to a birth cohort of women. Using a cohort perspective, however, has the disadvantages mentioned in Section 8.8 of data availability and the lack of up-to-dateness of the resulting measures. Because of the usefulness of PPRs, a number of demographers have explored the possibility of estimating PPRs on a period basis: that is, of working out PPRs which apply to particular time periods. The idea of calculating period PPRs was first suggested by the French demographer Louis Henry in the 1950s, and independently rediscovered by Feeney and Yu (1987) and Ní Bhrolcháin (1987).

Before describing the possible approaches, it is worthwhile considering the data sources which might be used for this purpose. It turns out that what is required is a single census or survey in which women are asked to give the dates of birth of all the children they have so far borne, and (ideally, but not necessarily) the date of their marriage. There are increasing numbers of such surveys available for national populations. They are termed retrospective surveys, because the women interviewed are asked to look back over their lives and recall the dates of past events. Note that the data required are obtained from a census or survey conducted at a single point in time.

Given data of this kind, there are two possible approaches to calculating period PPRs.

TRUE PARITY COHORTS

The first approach has been called by Ní Bhrolcháin (1987) the 'true parity cohort' approach. In this method, the survey data are used to identify all the births of a given order occurring to the women in the sample during a specified time period (let us call this the 'index period'). The subset of women who experience a birth of the specified order during the index period are then followed up to the date of the survey to see if they have a subsequent child before the survey. The date of birth of the subsequent child is then noted.

Suppose that the index period is a particular calendar year. Suppose, further, that the date of birth of the subsequent child is measured in years since the index year. We work out a series of proportions, which we can call q_x^*s (as they are analogous to the q_xs used in the analysis of mortality) using the following formula:

$$q_x^* = \frac{\text{Number of women who had a } j\text{th birth in the index year and who have their } (j+1)\text{th birth in year } x}{\text{Total number of women who had a } j\text{th birth in the index year} - \text{Number of these women who have already had their } (j+1)\text{th birth before the start of year } x}.$$

The q_xs in the original life table denoted the proportion of those still alive at the beginning of the current year who die during the year. These q_x^*s are exactly analogous: instead of age in the life table, we have 'duration since previous birth'; instead of death in the life table we have 'birth of subsequent child'.

The q_x^*s can be calculated for values of x ranging from 0 (q_0^* denoting the proportion of women who had their jth birth in the index year and also had their $(j+1)$th birth in that year) up to whatever maximum value is considered appropriate. In practice, however, in all populations the q_x^*s will become very small for values of x greater than 10.

The period PPR from the jth to the $(j+1)$th birth is then calculated as

$$a_j = 1 - (1 - q_0^*)(1 - q_1^*)(1 - q_2^*) \cdots . \tag{9.5}$$

The progression to the first birth poses an additional problem, since it is necessary to define a preceding event which can be dated. There are various possible choices: a woman's marriage, the date of a woman's own birth, or the date of a woman's 15th birthday (where it can be assumed that no children are born to women younger than 15 years).

Finally, it is conventional to omit from the analysis multiple births. That is, births in the index year which were the first born of twins, or the first and second born of triplets, are not followed. However, an index birth which represents the last born of a set of multiple births can be included. This convention is adopted because the progression from, say, the first born of a set of twins to the second born of the same set is not a matter of choice for the woman and it is mainly to take account of the fact that fertility is something about which people make choices that PPRs are used.

The index period must be a reasonable length of time before the survey in order to leave enough time for the women who had a *j*th child during the index period to have their $(j + 1)$th child before the survey. In other words, it is important to make sure that most of the women who are going to have another child have that child before the survey. If the index period is too close to the survey date the PPR will be underestimated, because a proportion of women who have not had their $(j + 1)$th child by the date of the survey will have that child afterwards. Generally speaking, in high-fertility populations, a period of five years will be sufficient to enable acceptable estimates of the PPRs to be obtained (Brass, 1996). In low-fertility populations, however, a longer period (say, ten years) may be needed for the estimated PPRs to be of an acceptable level of accuracy.

This need to allow enough time to elapse for women to have a chance to have another birth creates a problem with retrospective surveys because of censoring. Censoring has already been discussed in the context of mortality analysis in Sections 5.7 and 6.7. Censoring in retrospective surveys of fertility occurs because some women that are interviewed are too young to have completed their fertility. It may be illustrated using a Lexis chart (Figure 9.1).

Figure 9.1 illustrates a survey of women aged 15–49 years last birthday *at the time of the survey*. Most of these women will not have completed their childbearing (indeed, if the childbearing age range is considered to be 15–50 years it could be argued that all of them might possibly have more children). Some of the women in the survey may not

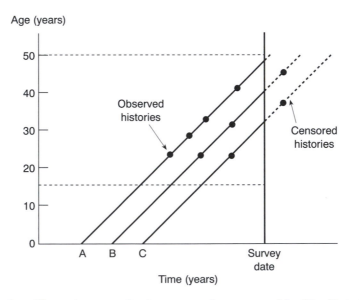

Figure 9.1 Lexis chart illustrating censoring in retrospective surveys of fertility. Dots denote births. Woman A has had four births before the survey. Woman B has had two births before the survey, but will have another child after the survey. Woman C has had one child before the survey, and will have another child after. The dates of birth of children born after the survey will not be available from the survey data

even have started their childbearing. Thus most (and arguably all) women's childbearing lives are censored.

There is a further issue to consider. In order to obtain a large enough sample of births of order j, the index period may have to be several years wide. This potentially creates a problem because the women who had their jth child at the beginning of the index period will be observed for longer before the survey than women who had their jth child at the end of the index period. The problem is essentially one of selectivity, and is discussed further in Chapter 11. One way of overcoming this is to fix the period of observation for all women to be that between the end of the index period and the survey date and to compute the proportion of women who have their jth child during the index period and go on to have their $(j + 1)$th child within this period of observation. For example, if the index period were from exactly five years before the survey to exactly ten years before the survey, the period of observation would be fixed at five years, and every woman would be observed for a period of five years following her jth birth (Figure 9.2).

Finally, it should be noted that if interest in the analysis focuses solely on the PPRs themselves, and not on the q_x^* values, then the PPRs for a given true parity cohort can be estimated more quickly using the following formula:

$$\begin{array}{c} \text{true parity cohort PPR} \\ \text{from the } j\text{th to the} \\ (j + 1)\text{th birth} \end{array} = \frac{\sum_{x=0}^{k} \begin{array}{c}\text{number of women having their } (j + 1)\text{th birth} \\ \text{in year } x, \text{ who had a } j\text{th birth in the index year}\end{array}}{\begin{array}{c}\text{number of women who had a } j\text{th birth} \\ \text{in the index year}\end{array}},$$

where x is measured in years since the index year, and k is the number of years for which the true parity cohort is followed.

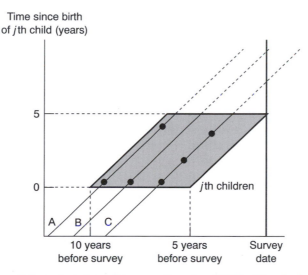

Figure 9.2 Illustration of the calculation of true parity cohort PPRs. The shaded area denotes the period of observation. All women who had a jth child between five and ten years before the survey date are observed for a period of five years. In the example shown, women A and C have a $(j + 1)$th child within five years (indeed woman C also has a $(j + 2)$th child); but woman B does not have a $(j + 1)$th child within five years. Note that the vertical axis on this chart does *not* indicate a woman's age

SYNTHETIC PARITY COHORTS

Because of the need to allow time for the $(j + 1)$th births to occur, the true parity cohort procedure produces PPRs which necessarily apply to quite a long period before the survey. Most of the births which are included in the calculation of the true parity cohort PPRs will usually have occurred several years before the census or survey. Is there any way in which PPRs which are even more up to date than these can be estimated? Yes, there is. What is done is to calculate PPRs based on the $(j + 1)$th births occurring in a particular year to women who had their *j*th births in a range of previous years. The following description of the method is adapted from Feeney and Yu (1987) and Ní Bhrolcháin (1987).

Consider the PPR from the first to the second birth. The first thing to do is to divide the time before the survey into periods of a given length. If the dates of birth of the children are recorded accurately enough, periods as short as one month could be used, but usually longer periods are used, and this exposition assumes that periods of one year are used. The method relies on tabulating the following information for each woman who had a first child: the year in which she had her first child; and the year in which she had her second child, if she had this child before the survey or census. We then work out a series of proportions, which we can call q_x^{**}s (as they are again analogous to the q_xs used in the analysis of mortality) using the following formula:

$$q_x^{**} = \frac{\text{Number of women who had their } j\text{th birth in the } x\text{th year before the current year and had their } (j+1)\text{th birth in the current year}}{\begin{array}{ll}\text{Total number of women who} & \text{Number of these women who have} \\ \text{had a } j\text{th birth in the } x\text{th year} - & \text{already had their } (j+1)\text{th birth} \\ \text{before the current year} & \text{before the start of the current year}\end{array}}.$$

The q_x^{**}s can be calculated for values of x ranging from 0 (that is, using women who had both their first birth and their second birth in the current year) up to whatever maximum value is considered appropriate. In practice, however, the q_x^{**}s will become very small for values of x greater than 10.

The period PPR from the *j*th to the $(j + 1)$th birth is then calculated as

$$a_j = 1 - (1 - q_0^{**})(1 - q_1^{**})(1 - q_2^{**})\cdots. \tag{9.6}$$

The key point to note about this approach to estimating period PPRs is that, whichever order of progression is being considered, and whatever the value of x, the numerators of the q_x^{**} calculations always refer to births in the same year. Therefore, the PPRs which result are all directly referring to fertility in the same year. This renders the PPRs as up to date as possible.

AN EXAMPLE

An example will illustrate the calculation of period PPRs using true and synthetic parity cohorts. The example is taken from the Tanzania Demographic and Health Survey of 1991–92. This was a survey of over 9000 women aged between 15 and 49 years last birthday. The survey collected information about each woman's date of birth, and about the dates of birth of all children they had borne before the survey – see Ngallaba *et al.* (1993) for more details. These data can be used to calculated true parity cohort and synthetic parity cohort PPRs from the first to the second birth.

The data are presented in Table 9.1. Following the recommendations of Feeney and Yu (1987), this table cross-classifies women who had a first birth in a given year according

Table 9.1 Data for calculation of period parity progression ratios, Tanzania

Year of birth of first child	Number of women having first child	Number of women having second child in relevant year																				
		1970	1971	1972	1973	1974	1975	1976	1977	1978	1979	1980	1981	1982	1983	1984	1985	1986	1987	1988	1989	1990
1970	99	1	9	38	24	11	6	2	0	1	1	1										
1971	119		0	9	47	37	7	6	3	1	0	0	0									
1972	121			0	17	51	26	9	4	2	2	0	0									
1973	123				0	11	55	30	5	7	1	1	0	1								
1974	159					2	12	68	40	40	10	7	2	3	1	1						
1975	110						0	10	40	30	6	6	1	1	1							
1976	161							0	13	42	12	10	6	2	0	2	2					
1977	137								3	12	66	42	29	16	6	6	1	2	0			
1978	149									0	7	55	54	14	7	1	2	2	1			
1979	162										1	17	61	40	19	8	2	3	1	0		
1980	180											4	5	67	45	23	8	4	2	0		
1981	207												1	12	70	62	23	10	8	3	5	
1982	185													2	11	62	51	22	8	4	1	3
1983	225														2	13	88	59	19	5	9	8
1984	239															0	25	68	76	28	10	6
1985	274																0	15	102	80	21	19
1986	271																	1	17	94	79	33
1987	314																		3	17	101	91
1988	342																			3	24	108
1989	358																				3	30
1990	367																					3

Source: Extracted from Tanzania Demographic and Health Survey, 1991–92, excluding women with inconsistent birth history data. A copy of these data may be obtained by applying to Demographic and Health Surveys, Macro International, Inc., Suite 300, 11785 Beltsville Drive, Calverton, MD 20705-3119, United States of America. Data may also be ordered via the Internet from **http://www.macroint.com/dhs/**.

to the year in which they had their second child, if they had a second child before the survey.

Using this table, the q_x^* values for true parity cohorts are obtained from each row. Thus, for example, for women who had their first birth in 1970, we have

$$q_0^* = \frac{\text{number of women who had their second birth in 1970}}{\text{number of women who had their first birth in 1970}}$$

$$= \frac{1}{99}$$

$$= 0.0101$$

and

$$q_1^* = \frac{\text{number of women who had their second birth in 1971}}{\substack{\text{number of women who had} \\ \text{their first birth in 1970}} - \substack{\text{number of women who had} \\ \text{their second birth in 1970}}}$$

$$= \frac{9}{99 - 1}$$

$$= 0.0918.$$

The other q_x^* values are obtained similarly, remembering that the calculations for each true parity cohort are based just on women who had their first births in a given year.

The values of q_x^{**}, representing synthetic parity cohorts, and relating to a given year, are obtained by dividing the numbers of second births in the column for that year by the relevant number of women in each row who had not yet had a second birth by the start of the year in question. Thus, for 1990 we have

$$q_0^{**} = \frac{\text{number of women who had their second birth in 1990}}{\text{number of women who had their first birth in 1990}}$$

$$= \frac{3}{367}$$

$$= 0.0082$$

and

$$q_1^{**} = \frac{\text{number of women who had their second birth in 1990}}{\substack{\text{number of women who had} \\ \text{their first birth in 1989}} - \substack{\text{number of women who had} \\ \text{their second birth in 1989}}}$$

$$= \frac{30}{358 - 3}$$

$$= 0.0845.$$

Notice that, when using synthetic parity cohorts, the q_x^{**} values are based, for each value of x, on a different set of women, classified according to the year in which they had their first births.

Once the q_x^* and q_x^{**} values have been calculated, the PPRs based on true and synthetic parity cohorts can be calculated using equations (9.5) and (9.6), respectively. The results for progression from the first to the second birth in Tanzania are shown in Table 9.2. In working out these PPRs, second births occurring more than ten years after the first birth have been ignored.

Table 9.2 True parity cohort and synthetic parity cohort PPRs from the first to the second birth in Tanzania

Index year	True parity cohort PPR	Year of second births	Synthetic parity cohort PPR
1970	0.939		
1971	0.924		
1972	0.950		
1973	0.911		
1974	0.931		
1975	0.882		
1976	0.963		
1977	0.978		
1978	0.960		
1979	0.963		
1980	0.889	1980	0.899
		1981	0.952
		1982	0.972
		1983	0.932
		1984	0.941
		1985	0.922
		1986	0.943
		1987	0.951
		1988	0.901
		1989	0.930
		1990	0.936

Source: Table 9.1.

Once the period PPRs have been calculated, an equation analogous to equation (9.4) can be used to calculate period TFRs from them.

9.5 Age-based and parity-based decomposition of total fertility

This chapter has shown how the fertility of a population (as expressed in its total fertility rate) may be decomposed either by the age of the women involved, or by their parity: that is, by the number of children they already have. By way of conclusion, some of the advantages and disadvantages of age-based and parity-based measures may be summarized.

AGE-BASED MEASURES

Age-specific fertility rates have the advantage of being simple to calculate and understand. Because they are non-decremental, they can also be summed to give a total fertility rate (either overall, or specific to a particular birth order). However, because they are non-decremental rates, the resulting TFRs, if calculated on a period basis, are sensitive to changes in the timing of births. This sensitivity arises because the exposed-to-risk with respect to age-based measures of fertility will vary in its composition between periods being compared. This is most obviously seen with respect to order-specific period TFRs. The true exposed-to-risk in these clearly depends on past order-specific fertility, which is not taken into account in conventional age-order-specific fertility rates.

As Feeney and Yu (1987) noted, age-based measures of fertility are most appropriate in high-fertility populations where contraception is rarely practised, for in such populations fertility varies principally with age, so that age-order-specific fertility rates are roughly constant over time, and the distortions which arise from using non-decremental rates are minimized. However, such populations are becoming rarer. Even in sub-Saharan Africa, where fertility is higher than in any other world region, there are a growing number of countries in which contraceptive use is increasing rapidly (Brass and Jolly, 1993).

PARITY-BASED MEASURES

We have shown that the total fertility rate may also be decomposed into its constituent PPRs. Feeney and Yu (1987, p. 100) maintain that

> two substantial arguments favour parity progression measures over age-specific birth rates. First, the decomposition of total fertility into birth order components is...a useful device for the study of fertility generally, because the birth order components represent family size, because they reflect the life cycle, and because the birth-order components computed from period parity progression ratios are clearly superior to the components calculated from age-order-specific birth rates.

There are therefore two principal advantages of the parity-based approach over the age-based approach. First, the parity-based approach controls for past order-specific fertility. Therefore, parity-based period measures are much less prone than are age-based period measures to temporal fluctuations caused by changes in the timing of births. Second, fertility is coming to reflect more and more the decisions made by people about the appropriate number of children to have. Parity-based measures explicitly acknowledge this fact, whereas age-based measures do not.

The disadvantages which parity-based measures have over age-specific fertility rates is that their calculation is more complex and their data requirements are heavier. However, as the number of large-scale surveys of fertility in both developed and developing countries increases, and computer technology advances, these disadvantages are being gradually reduced.

Further reading

Discussions of the use of parity progression ratios are contained in Brass (1989) and Ní Bhrolcháin (1992). A much fuller discussion of some of the issues raised in this chapter is provided in Feeney and Yu (1987). A good survey of various approaches to measuring fertility is Brass (1996).

Exercises

9.1   The data in Table 9E.1 give the numbers of births by order in the Seychelles in 1987 in five-year age groups, together with the mid-year population of women in each age group.
  (a)   Calculate age-order-specific fertility rates for birth orders 1–4.
  (b)   Calculate total fertility rates for birth orders 1–4.
  (c)   Calculate the total fertility rate for all birth orders combined.

Table 9E.1

Age group	Mid-year female population	Number of births in 1987 by order						
		1	2	3	4	5	6–9	10+
15–19	3664	204	27	3	0	0	0	0
20–24	3726	280	216	82	21	5	0	0
25–29	3062	77	145	135	80	34	11	0
30–34	2117	23	38	61	47	35	29	1
35–39	1490	3	20	12	19	7	34	1
40–44	1227	0	2	3	3	2	10	3
45–49	1330	0	0	0	0	1	0	1

Source: Republic of Seychelles (1991).

9.2   Describe circumstances in which you might expect the period TFR for first births to exceed 1.0.

9.3   Table 9E.2 gives parity progression ratios for a number of recent birth cohorts in England and Wales. Assuming that no woman in any of these birth cohorts had a fifth child, calculate:
   (a)   the proportion of women in each birth cohort who had exactly 0, 1, 2, 3 and 4 children;
   (b)   the total fertility rate for women in each birth cohort.

9.4   Using the results of Exercise 9.3, discuss recent fertility trends in England and Wales.

9.5   Table 9E.3 uses the same Tanzanian data as Table 9.1, and is constructed in a similar way, but shows progression from the fourth to the fifth birth. Use it to construct true parity cohort PPRs for women having their fourth births in the years 1970–80, and synthetic parity cohort PPRs from the fourth to the fifth birth relating to the years 1980–90.

9.6   In the data set used in Exercise 9.5, the number of women having fourth births in each year during the early 1970s is rather small. What problems might this cause for the interpretation of the true parity cohort PPRs, and how might these be overcome?

Table 9E.2

Calendar years of birth	Parity progression ratios			
	0–1	1–2	2–3	3–4
1931–33	0.861	0.804	0.555	0.518
1934–36	0.885	0.828	0.555	0.489
1937–39	0.886	0.847	0.543	0.455
1940–42	0.890	0.857	0.516	0.416
1943–45	0.892	0.854	0.458	0.378
1946–48	0.885	0.849	0.418	0.333

Source: Brass (1989, p. 23).

Table 9E.3

Year of birth of fourth child	Number of women having fourth child	Number of women having fifth child in relevant year																				
		1970	1971	1972	1973	1974	1975	1976	1977	1978	1979	1980	1981	1982	1983	1984	1985	1986	1987	1988	1989	1990
1970	34	1	3	17	6	0	2	1	1	0	0	0										
1971	28		0	3	12	11	0	0	0	0	0	0	0									
1972	41			0	4	14	12	6	3	0	1	0	0	0								
1973	43				1	7	18	9	2	2	2	1	1	2	0							
1974	53					0	4	20	18	5	2	0	2	1	2	0						
1975	84						2	6	33	25	6	3	1	2	2	1	0					
1976	77							2	6	42	16	3	4	1	2	1	0	1				
1977	74								0	5	27	22	11	2	2	0	1	1	1			
1978	90									3	7	23	18	16	3	2	0	0	0	2		
1979	103										3	10	34	27	9	4	3	0	1	2	0	
1980	99											2	6	40	28	9	3	0	2	0	2	1
1981	93												2	6	34	23	9	7	2	3	3	0
1982	96													2	6	32	35	6	6	2	1	1
1983	119														0	6	43	32	11	1	7	1
1984	108															1	12	38	30	7	6	4
1985	113																4	5	32	28	16	8
1986	125																	0	7	46	39	15
1987	126																		2	5	33	48
1988	120																			3	1	38
1989	146																				4	5
1990	127																					1

Source: Extracted from Tanzania Demographic and Health Survey, 1991–92, excluding women with inconsistent birth history data. A copy of these data may be obtained by applying to Demographic and Health Surveys, Macro International, Inc., Suite 300, 11785 Beltsville Drive, Calverton, MD 20705-3119, United States of America. Data may also be ordered via the Internet from **http://www.macroint.com/dhs/**.

10

The Determinants of Fertility

10.1 Introduction

Most of the variations in fertility which we can observe between different populations can be understood by looking at a relatively small number of *intermediate factors*. This chapter will be devoted to describing the effect of these factors. Demographers sometimes call them the *proximate determinants of fertility*. The model of the determinants of fertility that is used most often by demographers states that fertility is determined ultimately by social, economic and cultural factors, but that these operate through a number of intermediate factors, or proximate determinants of fertility (Davis and Blake, 1956; Bongaarts, 1978).

What are these intermediate factors or proximate determinants? There are, broadly speaking, three: *nuptiality*, specifically the age of women at marriage and the proportion of women ever marrying; *breastfeeding* (together with periods of abstinence from sexual intercourse after the birth of each child); and *birth control*, by which is meant the use of contraception and induced abortion. These factors have important effects on both the level of fertility and the distribution of births within the reproductive age range.

In this chapter attention is focused on the analysis of their effects on the level of fertility; in the next chapter we consider ways of analysing their effects on the distribution of births within the reproductive age range by considering the analysis of the intervals between births. The material in this chapter therefore provides a context within which to set the material in the next.

In Sections 10.2 and 10.3, the effects of nuptiality are examined. Section 10.2 makes the point that whether or not a woman is married is a principal determinant of her fertility. This may seem an obvious point, but there are a number of issues relating to the definition of marriage which need to be clarified. In Section 10.3 a number of measures of the fertility of married women are described. Sections 10.4 and 10.5 look at the way in which breast-feeding and contraceptive use affect fertility levels.

It will become clear as the discussion unfolds that a sensible way to look at the effect of these three proximate determinants is to view them as potentially reducing the portion of the childbearing age range for which a woman is at risk of becoming pregnant and bearing children. This is the approach taken by Bongaarts (1978), and Bongaarts and Potter (1983) who devised a method of quantifying their effects, which is summarized in Section 10.6. Section 10.7 summarizes the effects of the proximate determinants in real populations by an examination of some typical reproductive histories. Reproductive history data are examined further in Chapter 11.

10.2   The effects of marriage on fertility

We have seen already that fertility varies with age. After age, the next most important feature about a woman that determines her fertility is her marital status. In almost all populations, the fertility of married women is higher (usually much higher) than the fertility of single, widowed or divorced women. Because the majority of children are born to married women in most populations, much analysis of fertility only considers fertility within marriage, or *marital fertility*.

Now it might be objected that, in the 1990s, there are a good many births outside legal marriages in countries such as the United Kingdom. In fact, in the United Kingdom in the last few years, more than 30% of all births have taken place outside legal marriages. Thus only to consider births within legal marriages in analyses of fertility would seem to ignore a lot of fertility. There are two questions to consider here.

1   Is legal marriage a good definition of marriage for the purposes of analysing fertility differentials?
2   How representative is the United Kingdom of the world as a whole?

Let us consider the answers to these two questions.

The answer to the first question is, in short, 'no', because many couples live together, and bear children together, as if they were married, without ever going through a legal marriage ceremony. In England and Wales, for example, more than half of the illegitimate births these days are registered in the names of parents who live at the same address. In other words, they are born of cohabiting parents. In some other countries, a large proportion of couples (who may live together as husband and wife for many years) never formally marry. This is true, for example, in many Caribbean countries. It would appear, then, that in many populations, a strict legal definition of marriage is not really a very good one to use for the purposes of analysing marital fertility.

What we would like to know is not so much whether a woman is legally married, but whether or not she is in a steady sexual relationship – the kind of relationship in which childbearing is culturally and socially sanctioned. In other words, we want to include what are sometimes called 'common-law marriages' in our definition.

This is not easy. Censuses rarely collect the required information, and, in any case, people are often vague or untruthful in their responses to questions about cohabitation. Survey data are rather better at obtaining accurate responses.

Moving on to the second question, the United Kingdom has a higher proportion of children born outside legal marriages than most countries. Even though we treat cohabiting couples as 'married', the United Kingdom probably also has a higher proportion of children born to couples who are neither legally married nor living together as husband and wife than most countries. In many developing countries, for example, legal marriage is almost universal (this applies in most of Africa, and much of Asia), and childbearing outside legal marriage is quite rare.

Having clarified these issues, then, we can finally see how nuptiality reduces the proportion of her childbearing age range during which a woman is likely to have children. A woman who is not married has a much reduced risk of having sexual intercourse, and thus a much reduced risk of having children. This statement makes it clear why it is not the legal definition of marriage which matters, so much as the 'common-law' definition. Unfortunately, however, most national statistical agencies and other organizations which collect demographic data still rely on the legal definition.

At what point in the childbearing age range is the impact of nuptiality felt most keenly? It is most important at the younger ages, since the age at first marriage in most populations is greater than 15 years. Thus a very large proportion of women aged, say 15–19 years, are not married.

Of course, if divorce and widowhood are common in a population, then this will mean that many women spend parts of their childbearing age range in these marital statuses, which also reduces their risk of having sexual intercourse. However, the impact of divorce and widowhood on fertility is not so great as that of never being married, for two reasons. First, many women never become divorced or widowed, whereas all women will, at some stage in their lives, be in the state 'never married'. Second, divorced and widowed women tend to be older than never married women, and at older ages the physiological ability to have children is reduced.

10.3   Measures of fertility specific to marital status

So important is marital status as a determinant of fertility, that many of the fertility measures described in Chapters 8 and 9 are routinely calculated specific to marital status.

For example, recall the general fertility rate (GFR), which was described in Section 8.2. It was defined there as

$$GFR = \frac{\text{total number of births in a year}}{\text{mid-year population of women aged 15–49 years last birthday}}.$$

A general fertility rate specific to marital status is the *general legitimate fertility rate*, which is defined as

$$\text{general legitimate fertility rate} = \frac{\text{total number of legitimate births in a year}}{\substack{\text{mid-year population of married women aged} \\ \text{15–49 years last birthday}}},$$

where a *legitimate birth* is one occurring to a legally married woman. A strict legal definition of marriage is used to determine the exposed-to-risk in the denominator. Of course, the age range 15–44 years can also be used in the denominator.

We can also calculate a *general illegitimate fertility rate*, using only illegitimate births in the numerator, and non-married (that is, never married, widowed and divorced) women in the denominator.

Age-specific marital fertility rates (ASMFRs) are calculated in the same way as age-specific fertility rates, save that only legitimate births and married women are considered. Thus we have

$$\text{ASMFR}_x = \frac{\substack{\text{births in a year to married women aged } x \\ \text{last birthday at the time of the birth}}}{\substack{\text{mid-year population of married women} \\ \text{aged } x \text{ last birthday}}},$$

where ASMFR_x is the age-specific marital fertility rate at x last birthday.

Finally, a measure known as the *total marital fertility rate* (TMFR) is sometimes used. It is calculated using the formula

$$\text{TMFR} = \sum_{x=15}^{49} \text{ASMFR}_x.$$

The TMFR is the number of children a woman would end up with at exact age 50 years (the end of her childbearing age range) assuming that she marries at exact age 15 years, that she remains married all the time from her 15th to her 50th birthdays, and that she has, at each age x last birthday from ages 15 to 49 years, the $ASMFR_x$s used in the calculation. Clearly, it also supposes that the woman survives until her 50th birthday.

There is just one problem with using the TMFR to compare the fertility experiences of different populations. The age at marriage in many populations is higher than 15 years, and very few women are married at ages below about 20 years. Because of this, estimates of the ASMFRs for ages under 20 years are very imprecise. Moreover, the assumption that a woman marries at exact age 15 years does not represent the experience of many women (indeed, if the minimum legal age of marriage is greater than 15 years it may not represent the experience of *any* women). In such cases it makes more sense only to calculate ASMFRs for ages 20 to 49 years last birthday, and to calculate the TMFR using the formula

$$TMFR = \sum_{x=20}^{49} ASMFR_x,$$

and this is often done. Of course, five-year age groups can be used in the calculation, as with the total fertility rate for all women (see Exercise 10.1).

Age-order-specific fertility rates and parity progression ratios can also be calculated for married women only. Indeed, in the case of parity progression ratios, using only married women resolves the problem of how to measure the progression to the first birth, for a woman's marriage can be regarded as equivalent to a 'zeroth' birth, and the parity progression ratio from 0 to 1 births then measures the proportion of those women who ever marry who go on to have at least one child.

10.4   The effects of breastfeeding and abstinence from sexual intercourse after birth

BREASTFEEDING

The vast majority of infants in the world are breastfed for the first few months of life. In developed countries, breastfeeding frequently takes place only for a few months (fewer than six months is typical). In developing countries, however, many children are breastfed for much longer periods. For example, in Tanzania, the median age at which children cease to be fed from the breast is about 22 months (Hinde and Mturi, 1996). Similar figures characterize much of Africa and Asia.

Breastfeeding greatly reduces the chance that a woman will have a child because while she is breastfeeding, ovulation is inhibited. If a woman does not ovulate, then she is highly unlikely to get pregnant, even if she has regular sexual intercourse.

Therefore, if a woman breastfeeds each child for a period of, say, y years, then her ability to conceive will be much reduced for a total of ny years during her childbearing age range, where n is the number of children she has. In Africa, where n is, on average, about 6, and y is normally at least 1.5, the average woman has this reduced chance of having children for about nine years during her childbearing age range. Since the childbearing age range is 35 years, breastfeeding means that for more than a quarter of her childbearing years the average African woman will be quite unlikely to get pregnant.

The effect of breastfeeding is felt throughout the childbearing years. Its effect is to lengthen the intervals between successive births and thereby to reduce the level of fertility.

ABSTINENCE FROM SEXUAL INTERCOURSE AFTER THE BIRTH OF EACH CHILD

In most populations, there is a period after the birth of each child during which a woman does not have sexual intercourse. In developed countries, this period of *postpartum abstinence* is quite short (sometimes only a week or two). In developing countries, however, it can be much longer. Clearly, if a woman does not have sexual intercourse, she will not get pregnant.

However, the duration of the period of abstinence from sexual intercourse is usually less than the duration of breastfeeding. Thus the effect of abstinence from sexual intercourse after the birth of each child over and above the effect of breastfeeding is not usually very great.

10.5   Birth control

Birth control includes the use of contraception and induced abortion.

CONTRACEPTION

Modern contraceptive methods are very effective at preventing fertility. In developed countries, the vast majority of women use contraception at some stage during their child-bearing years. In developing countries, however, the use of contraception is often much rarer. The effect of contraception in preventing childbearing is seen throughout the child-bearing years, but tends to be more important at the older ages within this age range (at ages 30 years and over). This is because in most populations where contraception is common, women typically want only a few children. After they get married, they tend to have these children quite quickly. Once they have had the number they want, they use contraception to prevent further births.

We should not forget, though, that contraception may also be used to space births more widely, and thus it can affect the intervals between births.

ABORTION

Induced abortion is quite widely used to prevent births. In some senses, it is a substitute for the use of contraception. However, the effect of induced abortion in preventing births is less important than the effect of contraception. This is because of legal restrictions on induced abortions in many countries, together with the fact that abortion is seen by many as morally undesirable.

10.6   Quantifying the effects of the proximate determinants of fertility

Bongaarts (1978) and Bongaarts and Potter (1983) have proposed a simple but ingenious method of quantifying the relative effects of the proximate determinants of fertility in a given population. Their model is based on the view that the effect of each of the proximate determinants – marriage, breastfeeding and postpartum abstinence, and contraception and induced abortion – is to reduce fertility in a given population from some hypothetical level which might be achieved. This hypothetical level Bongaarts and Potter (1983) refer to as the

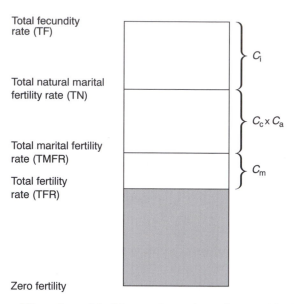

Total fecundity rate (TF)

Total natural marital fertility rate (TN)

Total marital fertility rate (TMFR)

Total fertility rate (TFR)

Zero fertility

C_i

$C_c \times C_a$

C_m

Figure 10.1 Bongaarts and Potter's model of the proximate determinants of fertility. The shaded area denotes the observed fertility. The unshaded area denotes potential fertility which is inhibited by non- or late marriage, contraception, induced abortion and postpartum infecundability. Source: Adapted from Bongaarts and Potter (1983, Fig. 4.1, p. 79).

total fecundity rate. It is defined as the average number of children which each woman in a population would have if marriage were universal in the childbearing age range, if breast-feeding and postpartum abstinence from sexual intercourse did not occur, and if no contraception and induced abortion were practised. The total fecundity rate is thus determined largely by physiological factors, and does not vary greatly between human populations, typically ranging between 13 and 17.

Of course, the fertility actually observed in a given population reflects the extent to which the proximate determinants actually cause fertility in that population to diverge from the total fecundity rate (TF). Suppose that the observed fertility is measured by the period total fertility rate (TFR). Bongaarts and Potter (1983) then define two further measures: the total marital fertility rate (TMFR), which is defined in the same way as described in Section 10.3 above, and the *total natural marital fertility rate* (TN), which is that fertility which would be achieved in the population with the TMFR in the absence of contraception and induced abortion. The four rates are shown in Figure 10.1.

It is clear that the difference between the TF and TN is accounted for by the prevalence of breastfeeding and postpartum abstinence in a population. Bongaarts and Potter refer to their joint effects by the term *postpartum infecundability*. The difference between the TN and TMFR is accounted for by the prevalence of contraception and induced abortion. Finally, the difference between the TMFR and TFR measures the fertility-inhibiting effect of delayed marriage, non-marriage and marital disruption.

The model proposes that four indices be calculated which quantify these fertility-inhibiting effects. They are defined as follows (Bongaarts and Potter, 1983, p. 80):

- C_m is an index of marriage, ranging in value from 1 if all women of reproductive age are married, to 0 in the absence of marriage;

- C_c is an index of contraception, ranging in value from 1 if no contraception is used in the population to 0 if all women use perfectly effective contraception;
- C_a is an index of induced abortion, ranging in value from 1 if abortion does not occur to 0 if all pregnancies are aborted;
- C_i is an index of postpartum infecundability, ranging in value from 1 if there is no breastfeeding or postpartum abstinence, to 0 in the (rather improbable) case that the duration of infecundability is infinite.

The relationship between these four indices and the four fertility rates listed above is as follows:

$$\text{TFR} = \text{TMFR} \times C_m, \tag{10.1}$$

$$\text{TMFR} = \text{TN} \times C_c \times C_a, \tag{10.2}$$

$$\text{TN} = \text{TF} \times C_i. \tag{10.3}$$

Substituting from equation (10.3) into equation (10.2) and then into equation (10.1) produces

$$\text{TFR} = \text{TF} \times C_i \times C_c \times C_a \times C_m.$$

ESTIMATION OF THE FERTILITY INDICES

In order to use Bongaarts and Potter's model to analyse the proximate determinants of fertility in a given population, it is necessary to be able to estimate the four fertility indices C_m, C_c, C_a and C_i from data. Bongaarts and Potter propose the following formulae.

First, the index C_m is estimated using a weighted average of the proportions currently married by age, where the weights are the age-specific marital fertility rates at each age. Thus

$$C_m = \frac{\sum_{x=15}^{x=49}(\text{ASMFR}_x \times \pi_x)}{\sum_{x=15}^{x=49}(\text{ASMFR}_x)}, \tag{10.4}$$

where the π_x are the proportions currently married at each age. Five-year age groups are often used in equation (10.4) instead of single years of age. The use of ASMFRs as weights ensures that the index C_m accurately captures the fertility-inhibiting effects of late marriage and non-marriage, as it gives greater weight to late marriage and non-marriage at those ages where fertility within marriage is highest.

The index C_c is calculated using the formula

$$C_c = 1 - 1.08ue, \tag{10.5}$$

where u is the proportion of married women of reproductive age currently using contraception, and e is the average use-effectiveness of contraception. The value of e depends on what is known as the contraceptive *method mix*: for a given proportion of married women using contraception, e will be higher the more effective are the methods being used. If, for a given contraceptive method, m, the use-effectiveness of that method is e_m, and a proportion u_m of currently married women are using that method, then

$$e = \frac{\sum u_m e_m}{\sum u_m},$$

where the summations are over all the different contraceptive methods being used. Clearly, $\sum u_m = u$, so

$$e = \frac{\sum u_m e_m}{u},$$ (10.6)

and substituting from equation (10.6) into equation (10.5) produces

$$C_c = 1 - 1.08 \sum u_m e_m.$$

Bongaarts and Potter (1983, p. 84) suggest that suitable values for e_m for different contraceptive methods are 0.9 for the pill, 0.95 for intra-uterine devices, 1.00 for sterilization, 0.8 for periodic abstinence (for example, the rhythm method), 0.9 for the condom, and 0.7 for other traditional methods, such as withdrawal.

The index C_i is estimated using the formula

$$C_i = \frac{20}{18.5 + i},$$

where i is the average duration (in months) of postpartum infecundability produced by breastfeeding and postpartum abstinence. This formula takes account of the fact that even when breastfeeding and postpartum abstinence do not occur in a population, a short period of postpartum infecundability (lasting about 1.5 months) will occur for physiological reasons. The value of i is not the same as the average duration of breastfeeding or abstinence (whichever average is the longer), but is closely related to this average. Since the average duration of breastfeeding is usually longer than the average duration of postpartum abstinence, Bongaarts and Potter (1983, p. 25) provide an equation for estimating i from a knowledge of the average duration of breastfeeding, d_b:

$$i = 1.753 \exp(0.1396 d_b - 0.001872 d_b^2).$$

Finally, the index C_a may be estimated using the formula

$$C_a = \frac{\text{TFR}}{\text{TFR} + 0.4(1 + u)\text{TA}},$$

where TA is the total abortion rate, or the average number of induced abortions a woman would have in her life (Bongaarts and Potter, 1983, pp. 85–86). Unfortunately, TA is rarely known, especially for the populations of developing countries. In such cases, it is common to assume that C_a is equal to 1.0. Clearly this is unrealistic, although some justification for it can be adduced by noting that in many developing countries induced abortion is illegal (and therefore possibly rather uncommon) and that the fertility-inhibiting effect of each induced abortion is relatively small, especially where the practice of efficient contraception is rare.

Finally, since the total fecundity rate does not vary very much between populations, it is sometimes reasonable, and might be helpful, to assume a value for TF. Bongaarts and Potter propose the value of 15.3.

ILLUSTRATION

The use of the Bongaarts–Potter model may be illustrated using data from the Tanzania Demographic and Health Survey, 1991–92 (Mturi and Hinde, 1994). The reported total fertility rate in Tanzania in the three years preceding this survey was 6.12. Mturi and

Hinde calculated the values of the three indices C_m, C_c and C_i as being 0.756, 0.908 and 0.587, respectively. They assumed that C_a was equal to 1.00. The estimated total fecundity rate was obtained as

$$TF = TFR/C_m C_c C_i$$
$$= 6.12/(0.756 \times 0.908 \times 0.587)$$
$$= 15.19.$$

The total marital fertility rate (TMFR) is equal to $TFR/0.756 = 6.12/0.756 = 8.10$. The total natural marital fertility rate is equal to $TMFR/0.908 = 8.10/0.908 = 8.92$.

Therefore, of the difference between the TFR and TF, the proportion accounted for by the fact that not all women in the childbearing ages are married, is equal to $(8.10 - 6.12)/(15.19 - 6.12) = 22\%$. The proportion accounted for by the use of contraception is $(8.92 - 8.10)/(15.19 - 6.12)$, or 9%. The remaining 69% of the difference is accounted for by the long periods of breastfeeding in Tanzania, which lead to lengthy intervals after each birth during which women are at a much reduced risk of conceiving.

10.7 Reproductive histories

We can summarize the effect of the major determinants of fertility in real populations by looking at some typical reproductive histories. A woman's reproductive history is a potted biography which includes dates relevant to her childbearing career, typically her date of birth, the date of her first marriage, the dates of birth of her children, and so on (Figure 10.2). A full reproductive history would also include the date(s) of death of her husband(s), the dates of any divorces she had, and the dates of any remarriages. It would also include the dates in which she entered steady sexual relationships which were not legally recognized as marriages, and the dates on which these relationships ended. Clearly, obtaining data on a full reproductive history is very difficult, and in practice demographers have to rely on only a subset of the information being available.

The major characteristics of the reproductive histories of average women in different populations can be summarized by looking at four different societal groups. In modern developed countries, marriage typically takes place in a woman's early twenties. Thereafter, a small number (usually no more than three) of children are born quite soon, and fairly close together. Once the children have been born women use contraception to prevent

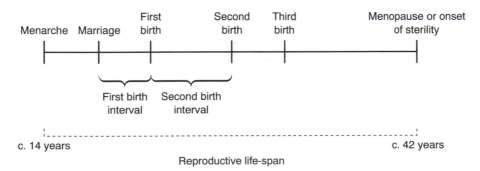

Figure 10.2 Illustration of a reproductive history

further births. The reproductive history therefore has a very long open birth interval (see Section 11.2).

In north-west Europe before about 1900, marriage was typically quite late – at about 25 years or older (Hajnal, 1965). Thereafter, women had children quite regularly. Contraceptive use was rare. The result was that five or six children would be born, with the last arriving on or around the woman's 40th birthday.

In developing countries, the age at first marriage is young, often under 20 years. Contraceptive use is much less prevalent than in developed countries, but births are spaced widely apart because of long periods of breastfeeding. The outcome is that six or seven children are born, with the last arriving on or around the woman's 40th birthday.

Finally, we return to the Hutterites, a religious sect in North America (see Section 8.4). Hutterite women who married between the two world wars have the highest fertility ever reliably recorded in a large population. Although their age at marriage was not very young, their limited practice of breastfeeding and their avoidance of contraception led to about nine or ten children being born to the average woman.

It is, therefore, clear that the three major proximate determinants of fertility have very different relative impacts in different populations. In modern developed populations, birth control is a much more important inhibitor of fertility than the other two; in most contemporary developing countries, postpartum infecundability is still the most important factor inhibiting fertility; in historical European populations, late and non-universal marriage was the main reason why fertility fell well below its biological potential.

Further reading

Readers who are interested in the details of the Bongaarts–Potter model should consult Bongaarts and Potter (1983). A succinct description is provided in Bongaarts (1978). There is a wealth of literature on the determinants of fertility. A good collection of papers is to be found in Bulatao and Lee (1983).

Exercises

10.1   The data in Table 10E.1 show the numbers of married women by age in England and Wales in 1991, together with the number of first, second, third, and fourth and higher-order live births recorded to married women in that year.

Table 10E.1

Age group	Number of married women (thousands)	Number of births by order (thousands)			
		1	2	3	4 or higher
Under 20	26.1	6.7	2.0	0.2	0.0
20–24	431.4	51.2	32.8	9.4	2.3
25–29	1108.6	84.5	73.9	26.8	11.1
30–34	1299.4	40.2	53.0	27.5	14.8
35–39	1281.7	9.7	14.7	10.5	8.9
40–49	1467.3	1.3	1.9	1.8	2.7

Source: *Population Trends* 87 (1997), p. 53; Office of Population Censuses and Surveys (1993a, p. 64).

Table 10E.2

Year	Mid-year female population aged 16–44 (thousands)			Births (thousands)	
	Never married	Married	Widowed and divorced	Total	To married women
1971	2434	6419	200	783.2	717.5
1981	3196	6235	523	634.5	553.5
1991	4093	5780	829	699.2	487.9

Source: *Population Trends* 87 (1997), pp. 49 and 53.

(a) Calculate age-order-specific marital fertility rates for 1991.
(b) Calculate total marital fertility rates for each birth order.
(c) Calculate the total marital fertility rate for 1991.
(d) Comment on your results.

10.2 Table 10E.2 shows the female population of England and Wales classified by marital status in various years, and the number of live births classified by marital status in the same years. Use these data to calculate general legitimate fertility rates and general illegitimate fertility rates for the years in question. Note that the age range 16–44 years should be used in the denominator.

10.3 In nineteenth-century England it was common for unmarried women who became pregnant to get married prior to the birth. In rural areas, as many as 40% of brides were pregnant when they got married. Discuss the implications of high rates of bridal pregnancy like this for:
(a) the use of age-specific fertility rates to measure fertility;
(b) the use of age-specific marital fertility rates to measure fertility;
(c) the usefulness of the total marital fertility rate as a measure of fertility.

10.4 The data in Tables 10E.3 and 10E.4 are taken from the Demographic and Health Survey of Jordan in 1990. Use them to estimate the indices C_i, C_c and C_m in the Bongaarts–Potter model of the proximate determinants of fertility, and hence to estimate the TMFR, TN and TF. Note that the average length of postpartum infecundability in Jordan, according to the Demographic and Health Survey data, is about eight months (that is, $i = 8$). Assume that there is no abortion in Jordan.

Table 10E.3

Age group	Age-specific marital fertility rate (per woman)	Proportion of women currently married
15–19	0.471	0.104
20–24	0.492	0.445
25–29	0.416	0.712
30–34	0.307	0.860
35–39	0.211	0.893
40–44	0.087	0.908
45–49	0.022	0.883

Source: Zou'bi *et al.* (1992, pp. 31, 56).

Table 10E.4

Method m	Proportion of women using u_m	Effectiveness e_m
Pill	0.05	0.90
Female sterilization	0.06	1.00
Intra-uterine device	0.15	0.95
Periodic abstinence	0.04	0.80
Other modern methods	0.01	0.90
Other traditional methods	0.08	0.70

Sources: Zou'bi *et al.* (1992, p. 38); Bongaarts and Potter (1983).

11

Birth Interval Analysis

11.1 Introduction

In Chapter 10, it was seen how the major proximate determinants or intermediate factors which affect fertility – nuptiality, breastfeeding and birth control – do so by influencing women's reproductive histories. This suggests that it might be helpful to analyse these reproductive histories directly. This analysis is the theme of this chapter. The analysis of reproductive histories is a complex and somewhat daunting undertaking, and this chapter can only serve as an introduction to some of the important issues which arise. Readers who are interested in pursuing the topic in more detail are referred to the suggestions for further reading given at the end of this chapter.

Section 11.2 describes the place of birth intervals within reproductive histories. It considers two important issues which arise immediately: the question of births before marriage, and the fact that reproductive histories contain birth intervals of two distinct types, usually referred to as 'open' and 'closed' intervals. The second of these issues has assumed greater importance in recent years, with the advent of large-scale retrospective surveys in which reproductive history data have been collected. The nature of these data and the problems they pose for birth interval analysis are the subject of Section 11.3. In particular, the analytical problems raised by censoring and selection are discussed.

After a brief discussion of the components of birth intervals in Section 11.4, the use of traditional life table analysis to study birth intervals is described in Section 11.5. In Sections 11.6–11.9 this theme is extended by looking at the use of survival analysis methods in the context of birth interval analysis. Section 11.6 shows how survival analysis can be applied to a wide variety of longitudinal data. Section 11.7 discusses the incorporation of covariates into survival analysis in the context of the proportional hazards model. Section 11.8 describes the estimation of the parameters in a proportional hazards model, and Section 11.9 illustrates extensions to the survival analysis model which are often of use in the analysis of birth interval data.

11.2 Birth intervals

Look again at Figure 10.2, which shows the reproductive history of a typical woman. This reproductive history can be viewed as a series of events. These events are, in chronological order:

- *menarche*, or the onset of the ability to conceive and bear children (this happens between the ages of about 11 and 18 for most women);
- *marriage*, or, using the 'common-law' definition we considered in the previous chapter, entry into a stable sexual union;
- the *births*, in order (note that multiple births are treated here as single events);
- *menopause*, or the onset of sterility (this happens between the ages of about 35 and 50 for most women).

We might note in passing that it is possible for a woman to get married before menarche – indeed, it happens not infrequently in a few societies. Nevertheless, in global terms, the proportion of women who get married prior to menarche is very small.

A *birth interval* is the length of time elapsing between each birth and the event which preceded it. Each birth interval is given a number relating to the birth which ends the interval. Thus a woman's *first birth interval* is the length of time elapsing between her marriage and her first birth, her *second birth interval* is the length of time elapsing between her first birth and her second birth, and so on.

This seems straightforward enough, but there are at least two difficulties with this definition.

1   Some women have children before they marry. For these women, the length of the first birth interval according to the definition just given will be negative. Note that this might be the case whether or not the legal definition of marriage is used.
2   There is a period of time after the last birth (which for many women, especially in developed countries, is very long), and which, according to the definition just given, is not part of any birth interval.

Let us consider these in turn.

BIRTHS BEFORE MARRIAGE

The first difficulty (women having children before they marry) is sometimes circumvented in the analysis of real populations by only analysing the birth histories of women who marry before the birth of their first child. (Whether a strict legal definition of marriage or a 'common-law' definition is used depends on factors such as the availability of data and the frequency of premarital or extramarital cohabitation in the population under investigation.)

Only analysing the birth histories of women who marry before the birth of their first child can effectively eliminate from the analysis quite a large proportion of women. This is especially true in developed countries during the last 20 years if a legal definition of marriage is used. It is much less true of many developing countries. One could, of course, ignore births before marriage when drawing up a woman's reproductive history, treat the first birth after marriage as if it were a woman's first ever birth, and renumber all successive births accordingly. This would increase the fraction of women whose reproductive histories could be included in the analysis. This increase, however, is bought at the price of knowing that not all the events we are analysing are alike. We might ask whether the interval between marriage and the first birth after marriage for a woman who did not have a child prior to her marriage will be determined by the same factors as the corresponding interval for a woman who did have a child (or children) before her marriage.

Alternatively, we could simply ignore marriage (whether legally recognized or not). This raises two issues. First, the first birth interval will need to be redefined, as it can no longer start with marriage. There are various possible approaches. We could start the first birth interval at menarche. However, data about the onset of menarche are rather rare in practice. Alternatively, we could start the first birth interval at an age which is just below the youngest age at which women in the population we are studying start to bear children.

The second issue is that by ignoring marriage we are inevitably conflating the experience of women who married at different ages. If the reproductive experience of women who married at a young age is very different from those who married when they were much older, there is a danger that the results obtained by ignoring this heterogeneity will represent no more than a very few of the women in the actual population, being a kind of rather artificial 'average'.

There is no ideal solution to the problem of births before marriage. As a general guide, ignoring marriage is often a workable solution in populations where the age at marriage is roughly the same for a large proportion of women, and in which the vast majority of women marry. Such is the case in many developing countries. The unfortunate thing, of course, is that it is in precisely these populations that the proportion of births before marriage is lowest, and so the problem is least acute. In populations where ages at marriage vary widely, and in which a relatively high proportion of births take place before marriage, a simple solution to the problem does not exist. However, the use of survival analysis, which is described later in this chapter, offers a more complex but flexible approach.

OPEN AND CLOSED BIRTH INTERVALS

The second difficulty with the definition of birth intervals is caused by the fact that there is a period of time after the birth of a woman's last child during which she is still capable of bearing children but does not.

Since a primary aim of the analysis of birth intervals is to try to understand the determinants of fertility, the fact that a woman does not have another child is important, since it may reveal behavioural patterns such as the use of contraception. Birth control is widely used in many populations to prevent a further birth once a woman has had the number of children she and her partner want.

The time which elapses after the birth of a woman's last child is called the *open birth interval*. Open birth intervals do not end with a birth. The other birth intervals, which do end with a birth, are known as *closed birth intervals*. For any analysis of birth intervals to contribute fully to our understanding of fertility, open birth intervals need to be included.

11.3   Data for the analysis of birth intervals

The problem of open birth intervals is even more important than suggested in Section 11.2. This is because of the nature of a large proportion of the reproductive history data which demographers now have available.

Because censuses and vital registration do not yield sufficient information about individual women for reproductive histories to be constructed, relevant data are usually derived from surveys. As described in Section 1.6, there are two types of survey data which might be used: prospective studies and retrospective surveys.

PROSPECTIVE STUDIES

These follow a sample – ideally a randomly drawn sample – of women (for example, a birth cohort) through their lives, and note down the relevant events as and when they happen. An example is the British Office for National Statistics Longitudinal Study of a 1% sample of the population of England and Wales.

The Longitudinal Study includes data on a sample of people born on one of four dates in the year. (For reasons of confidentiality, these dates are not known to the public; indeed, they are known only to a handful of people.) Information about these people, taken from the population censuses of 1971, 1981 and 1991, is included, as are details of the births of children to the women in the Longitudinal Study, as and when they occur. Data on deaths and migration are also linked to the census information.

Prospective studies are very expensive to undertake, as continuous monitoring of events is needed over a long period of time. For this reason, retrospective surveys are often used to gather data on reproductive histories.

RETROSPECTIVE SURVEYS

These are questionnaire surveys done at a particular point in time, in which (ideally) a random sample of women are interviewed. The women in the sample are asked to give details of their past lives, including the dates of their marriage(s) and the dates of birth of all their children so far. Other information might be collected, such as the dates of any divorces they have experienced, the date of death of any husband who has died, contraceptive use, and other social and economic characteristics of the women.

Because they can be carried out at a single point in time, retrospective surveys are cheaper and easier to administer than prospective studies. There are a great many examples which can be used for the analysis of birth intervals, of which a few are listed below.

1   The 1980 Women and Employment Survey of over 5000 British women aged 16–59 years in 1980.
2   The Demographic and Health Surveys in developing countries. There are now more than 40 of these, which have been undertaken in Africa, Asia and Latin America since 1985. They interview a sample of several thousand women aged (typically) between 15 and 49 years in each country.
3   The World Fertility Survey, which comprised a series of surveys similar in form to the Demographic and Health Surveys, but carried out during the 1970s, and including some developed countries as well as developing countries. Many of the methods now used by demographers to analyse birth intervals were developed in order to analyse World Fertility Survey data (see Hobcraft and Rodríguez, 1980).

THE PROBLEM OF CENSORING

Both retrospective surveys and prospective studies suffer from the problem of censoring. The cause of the problem differs for the two types of survey.

Prospective studies

Censoring in these usually occurs because some women drop out of the study before their reproductive histories are complete (perhaps because they emigrate, or because they die, or because they become fed up with providing information to those carrying out the survey). This, of course, is essentially the same problem of censoring which was described in Chapters 5 and 6 in the context of mortality investigations. Women in prospective studies of fertility who die, or who emigrate, are analogous to withdrawals from mortality investigations carried out by life insurance companies.

Retrospective surveys

Censoring in retrospective surveys occurs because some women who are interviewed are too young to have completed their fertility. This phenomenon was mentioned in Section 9.4 (see Figure 9.1). Most women in retrospective surveys will not have completed their childbearing; indeed, some women will probably not yet have had any children. Thus, most women's reproductive histories are censored. This does not mean that all the birth intervals observed in retrospective surveys are open intervals. In fact, every woman can have one, and only one, open birth interval. Retrospective surveys will contain data on many closed birth intervals.

THE PROBLEM OF SELECTION

There is a further difficulty with retrospective surveys. Suppose we wished to analyse the length of the interval preceding the third birth for each woman. The women in the survey can be divided into three groups:

1   Women who have had a third birth by the time of the survey (for example, woman A in Figure 11.1). Such women will have closed third birth intervals.
2   Women who have had a second birth by the time of the survey but who have not had a third birth (for example, woman B in Figure 11.1). Such women will have open third birth intervals.

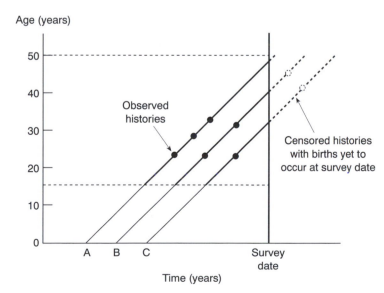

Figure 11.1 Retrospective survey data and the analysis of birth intervals (dots denote births)

3 Women who have not had a second birth by the time of the survey (for example, woman C in Figure 11.1). Such women will not have third birth intervals at all.

The last category of women will not be included in the analysis of third birth intervals. Of course, some of them may well go on to have second (and third) children, but we cannot observe this. The problem with not including them is that it may bias the sample. In other words, the women who are in categories **1** and **2** above may not be representative of all women. For example, women who marry and have their children at a young age are more likely to be in categories **1** or **2** than women who marry and have children when they are rather older.

This phenomenon is known as *selection* (or sometimes *selectivity*). The women whom we do include in the analysis are selected for having certain characteristics – in this case early childbearing. If we analyse these women as if they were a random sample from the population, as was the survey as a whole, and try to make inferences from our results about the population from which the sample was taken, we might reach erroneous conclusions about the determinants of fertility in the population as a whole.

Selection might also be a problem in prospective studies if women who withdraw from the study are different in some way from women who do not (see Section 11.6).

11.4 The components of birth intervals

The final preliminary we need to examine before looking at the analysis of birth intervals concerns the components of birth intervals. Figure 11.2 shows that all closed birth intervals except the first interval (between marriage and the first birth) have three separate components.

1 There is a time after the birth which began the interval (known usually as the *index birth*) during which the woman cannot become pregnant because of abstinence from sexual intercourse or a lack of ovulation caused by breastfeeding. This interval is known as the *postpartum infecundable interval*.
2 Once the woman begins to ovulate again, she can, in principle, become pregnant. Of course, whether she does conceive will depend on many factors, of which the chief one is the use of contraception. The length of time which elapses between the end of the postpartum infecundable interval and the woman becoming pregnant is known as the *conception wait*, or sometimes the *waiting time to conception*.
3 Between the conception and the birth which ends the birth interval is a period of *gestation*.

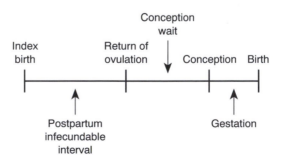

Figure 11.2 Components of birth intervals

The duration of the first of these components is determined mainly by the duration of abstinence after the birth of the index child, and/or the length of time for which that child is breastfed. The duration of the second of these components is determined mainly by contraceptive use. Thus the duration of both components **1** and **2** is determined by behavioural factors which may vary from population to population, and from woman to woman within a population.

The duration of gestation, however, is always about nine months, and does not vary from population to population, or (very much) from woman to woman. Thus it is not very interesting to analyse this duration if we are wanting to find out about the determinants of fertility.

In practice, therefore, closed birth intervals are often modified to relate to the time between the index birth and the conception of the next birth. The date of conception of the next birth is obtained by subtracting nine months from the reported date of the birth which ends the birth interval.

Open birth intervals do not have component **3** in the list above (gestation). In retrospective surveys they may not have component **2** either, if the survey takes place very shortly after the index birth. This creates a further complication with retrospective survey data. Some women may be pregnant at the time of the survey (see Figure 11.3). This creates a problem for the analysis if we are going to use the conception of the next birth as the event which ends a birth interval, for we cannot calculate the date of conception by subtracting nine months from the date of births which have not occurred by the survey date (since we do not know the dates of such births – or indeed whether they will occur at all). If we do not do something to sort this out, we shall misclassify the birth intervals of women who are pregnant at the time of the survey as open intervals, whereas in fact they are really closed intervals.

One solution which is adopted in practical work is to imagine that the survey took place nine months before it actually did. We then ignore all the data which relate to the last nine months before the survey (the shaded area in Figure 11.3), except the data on births within nine months of the survey date, which we use to determine the dates of conception of these

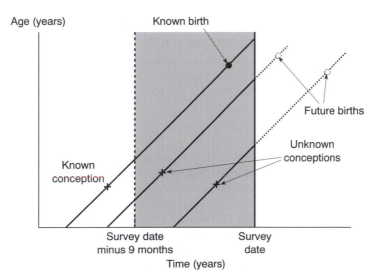

Figure 11.3 Lexis chart illustrating the problem of pregnancy at the time of the survey in retrospective surveys

babies. If we do this, we can work out whether or not each woman was pregnant at a point nine months before the actual survey date. We then proceed with the analysis using the data up to the date nine months before the survey date.

The first birth interval is also different from the others, in that it does not have component **1**, the postpartum infecundable interval. Because of this, first birth intervals are often analysed separately from the rest.

Finally, there is a problem with pregnancies which do not result in a live birth. These will not usually be reported in surveys, but they will clearly tend to lengthen birth intervals. It is difficult to solve this problem in practical work, since most surveys do not ask about miscarriages and abortions. In such a case, demographers simply acknowledge their awareness of the problem, and live with it. The proportion of conceptions which do not result in a live birth is quite high; however, most conceptions which do not lead to a live birth arise because of miscarriages very early in pregnancy, and these do not tend to lengthen birth intervals by very much.

11.5   Life table analysis

In Chapter 9, it was shown how life table analysis could be used to calculate period parity progression ratios. It should by now be clear that the same kind of analysis can be used to analyse birth intervals. In the conventional life table (described in Chapter 4), we follow a group of people all born during a particular period, and construct a life table based on age-specific death rates for this birth cohort. An analogous procedure can be applied to birth intervals by following a group of women who all had a birth of a particular order during a certain period, and constructing a life table based on the proportions who conceive their next child at given durations since this index birth. (In the case of first birth intervals, marriage or the attainment of some age immediately prior to childbearing can be used as the initial 'event'.) We therefore have 'duration-specific conception rates' instead of age-specific death rates. Censoring can be handled in this analysis quite easily using the method described in Section 5.7.

Life tables of this type are often employed in birth interval analysis, but there are a number of difficulties with them. These arise chiefly from the heterogeneity among the women typically included in the analysis of any particular birth interval. In earlier sections of this chapter many possible sources of such heterogeneity were mentioned. For example, women who had their first birth prior to marriage may have subsequent fertility behaviour which is different from that of other women. Fertility also varies considerably with the age of the woman. Comparison of populations, or of subgroups within the same population, even for the same birth interval, may therefore be seriously confounded if the age composition of the women included in the analysis differs for different populations. This difficulty is compounded by the selectivity issue discussed in Section 11.3.

An obvious way to circumvent problems posed by heterogeneity is to stratify the women in the sample on the basis of confounding factors such as age or premarital childbearing, and calculate separate life tables for the same birth interval for each of the resulting sub-groups. Unfortunately, when using survey data, this potential solution tends to result in very small numbers of women in some strata, with the result that the duration-specific conception rates have to be based on extremely small numbers.

There is, however, a solution to this problem which is becoming adopted increasingly widely by demographers. The solution involves survival analysis, which, as we showed in Chapter 6, is an alternative way of looking at the life table. In the remaining sections of this chapter, the application of survival analysis to birth intervals is described. Some of the material is quite difficult. In particular, Section 11.8 may be omitted by readers unfamiliar with calculus and daunted by algebra.

11.6   Survival analysis

In Chapter 6 we outlined a situation in which we followed a cohort of people from their 40th birthday until the first of three possible events occurred: death before their 41st birthday; withdrawal from the investigation while still alive before their 41st birthday; or their 41st birthday. All the people in the investigation could be classified according to which of these three things happened to them. A person who withdrew from the investigation while still alive, or who remained under observation until his/her 41st birthday, was considered to have his/her life censored at either the date of withdrawal or the date of his/her 41st birthday. We also showed that we could apply exactly the same method to analyse mortality between any two exact ages x and $x + n$. In this section this scenario is generalized to illustrate how survival analysis may be used to analyse and model reproductive histories and, in particular, birth intervals.

The critical features of the scenario just outlined are as follows:

1   There is a well-defined *start event* which denotes the point at which we begin to observe each person. In the case of mortality investigations, this start event was, typically, each person's xth birthday.
2   There is a well-defined *terminal event*. In the case of mortality investigations, this terminal event was death.
3   We have a means of measuring, for each person, the *duration since the start event*. We can call this means of measuring a *metric of time*. In the case of mortality investigations in which the start event is birth, the metric of time is simply a person's age. If the start event is a person's xth birthday (where $x > 0$), then the metric of time is age minus x.
4   For people who do experience the terminal event during the period of the investigation, we measure (using our metric of time) the duration between the start event and the terminal event.
5   For people who do not experience the terminal event during the period of the investigation, we measure (using our metric of time) the duration between the start event and some later time which we know to be before they experience the terminal event. This later time is the time at which they are censored.

There is, in fact, another important feature, which was not mentioned in Chapter 6. This is that *censoring is uninformative*. By 'uninformative', we mean that the fact that a person's history is censored tells us nothing about when he/she is likely to experience the terminal event, but only that he/she has not experienced it at the duration when his/her history is censored. This issue deserves a little discussion in the context of birth interval analysis.

Data for analysing birth intervals come typically from either retrospective survey data or prospective studies. In Section 11.3 it was shown that the cause of censoring in retrospective history data is the occurrence of the survey itself. Since this event occurs to all women in the survey, and at a time which is determined by the people arranging and financing the survey, not by any of the women who are being surveyed, it may be assumed that any censoring which arises is unrelated to the fertility behaviour of any of the surveyed women. Thus in retrospective surveys, the assumption that censoring is uninformative seems reasonable.

In prospective studies, however, censoring occurs because of decisions taken by individual members of the cohort being studied (for example, their decision to emigrate or not to participate further in the survey). It may well be that women whose histories are censored differ in their fertility behaviour from those who are not censored.

Unfortunately, there is no easy way of testing the assumption of uninformative censoring in prospective studies. Some idea of possible differences between those women who are lost to follow-up and those who are not might be obtained by looking at the composition of the two groups with respect to background factors (age, social class, age at marriage, and so on), but this is only an approximate test.

Let us return to our reproductive history data. In the remainder of this chapter, we are going to suppose that we have data from a retrospective survey in which questions have been asked about the reproductive histories of a random sample of *n* women aged 15–49 years last birthday at the time of the survey. (This might seem a very specific supposition, but this is by far the most common form of reproductive history data.)

Suppose that we are interested in analysing the length of the second birth interval (that is, the duration between the first birth and the conception which leads to the second birth). Figure 11.4 depicts the data for four women in the sample. Woman A had her first child when she was exactly 35 years of age, and her second child when she was exactly 40 years of age. Thus the interval between her first birth and the conception of her second child was $4\frac{1}{4}$ years (since conception occurs nine months before birth). Woman B had four children before the survey date, the first when she was exactly 20 years of age and the second when she was exactly $22\frac{3}{4}$ years of age. Thus the interval between her first birth and the conception of her second child was two years. Woman C only had one child prior to the survey. Thus her second birth interval is open. However, since we know that woman C was aged exactly 25 years when she had her first child, and we know that she was aged exactly 33 years at the time

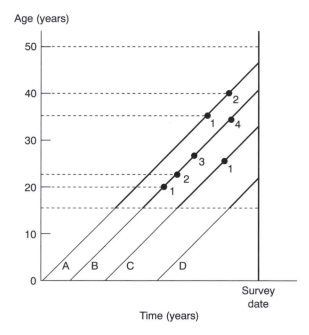

Figure 11.4 Lexis chart illustrating reproductive history data (dots denote births)

of the survey, the interval between her first birth and her second conception was at least $7\frac{1}{4}$ years (remember that she might have been pregnant at the time of the survey).

Woman D in Figure 11.4 does not have a first birth prior to the survey date, and therefore we cannot include her in the analysis of the second birth interval. The fact that we exclude certain women from the analysis raises the possibility of non-random selection (see Section 11.3).

Look back at the list of critical features of the survival analysis scenario. Does the second birth interval scenario have these features? Consider them in turn.

1   Well-defined start event. The birth of the first child is the start event. It is well defined. We know when it happened for each woman who had a first birth prior to the survey date.
2   Well-defined terminal event. The terminal event is the conception of the second child. This is well defined provided that we only consider conceptions occurring more than nine months before the survey date.
3   Metric of time. The metric of time is the duration since a woman's first birth.
4   Duration until terminal event. For women who conceived their second child more than nine months before the survey date, the length of the second birth interval may be calculated either by subtracting the age of each woman at the birth of her first child from her age when she conceived her second child, or by subtracting the date at which a woman bore her first child from the date when she conceived her second child. Either of these will produce a measure of the duration since a woman's first birth.
5   Duration until censoring. For women who did not conceive their second child more than nine months before the survey date, the length of the second birth interval (in years) is given by

$$\begin{array}{c}\text{length of second} \\ \text{birth interval}\end{array} > \begin{array}{c}\text{survey} \\ \text{date}\end{array} - 0.75 - \begin{array}{c}\text{date of birth} \\ \text{of first child}\end{array}.$$

Thus we have a survival analysis scenario. It is illustrated in Figure 11.5, in which the second birth intervals of women A, B and C from our example have been depicted, using duration since first birth on the horizontal axis, rather than age.

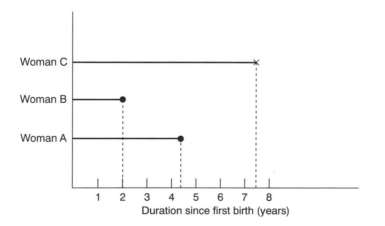

Figure 11.5 Chart showing the second birth intervals of women A, B and C. Dots denote conceptions of second children; the cross denotes a censored observation. Source: Figure 11.4

We are, therefore, going to analyse the distribution of the random variable 'duration between first birth and conception of second child', which we denote as T. The random variable T has a survivor function, $S(t)$, a probability density function $f(t)$, and a hazard function $h(t)$, defined in exactly the same way as they were with the analysis of mortality data.

We are interested in estimating these functions from the data we have. The procedure is exactly the same as before. We use the method of maximum likelihood, which consists of the three stages described in Section 6.8.

11.7   Covariates

We now introduce an additional feature of the analysis. Samples of women used in the analysis of birth intervals are, as has been seen, typically rather heterogeneous. This heterogeneity confounds the comparison of different populations. Moreover, in Chapter 10 it was shown how various determinants of fertility affect reproductive histories, and therefore birth intervals.

Analysts are, therefore, often interested in both controlling for confounding factors (such as age) and measuring the effect of certain variables or factors on the lengths of the birth intervals (such factors might include breastfeeding behaviour, use or non-use of contraception, marital status, or social and economic variables). We call these factors or variables *covariates*. How may they be incorporated?

Recall that in the survival analysis of mortality, the hazard was assumed to be constant at some unknown value λ. We estimated λ using data on all the people in the investigation.

We can incorporate a covariate into our analysis of the second birth interval by supposing that the hazard of conceiving a second child *is different for women who have different values of the covariate*. For example, consider the covariate *age at first birth*. We suppose that the hazard of conceiving a second child depends on a woman's age at the birth of her first child. This, after all, is likely to be the case, for it is known that a woman's ability to become pregnant varies with her age for biological reasons.

We need, therefore, to specify the hazard as a function of the duration t since a woman's first birth and her age at the first birth. This may be done by assuming that the hazard $h(t)$ is given by the equation

$$h(t) = \lambda \exp(\beta a), \tag{11.1}$$

where a denotes age at first birth, and λ and β are unknown parameters. Equation (11.1) says that the hazard of a woman conceiving her second child is constant over all durations since her first birth, but that the value of this constant varies according to her age when she had her first birth. For a woman who had her first birth at age 20 years, the hazard is equal to $\lambda \exp(20\beta)$. For a woman who had her first birth at age 25 years, the hazard is equal to $\lambda \exp(25\beta)$, and so on. For a woman who had her first birth at age zero, the hazard is equal to $\lambda \exp(0\beta)$, which is just λ.

This seems fine, but there is a slight difficulty. Women do not give birth at age zero, so the parameter λ seems to be purely hypothetical. We can avoid this problem by replacing the covariate a in equation (11.1) by a covariate a^*, where a^* is, say, age at first birth minus 15 years. If we do this, we can rewrite equation (11.1) as

$$h(t) = \lambda^* \exp(\beta a^*). \tag{11.1a}$$

The value of a^* for a woman who had her first birth at exact age 20 years would be equal to $20 - 15 = 5$. The parameter λ^* in equation (11.1a) measures the hazard of conceiving a second birth for a woman who had her first child when she was aged exactly 15 years. This is so, because a^* is then $15 - 15 = 0$, so the hazard for a woman who had her first child at exactly 15 years is equal to $\lambda^* \exp(0\beta)$, which is just λ^*. The parameter λ^* is not hypothetical, since some women do, indeed, have children at age 15 years.

Another feature of both equations (11.1) and (11.1a) is very important. They posit that the effect of age at first birth on the hazard is *multiplicative*. In equation (11.1a), for example, the hazard of conception (of a second child) for a woman who has her first birth at exact age a years is equal to the hazard of conception (of a second child) for a woman who has her first birth at age 15 years, λ^*, multiplied by the quantity $\exp[\beta(a - 15)]$. This implies that the hazards for two women of different ages are *proportional*.

To see this, consider two women, one who had her first child at exact age 25 years (call her woman K) and one who had her first child at exact age 30 years (call her woman L). We have, using equation (11.1a),

$$\text{hazard of conception for woman K} = \lambda^* \exp[\beta(25 - 15)]$$

$$= \lambda^* \exp(10\beta)$$

and

$$\text{hazard of conception for woman L} = \lambda^* \exp[\beta(30 - 15)]$$

$$= \lambda^* \exp(15\beta).$$

Thus

$$\frac{\text{hazard of conception for woman L}}{\text{hazard of conception for woman K}} = \frac{\lambda^* \exp(15\beta)}{\lambda^* \exp(10\beta)}$$

$$= \frac{\lambda^* \exp[(10\beta + 5\beta)]}{\lambda^* \exp(10\beta)}$$

$$= \frac{\lambda^* \exp(10\beta) \exp(5\beta)}{\lambda^* \exp(10\beta)}$$

$$= \exp(5\beta).$$

In other words, the effect of age is to multiply the hazard by a constant factor, where, by the word 'constant', we mean a factor which does not depend on the duration since the start event.

11.8   Estimating the parameters

All we now need to do to measure the effect of a woman's age at first birth on the hazard of her getting pregnant with her second child is to estimate β. Of course, to measure the hazard itself we also need to estimate λ (or λ^*).

First, we write down the likelihood as a function of the parameters. If we define t_i to be the duration between the first birth and the point at which a woman either conceives her second child or is censored, and define a variable δ_i, such that if woman i conceives her second child then δ_i is equal to 1, and if woman i does not conceive her second child by the date nine months before the survey date then δ_i is equal to zero, then the likelihood, L, for the whole sample of n women, is given by the equation

$$L = \prod_{i=1}^{n} [f(t_i)]^{\delta_i} [S(t_i)]^{1-\delta_i}. \tag{11.2}$$

(Compare this with equation (6.12).)

We now need to express L as a function of the unknown parameters β and λ (we shall use λ rather than λ^* from now on for simplicity – it makes no difference to the exposition).

First, we noted in Chapter 6 that

$$S(t) = \exp\left[-\int_0^t h(u)\,du\right]. \tag{11.3}$$

If $h(u)$ is equal to $\lambda \exp(\beta a)$, then equation (11.3) becomes

$$S(t) = \exp\left[-\int_0^t \lambda \exp(\beta a)\,du\right].$$

Performing the integration gives us

$$S(t) = e^{-\lambda t \exp(\beta a)}. \tag{11.4}$$

We also know from Chapter 6 that

$$f(t) = h(t)S(t). \tag{11.5}$$

If $h(t)$ is equal to $\lambda \exp(\beta a)$, and $S(t)$ is equal to $e^{-\lambda t \exp(\beta a)}$, then, substituting these values into equation (11.5), we obtain

$$f(t) = \lambda \exp(\beta a)[e^{-\lambda t \exp(\beta a)}]. \tag{11.6}$$

Using equations (11.4) and (11.6) to substitute into equation (11.2) expressions for $f(t_i)$ and $S(t_i)$ in terms of λ and β produces the following equation for the likelihood:

$$L = \prod_{i=1}^{n} (\lambda \exp(\beta a_i)[e^{-\lambda t_i \exp(\beta a_i)}])^{\delta_i} (e^{-\lambda t_i \exp(\beta a_i)})^{1-\delta_i}, \tag{11.7}$$

where a_i is the age at first birth of woman i.

Equation (11.7) looks complex. It may, however, be simplified by noting that it can be written

$$L = \prod_{i=1}^{n} [\lambda \exp(\beta a_i)]^{\delta_i} [e^{-\lambda t_i \exp(\beta a_i)}]^{\delta_i} [e^{-\lambda t_i \exp(\beta a_i)}]^{1-\delta_i}.$$

Since $\delta_i + (1 - \delta_i)$ is equal to 1, this simplifies to

$$L = \prod_{i=1}^{n} [\lambda \exp(\beta a_i)]^{\delta_i} [e^{-\lambda t_i \exp(\beta a_i)}]. \tag{11.8}$$

We can then take natural logarithms of equation (11.8) to obtain

$$\ln L = \sum_{i=1}^{n} \delta_i \ln[\lambda \exp(\beta a_i)] + \sum_{i=1}^{n} \ln[e^{-\lambda t_i \exp(\beta a_i)}]. \tag{11.9}$$

Since $\ln(e^z)$ is simply equal to z, the second term on the right-hand side of equation (11.9) becomes simply $\sum_{i=1}^{n} -\lambda t_i \exp(\beta a_i)$, so we have

$$\ln L = \sum_{i=1}^{n} \delta_i \ln[\lambda \exp(\beta a_i)] - \sum_{i=1}^{n} \lambda t_i \exp(\beta a_i). \tag{11.10}$$

Moreover, since $\ln[\lambda \exp(\beta a_i)]$ is equal to $\ln \lambda + \ln[\exp(\beta a_i)]$, which is equal to $\ln \lambda + \beta a_i$, equation (11.10) becomes

$$\ln L = \sum_{i=1}^{n} \delta_i \ln \lambda + \sum_{i=1}^{n} \delta_i \beta a_i - \sum_{i=1}^{n} \lambda t_i \exp(\beta a_i). \tag{11.11}$$

Equation (11.11) can be partially differentiated with respect to λ and β to obtain

$$\frac{\partial(\ln L)}{\partial \lambda} = \frac{\sum_{i=1}^{n} \delta_i}{\lambda} - \sum_{i=1}^{n} t_i \exp(\beta a_i)$$

and

$$\frac{\partial(\ln L)}{\partial \beta} = \sum_{i=1}^{n} \delta_i a_i - \sum_{i=1}^{n} \lambda t_i a_i \exp(\beta a_i).$$

For the partial derivatives to be zero, we must therefore have

$$\frac{\sum_{i=1}^{n} \delta_i}{\lambda} = \sum_{i=1}^{n} t_i \exp(\beta a_i) \tag{11.12}$$

and

$$\sum_{i=1}^{n} \delta_i a_i = \sum_{i=1}^{n} \lambda t_i a_i \exp(\beta a_i). \tag{11.13}$$

Solving the simultaneous equations (11.12) and (11.13) for λ and β is not straightforward, but it can be done iteratively. (Note that the second-order derivatives of $\ln L$ with respect to λ and β are both negative, so we do have maxima in both cases.)

This method can be applied to any number of covariates simultaneously (the algebra becomes more complex, but computer programs to solve the likelihood equations are widely available). Moreover, to make the procedure more accurate, each birth interval may be broken down into smaller-duration segments, as was done in the case of mortality by analysing each year of age separately, and a separate hazard estimated for each segment. The overall life table for any birth interval can then be constructed from the results using the method described in Section 6.9.

11.9 Extensions using survival analysis

We have seen how we can now estimate not just the hazard of conception but also the effect of covariates on the hazard. The model we have outlined is very simple, but it can be made more complicated. This can be done in several ways.

One way is to allow the hazard to depend on the duration since the first birth. This is sensible, since it seems likely that a woman's chance of becoming pregnant with her second child will vary with the duration since her first birth. This may happen because women have views about what constitutes a desirable birth interval. We could, for example, specify the hazard as follows:

$$h(t) = (\lambda_0 + \lambda_1 t)\exp(\beta a).$$

We now have three parameters to estimate: λ_0, λ_1 and β.

It is also possible to break a birth interval down into segments, and estimate the hazard of conception within each segment, allowing the hazard to be constant within each segment but to vary between segments. This is, in effect, just what was done in Chapter 6 when using survival analysis to estimate a life table. A person's life was divided into years of age, and the hazard of death was assumed to be constant within each year of age.

We can also allow the effect of a covariate to depend on the duration since the first birth. This might be achieved by specifying the hazard function as follows:

$$h(t) = \lambda \exp[a(\beta_0 + \beta_1 t)].$$

Here again there are three parameters to estimate: λ, β_0 and β_1. Alternatively, it can be achieved by dividing a birth interval into segments, and allowing the value of a covariate to change between segments.

This extension is especially valuable since it enables us to measure the effects of breast-feeding and contraception on the length of birth intervals. Clearly, a woman will stop breastfeeding the index child part of the way through a birth interval. Similarly, a woman may use contraception for part of a birth interval, and then stop using it once she decides that she wants to have another baby. To take account of this, we can define covariates which take the value 1 if a woman is breastfeeding (or using contraception) and 0 if she is not breastfeeding (or using contraception). The actual values of these covariates will vary over time for the same woman. In particular, if a woman's birth interval is divided into segments, they will typically change between segments.

Whatever we do, the method of maximum likelihood can be used to estimate the parameters. The algebra can become more complicated, however, and we often have to use iterative methods.

Further reading

A classic work on the analysis of birth intervals from retrospective survey data is Hobcraft and Rodríguez (1980). Werner (1988a; 1988b) describes illustrative analyses of the reproductive histories of the women in the Office for National Statistics Longitudinal Study. On the range of applications of life table analysis, see Cox (1975b).

Readers interested in other applications of survival analysis may like to consult Allison (1984), Yamaguchi (1991) and Blossfeld et al. (1989). These texts are all very practical in their orientation, and describe the use of various widely available computer packages to perform survival analysis.

Exercises

11.1   The data in Table 11E.1 come from the 1980 Women and Employment Survey of British women. This was a retrospective survey, in which over 5000 women aged 16–59 years in 1980 were interviewed, and details of their reproductive histories collected. The table gives the distributions of the length of the interval between the first and the second birth (in other words, the numbers of women who had second birth intervals of different lengths) for three of the birth cohorts included in the

Table 11E.1

Length of birth interval (months)	Number of women		
	1930–39 birth cohort	1940–49 birth cohort	1950–64 birth cohort
18 or fewer	116	138	45
19–24	99	136	47
25–30	111	138	45
31–36	104	126	38
37–60	155	173	42
61 and over	80	49	6
Did not have a second birth before the survey	125	175	254

Source: Derived from the Women and Employment Survey, 1980.

survey. The table only relates to women who had their first birth when they were aged 20–29 years.

(a) Draw a Lexis chart to illustrate the survey, and the sample of women for whom data are presented in Table 11E.1.

(b) Suggest reasons for the differences in the distributions of birth interval length among the three birth cohorts.

11.2 A population of women has a hazard of conception, $h(x)$, which does not vary with the number of months since their previous birth – that is,

$$h(x) = \lambda.$$

If the expected remaining waiting time to conception among those who have not yet conceived three months after their previous birth is five months, calculate λ.

11.3 Describe how you might use data from retrospective surveys to investigate the following:

(a) the effects of social and economic factors on birth interval length;

(b) the effect of breastfeeding on birth interval length.

11.4 It has been suggested that waiting times between marriage and the conception of their first postmarital birth for women who were not pregnant when they got married follow a *Weibull* distribution. In this distribution, the survivor function $S(x)$ is given by the formula

$$S(x) = \exp[-(\lambda x)^{\beta}].$$

where λ and β are parameters.

(a) Derive an expression for the hazard of conception, $h(x)$, in terms of λ and β.

(b) In general, it is the case that the hazard of conception of the first child decreases monotonically with duration since marriage. What values of β will give rise to a hazard which decreases monotonically with duration x?

12

Population Growth

12.1 Introduction

So far in this book, mortality and fertility have been examined separately. The time has now come to look at how they act together to determine population growth.

In Chapter 1, the basic demographic equation was introduced. This equation says that, if we consider the population, P_t, of a country at some time t, then the size of this population one year later, P_{t+1}, is given by

$$P_{t+1} = P_t + B_t - D_t + I_t - E_t,$$

where B_t and D_t are respectively the numbers of births and deaths occurring in the population between times t and $t+1$, and I_t and E_t are respectively the numbers of immigrants and emigrants during the same period.

The quantity $B_t - D_t$, known as natural increase, is clearly dependent upon a population's fertility (which determines B_t) and mortality (which determines D_t).

In this chapter and the next we shall be looking at the relationships between mortality, fertility and population growth. Migration will not be considered. Indeed, we shall assume that migration is negligible. The assumption that migration is negligible is, of course, unrealistic for most populations. However, it may be justified on two grounds.

1 For national populations, migration (especially net migration) is not normally very great in comparison to the size of the population. For example, in England and Wales, net migration during the past ten years has rarely been more than 50 000 people per year, against a total population of around 50 million.
2 By ignoring migration we can construct elegant mathematical models which relate population growth to fertility and mortality. These models prove to be very useful in a variety of applications.

The analysis of migration forms the subject of Chapter 15.

This chapter proceeds as follows. Section 12.2 discusses the relationship between fertility and population growth, and introduces an alternative measure of fertility which is more closely related to population growth: this measure is known as the gross reproduction rate. In Section 12.3, the joint effects of fertility and mortality on population growth are analysed by considering a measure known as the net reproduction rate. Section 12.4 discusses geometric and exponential population growth. Finally, Section 12.5 shows how the exponential rate of population growth is related to the net reproduction rate, and hence to fertility and mortality.

12.2   Fertility and population growth

In Section 8.5, a measure known as the total fertility rate (TFR) was introduced. The TFR is obtained by summing the age-specific fertility rates (ASFRs) over the ages from 15 to 49 years last birthday. It is a measure of the total number of children a woman will have in her life (provided that she survives until age 50 years). If we denote the ASFR at age x last birthday by the symbol f_x, we have

$$\text{TFR} = \sum_{x=15}^{49} f_x.$$

The TFR gives us an indication of whether or not a population is growing. It does this because we can suppose that if the average woman has more than two children in her life, the population will grow. So, if the TFR in a population exceeds 2.0, we can suppose that that population will, as time goes on, increase in size. If the TFR falls below 2.0, the population will decline.

However, the TFR, as a measure of population growth, neglects two important additional factors which, in practice, affect population growth: the sex ratio of births, and mortality.

THE SEX RATIO OF BIRTHS

Population growth depends much more on the number of girls that are born than on the number of boys. Thus the *sex ratio of births* is important. Just to make the point using a trivial example, imagine a population in which every woman had three children, but that they were all boys. Such a population would become extinct within one generation, despite having a TFR greater than 2.0.

The sex ratio of births is taken into account by the *gross reproduction rate* (GRR). The GRR is calculated using the same principle as the TFR, except that only female births are counted. Thus the GRR is the total number of daughters a woman would have in her life (assuming that she survives until age 50 years). The formula for the GRR is

$$\text{GRR} = \sum_{x=15}^{49} f_x^{\text{d}},$$

where we use f_x^{d} rather than f_x to denote the fact that we are only counting daughters: f_x^{d} is the age-specific fertility rate at age x last birthday for daughters only. The sex ratio of births in most populations of European descent is about 106 boys to every 100 girls. In populations of African descent it is nearer to 103 boys to every 100 girls. In some Asian populations it is greater than 106 boys to every 100 girls. Thus the GRR is, typically, just under half the TFR.

MORTALITY

Some of the children born will not survive to have children themselves. We know that age-specific death rates in infancy and childhood are quite high, especially in developing countries (where, typically, only nine out of every ten children born survive until their first birthday). Children who die in infancy will not contribute to the growth of the population in the long run, since they will have no children themselves.

What matters for the rate of population growth, therefore, is the number of daughters a woman has who survive to have children themselves. This fact is explicitly taken into account by a measure known as the net reproduction rate.

12.3   The net reproduction rate

The *net reproduction rate* (NRR) is simply the gross reproduction rate adjusted for mortality. It is a most important measure of the rate of population growth.

A good way to understand how the NRR works is to imagine a birth cohort of women. The original number of women in this birth cohort will be denoted by the symbol l_0. At each age x last birthday, they have daughters at an age-specific fertility rate for daughters only of f_x^d.

What is f_x^d measuring? It is measuring the average number of daughters that one woman has while she lives through the entire year of age between exact ages x and $x + 1$. (Remember that age-specific fertility rates like f_x^d assume that mortality is zero.) In other words, f_x^d measures the average number of daughters born per woman-year lived between exact ages x and $x + 1$, and we can write the following:

$$
\begin{array}{c}
\text{number of daughters born between} \\
\text{exact ages } x \text{ and } x + 1 \text{ to a} \\
\text{birth cohort of women}
\end{array}
= f_x^d \times
\begin{array}{c}
\text{number of woman-years lived} \\
\text{between exact ages } x \text{ and} \\
x + 1 \text{ by the birth cohort}
\end{array}
$$

Thus, if there is no mortality, l_0 women will have a total of $f_x^d l_0$ daughters as they live through the year of age between exact ages x and $x + 1$. Each woman will live exactly one woman-year between exact ages x and $x + 1$. Thus each of the l_0 women will have f_x^d daughters between exact ages x and $x + 1$.

To take mortality into account, all we need to do is to adjust the number of woman-years lived between exact ages x and $x + 1$. This adjustment is made using the life table function L_x, the very definition of which is the number of woman-years lived between exact ages x and $x + 1$ by a cohort of l_0 women. Thus, if the mortality experience of the birth cohort of women is described by a life table, in which L_x is the number of woman-years lived between exact ages x and $x + 1$ by l_0 women born, then we have

$$
\begin{array}{c}
\text{number of daughters born between exact ages} \\
x \text{ and } x + 1 \text{ to a birth cohort of women}
\end{array}
= f_x^d L_x.
$$

Summing these quantities over all ages x last birthday from 15 to 49 years gives the total number of daughters born to a birth cohort of l_0 women, who, at each age, experience an ASFR (for daughters only) of f_x^d, and whose mortality is described by the life table of which L_x is a function. In symbols, therefore,

$$
\begin{array}{c}
\text{total number of daughters born} \\
\text{to a birth cohort of } l_0 \text{ women} \\
\text{(after adjusting for mortality)}
\end{array}
= \sum_{x=15}^{49} f_x^d L_x.
$$

The average number of daughters born per woman is, therefore, just $\sum_{x=15}^{49} f_x^d L_x$ divided by l_0. This average number of daughters born per woman is the NRR.

Now, because l_0 was chosen arbitrarily when drawing up the life table, we could choose it to be equal to 1. If we do this, then we can write

$$NRR = \sum_{x=15}^{49} f_x^{\mathrm{d}} L_x.$$

It is important to remember that when we define the NRR using this equation, the L_x refers to the number of woman-years lived between exact ages x and $x + 1$ by a birth cohort of size 1. (Some people have difficulty with this idea, because they have the notion that people are indivisible. Although this notion may be useful in everyday life, it is, at times, unhelpful in demographic analysis.)

Note that, if we take $l_0 = 1.0$, L_x cannot be greater than 1.0. Therefore, the NRR will always be less than the GRR, unless there is no female mortality at ages under 50 years, in which case the two reproduction rates will have the same value.

The NRR is the total number of daughters that each member of a birth cohort of women produces, after allowing for the mortality of the birth cohort. If we consider the original birth cohort as one generation, then the NRR measures the *size of the next generation relative to the size of the present one*.

If the NRR exceeds 1.0, then the next generation will be bigger than the present one, and the population will grow. If the NRR is less than 1.0, then the next generation will be smaller than the present one, and the population will decline. If the NRR is equal to 1.0, then the next generation will be exactly the same size as the present one, and the population size will remain constant.

If fertility and mortality remain constant for a very long period of time (many generations), then the NRR will also remain the same, and the rate of population growth per generation will remain the same. Since the NRR is calculated using only the prevailing ASFRs, and the mortality rates which are part of the life table from which the L_x values are derived, this suggests that *a population with constant fertility and mortality rates will, in the long run, have a constant rate of growth*. In fact, this is true (see Chapter 14).

DATA FOR CALCULATING THE NET REPRODUCTION RATE

As we have defined it, the NRR is a cohort measure of population growth. Of course, we could calculate it using fertility and mortality data for a single calendar year (or for a particular period). However, the NRR calculated using data for a single calendar year will only be a good measure of the long-run growth of a population if the fertility and mortality rates for that calendar year are maintained for a long period of time.

Interpreting an NRR which has been calculated using data for a single calendar year as an indication of the rate of population growth is, therefore, a dangerous procedure. Consider, for example, what conclusions the Japanese would have reached about their population growth if they had calculated an NRR based only on data for 1966 (see Section 8.7). Since the TFR for the year 1966 was 1.6, then, unless there was a very remarkable sex ratio of births in that year, the GRR cannot have been greater than about 0.8, and the NRR would have been slightly less than this. Therefore, the NRR for Japan based on data for 1966 is less than 1.0. However, the NRR of most of the birth cohorts which contributed to the period fertility in 1966 is probably about 1.0. This is because the TFR for the years before and after 1966 was just over 2.0.

Incorrect interpretations of the NRR based on data for a particular period have been the cause of much unwarranted concern about population decline in the past.

12.4   Geometric and exponential growth

We have now introduced one measure of population growth, the net reproduction rate. In this section, we consider population growth more abstractly. We look at two models of population growth, the geometric growth model and the exponential growth model. We show that the exponential growth model is to be preferred as a description of how populations grow.

GEOMETRIC GROWTH

Consider a population at some point in time, and suppose that its size at that time is P_0. Suppose that its size one year later is P_1. The absolute change in the population size during the year is equal to $P_1 - P_0$, and the *rate of change*, which we shall denote by the symbol r, is defined by the equation

$$r = \frac{P_1 - P_0}{P_0}.$$

Rearranging this equation produces

$$P_0 r = P_1 - P_0, \qquad P_0 + P_0 r = P_1$$

and it can be seen that

$$P_1 = P_0(1 + r). \tag{12.1}$$

If the rate of change remains the same during the next year, then the population after one more year, P_2, will be given by the equation

$$P_2 = P_1(1 + r). \tag{12.2}$$

Substituting for P_1 from equation (12.1) into equation (12.2) gives

$$P_2 = P_0(1 + r)(1 + r)$$
$$= P_0(1 + r)^2.$$

Repeating this procedure for subsequent years produces the general formula

$$P_t = P_0(1 + r)^t. \tag{12.3}$$

If r is positive, we have population growth. If r is negative, we have population decline. Since decline is just negative growth, we can use the word 'growth' to describe population change in general.

   Equation (12.3) describes a process known as *geometric growth*. It shows the development of a population's size, assuming that the annual rate of growth is constant.

EXPONENTIAL GROWTH

It may be thought that geometric growth is quite a good description of how populations grow. Indeed it is, but we can do a little better. The geometric growth model in equation (12.3) assumes that the population change in each year, which involves people either being added to the population or taken away, happens at a single point in time. In other words, we have a *single annual increment*.

However, in reality, people are being born all the time, and they are dying all the time. Thus the population growth is a continuous process. How can we modify equation (12.3) to allow for this?

We have said that equation (12.3) implies a single annual increment to the population. (This is like the interest on a bank deposit being paid once a year.) Suppose that, instead of a single increment, we have two increments per year. Suppose further that the first of these increments occurs after six months, and the other at the end of the year (strictly speaking, just before the end of the year). Finally, suppose that the annual rate of growth remains the same.

If the annual rate of growth is r, then the two increments will each be at a rate of $\frac{1}{2}r$. Thus the population after six months, $P_{1/2}$, will be given by the equation

$$P_{1/2} = P_0(1 + \tfrac{1}{2}r).$$

The increment at the end of the year will also be at a rate of $\frac{1}{2}r$, but it will apply to the population after six months, $P_{1/2}$.

Thus, the population after one year, P_1, is given by the equation

$$
\begin{aligned}
P_1 &= P_{1/2}(1 + \tfrac{1}{2}r) \\
&= P_0(1 + \tfrac{1}{2}r)(1 + \tfrac{1}{2}r) \\
&= P_0(1 + \tfrac{1}{2}r)^2.
\end{aligned}
\tag{12.4}
$$

Equation (12.4) may be expanded to give

$$P_1 = P_0(1 + r + \tfrac{1}{4}r^2), \tag{12.5}$$

and comparing equation (12.5) with equation (12.1) reveals that adding the additional people in two increments produces a slightly larger population after one year than adding them in a single increment, assuming that the annual rate of growth, r, is held constant. The difference amounts to $\frac{1}{4}r^2$.

Of course, people are not added to populations just twice a year; they are added virtually continuously throughout the year. We can represent this by imagining that we have a annual growth rate r being achieved by a very large number of increments, each one of which is, of course, very small.

In general, if we have an increment j times a year, we can write a general form of equation (12.4) as follows:

$$P_1 = P_0\left[1 + \frac{r}{j}\right]^j. \tag{12.6}$$

Now, imagine that j becomes very large. For growth which is effectively continuous, we want to find out what happens to equation (12.6) as $j \Rightarrow \infty$. In symbols, we want to find

$$\lim_{j \Rightarrow \infty} P_0\left[1 + \frac{r}{j}\right]^j.$$

It can be shown that

$$\lim_{j \Rightarrow \infty}\left[1 + \frac{r}{j}\right]^j = e^r.$$

To show it, replace j in the expression $\lim_{j \Rightarrow \infty}[1 + (r/j)]^j$ by rM (that is, put $j = rM$). Then, since r is constant as far as the limiting process goes, $M \Rightarrow \infty$ with j, so we can write

$$\lim_{j \Rightarrow \infty} \left[1 + \frac{r}{j}\right]^j = \lim_{M \Rightarrow \infty} \left[1 + \frac{r}{rM}\right]^{rM}$$

$$= \lim_{M \Rightarrow \infty} \left[1 + \frac{1}{M}\right]^{rM}$$

$$= \lim_{M \Rightarrow \infty} \left(\left[1 + \frac{1}{M}\right]^M\right)^r$$

$$= e^r.$$

Thus, for a continuously increasing population, we can express equation (12.6) above in the form

$$P_1 = P_0 e^r,$$

and, if r is constant, we have

$$P_2 = P_1 e^r$$

$$= (P_0 e^r) e^r$$

$$= P_0 e^{2r}.$$

In general, therefore,

$$P_t = P_0 e^{rt}. \tag{12.7}$$

Equation (12.7) is of vital importance. It describes *exponential growth* (see Figure 12.1). Its importance for the analysis of population growth stems from the fact that it can be proved (see Chapter 13) that a population in which fertility and mortality are constant will be characterized by exponential growth. Equation (12.7), therefore, is our preferred model of population growth.

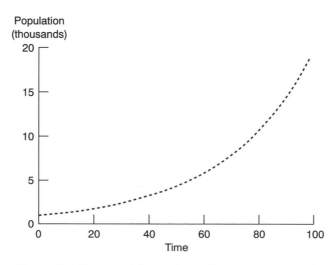

Figure 12.1 Exponential growth at 3% per year ($r = 0.03$)

12.5 The annual rate of growth and the net reproduction rate

Of course, equation (12.7) says nothing about fertility and mortality. It simply describes a rather abstract growth model. However, we can link the annual rate of population growth to the net reproduction rate.

Recall that the NRR is a measure of the population size in the next generation divided by the population size in the present generation. All we need to link the NRR to the annual rate of growth is to know how long a generation is. If we denote the *length of a generation* (in years) by the symbol g, then, if the size of the present generation is P_0, and the size of the next generation (that is, the generation alive in g years' time) is P_g, then

$$\text{NRR} = \frac{P_g}{P_0}.$$

But, if the population is growing exponentially, then we know from equation (12.7) that P_g is equal to $P_0 \, e^{rg}$. Thus, we can write

$$\text{NRR} = \frac{P_0 \, e^{rg}}{P_0} = e^{rg}.$$

Thus, we have expressed the net reproduction rate in terms of a constant annual rate of population growth.

Exercises

12.1 Table 12E.1 gives information on age-specific fertility rates in England and Wales in 1991, together with data on female mortality.
  (a) Assuming that 105 boys are born for every 100 girls, estimate the gross reproduction rate and the net reproduction rate.
  (b) What does your answer to part (a) tell you about population growth in England and Wales in the long run?

12.2 The mid-year population of Kenya was 18 million in 1982. Between 1970 and 1982 the average annual rate of growth was 4%. The World Bank estimated that, in mid-1990, Kenya's population was 26 million, and that by the middle of the year 2000 it will be 40 million (World Bank, 1984).
  (a) Assuming that the growth in the population of Kenya between 1982 and 1990, and between 1990 and 2000, is exponential, calculate the annual growth rates using the World Bank's estimates of the population.

Table 12E.1

Age group	Age-specific fertility rate	Female survivors to mid-point of age group per 10 000 women born
15–19	0.033	9903
20–24	0.090	9890
25–29	0.120	9871
30–34	0.087	9850
35–39	0.032	9817
40–44	0.006	9766
45–49	0.000	9685

Sources: *Population Trends* 87 (1997), p. 52; Office for National Statistics (1997a, p. 2).

(b)   Assume that the World Bank's estimate of 40 million in 2000 is correct. If Kenya's population continues to increase after 2000 at the same rate as the World Bank assumed it would increase between 1990 and 2000, when will it reach 80 million?

12.3 It is a well-known rule of thumb in demography that the length of time (in years) which a population takes to double in size may be approximately calculated using the formula

$$\text{doubling time} \cong \frac{70}{\text{growth rate in per cent per year}}.$$

This is known as the 'rule of 70'. Explain why this approximation works.

13

Models of Population Structure

13.1 Introduction

This chapter describes population structure and population dynamics. It begins by describing in Section 13.2 what demographers mean by the age structure of a population. Section 13.3 describes in general terms how the age structure of a population is related to its rate of growth, and hence to its fertility and mortality experience. A population's age structure is shown to be a dynamic, changing thing, revealing aspects of a population's demographic history, as illustrated in Section 13.4, and anticipating aspects of its demographic future.

Section 13.5 then introduces and defines two special types of population, which are called the stationary population and the stable population. These are simplified populations in which certain assumptions about fertility and mortality are met. Section 13.6 shows how the age structures of stationary and stable populations are related to their rate of growth, and hence to their fertility and mortality experience. These relationships can be expressed using relatively simple formulae.

The central propositions of the stable population model are that a population which has constant fertility and mortality for a long period will develop an age structure which does not change, in relative terms, over time (that is, there will be a constant proportion of people at each age), and this population will increase (or decrease) in size at a constant rate. Sections 13.7 and 13.8 include outline proofs of these propositions. The proofs require the use of calculus, and these sections can be omitted by readers who are not mathematically inclined. Finally, in Section 13.9, we return to the idea of the length of a generation, introduced in Section 12.5, and show how this quantity may be estimated from data on age-specific fertility and mortality.

13.2 The age and sex structure of a population

The *age and sex structure* of a population is simply the distribution of the population by age and sex. The terms *age and sex composition* and *age and sex distribution* are also used, and mean exactly the same thing.

Age and sex structures may be presented using absolute numbers or percentages. In the context of population dynamics, we usually deal with *percentage* age and sex structures, giving the percentage of the total population constituted by each age (and sex) group.

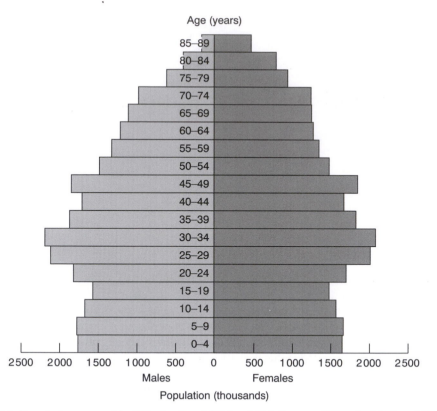

Age (years)

Figure 13.1 Population pyramid of England and Wales, 1995. Source: Office for National Statistics (1997b, p. 1)

Population pyramids are an elegant and useful way of presenting an age and sex distribution graphically. A pyramid looks like two ordinary histograms placed on their sides and back to back. The population pyramid for England and Wales in 1995 is shown in Figure 13.1. Population pyramids may be drawn using absolute numbers or percentages, and the horizontal scale should be labelled accordingly. When using percentages, the base for the percentages is the total population of both sexes combined (see Figure 13.2). Pyramids drawn using percentages are useful for comparing age and sex distributions, provided that the horizontal and vertical scales used in all the pyramids are the same.

THE DEPENDENCY RATIO

This is a frequently used summary measure of the age and sex structure. It is defined as the ratio of economically inactive to economically active persons. In other words, it measures the number of inactive people whom each economically active person has to support.

Since the economically inactive are mainly the young and the old, in practice the dependency ratio is calculated by dividing the total number of people under age a_1 and over age a_2 by the number of people between ages a_1 and a_2. Usually, a_1 is taken to be an age close to the school-leaving age, and a_2 an age close to the retirement age. In the United Kingdom, for example, $a_1 = 16$ years and $a_2 = 60$ or 65 years. The availability of data may determine

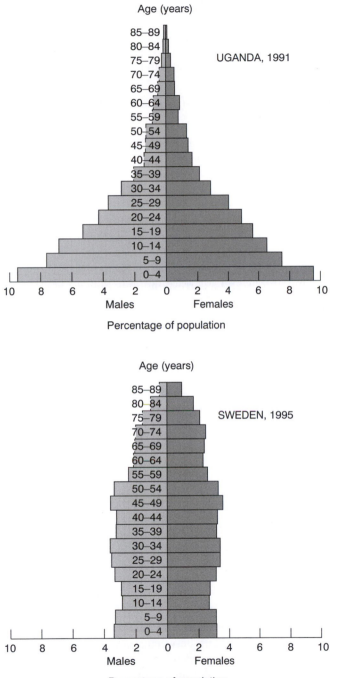

Figure 13.2 Population pyramids for Uganda (1991) and Sweden (1995). Sources: United Nations (1995b, pp. 190–191); Statistics Sweden (1997, p. 40)

the exact ages to be used in a particular population. In addition, it is possible to take a_2 to be different for males and females, if statutory retirement ages differ by sex.

An alternative measure of dependency is the proportion of the total population which is of working age (that is, aged between a_1 and a_2). This measure uses the same information as the dependency ratio, but is easier to understand.

13.3   The demographic determinants of the shape of the population pyramid

Fertility and mortality both affect the shape of the population pyramid. Their effects are rather different from one another, however.

FERTILITY

This has the most important effect in theory, and also often in practice. Its largely determines the width of the *base* of the population pyramid. High-fertility populations have pyramids which, when drawn using percentages, are wide at the base and narrow at the top (see the example of Uganda in Figure 13.2). Low-fertility populations tend to have a smaller percentage of children, and a correspondingly higher percentage of old people (see the example of Sweden in Figure 13.2).

MORTALITY

In general, mortality has a lesser effect than fertility on the shape of the population pyramid. We can distinguish two effects.

The effect of the overall level of mortality

This is quite small. In other words, populations which differ only in their level of mortality will have quite similar population pyramids (expressed as percentages of the population of each age).

The effect of the relative mortality levels of children and adults

This can be larger. If, for example, childhood mortality is high relative to adult mortality, the population pyramid will slope quite rapidly at young ages, but less rapidly at adult ages. Conversely, when adult mortality is high relative to that of children, the pyramid will slope more at adult ages, and there will be fewer adults aged over 50 years in the population.

THE RATE OF GROWTH

Since fertility and mortality together determine the rate of growth of a population, we can surmise that there will be a relationship between a population's growth rate and the shape of its population pyramid. This is indeed the case (see Figure 13.3). In general, a rapidly growing population will have a population pyramid which is wide at the base and which grows narrower rapidly as age increases. A population which is neither growing nor declining has a more 'rectangular' age pyramid, with only a slow decline with increasing age in the proportions in each age group. A population which is declining (because the number of births is fewer than the number of deaths) has a population pyramid which is narrower at the base than half-way up.

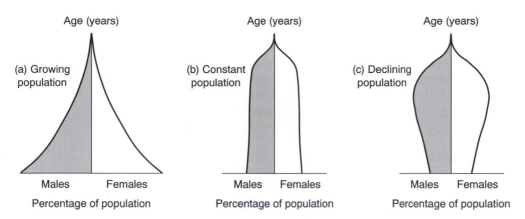

Figure 13.3 The relationship between a population's rate of growth and its age structure

13.4   The age and sex structure as a historical record

A country's age and sex structure is influenced by events in the past. Indeed, it can be seen as a kind of historical record. Events in the past which caused fertility and mortality to change will be 'remembered' by the age and sex structure long afterwards (although, as will be seen later, not for ever). For example, a temporary shortage of births caused by a war will leave a 'gap' in the population pyramid for many decades afterwards, where the missing children (who were not born during the war) should be. This 'gap' will gradually move up the population pyramid as the relevant birth cohort grows older. It will not disappear until everyone in the relevant birth cohort has died (and the 'gap' passes out of the top of the population pyramid).

This can be illustrated by examining the population pyramid of France in 1992, shown in Figure 13.4. The major features of this pyramid may be accounted for as follows:

1   There is a large 'gash' in the population pyramid at ages 72–77 years (marked 1 in Figure 13.4). This is the result of a large shortfall of births during the years 1914–19, when husbands and wives were separated by the military campaigns of the First World War.
2   The fact that there are fewer females in their fifties than there are in their sixties reflects a reduced number of births in the mid- and late 1930s when the small cohort born during the First World War were at their peak childbearing age. For males, the effect of changing fertility is offset by the greater mortality of males between the ages of 55 and 69 years.
3   There also appears to have been a deficit of births during the Second World War, which causes the indentation around age 50 years (marked 3 in Figure 13.4). This deficit of births is nowhere near as great as that during the First World War, but it occurred for the same reason.
4   There was a considerable boom in births immediately after the Second World War. Part of this boom is undoubtedly the result of demobilization (men and women being reunited after long periods spent apart during the war), but, since high fertility seems to have lasted for much longer than we should expect if demobilization were the only cause, there may be other factors at work.
5   A decline in fertility during the 1970s is reflected in the relatively small number of people aged 12–16 years.

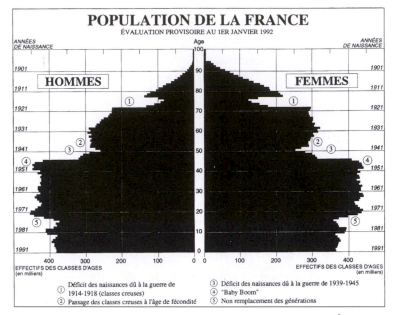

Figure 13.4 Population pyramid of France in 1992. Source: Institut National d'Études Démographiques (1992, p. 1114)

6   At ages above 65 years there are many more females alive than there are males. We would normally expect an excess of females at these ages because females generally live longer than males. In this case, the deficit is exaggerated, because of excess mortality of males during the Second World War.

13.5   Stationary and stable populations

Stationary and stable populations are special theoretical types of population which are widely used in demography. They are model populations which satisfy certain assumptions about fertility and mortality. Although the assumptions are clearly not realistic (in that few real populations satisfy them exactly), it turns out that the behaviour of many real populations may be approximated by that of stationary and stable populations.

STABLE POPULATION

Let us begin by defining what a stable population is. A *stable population* has the following features:

1   Its age-specific *fertility* rates (ASFRs) have been constant over time for a long period.
2   Its age-specific *death* rates (ASDRs) have been constant over time for a long period.
3   It is *closed to migration*. This means that there has been no migration into or out from the population for a long period. (Strictly speaking, a population can still be regarded as stable if there is migration, provided that the migration is such that it does not affect the age structure. In what follows, we ignore this rather unlikely possibility.)

It can be proved that if fertility and mortality remain constant for long enough, and migration is zero, then ultimately the population will develop a *constant rate of growth* and a *constant age structure* expressed in percentage terms. Thus a stable population may increase (or decrease) in size at a constant rate, but its age structure remains constant (that is, the percentage of the population at each age remains the same).

STATIONARY POPULATION

If the rates of fertility and mortality in a stable population are such that the rate of population growth is zero, the population is said to be *stationary*. A stationary population has an *unchanging size and an unchanging age structure*. That is, not only is the percentage of the population in each age group the same, but also the absolute number of people in each age group is the same.

It is common to regard stationary and stable populations as consisting purely of individuals of one sex. This sex is normally considered to be female. There is a rationale behind this, in that population growth is largely determined by the number of girls that are born. It is worth noting that the net reproduction rate, defined in Section 12.3, which is an important measure of the rate of population growth, is a purely 'female' measure.

13.6 Fertility, mortality and the age structure in stable populations

Let us denote the constant age-specific fertility rate at age x last birthday in the single-sex stable population by the symbol f_x, and let the proportion surviving to exact age x out of an original cohort of births be l_x (we use l_x because the proportion surviving to exact age x is the same as the l_x column of a life table which uses the constant age-specific death rates in the stable population, provided that we set l_0 equal to 1.0).

We now derive an expression for the number of people alive in the current year aged x last birthday in a stable population. We shall denote this number by the symbol A_x.

On average, the people alive and aged x last birthday are aged exactly $x + \frac{1}{2}$ years. Thus we can write

A_x = number of births $x + \frac{1}{2}$ years ago × proportion surviving to exact age $x + \frac{1}{2}$

   = births $x + \frac{1}{2}$ years ago × $l_{x+\frac{1}{2}}$.

Now, assuming that deaths are evenly distributed within each single year of age,

$$l_{x+\frac{1}{2}} = \tfrac{1}{2}(l_x + l_{x+1}).\tag{13.1}$$

The right-hand side of equation (13.1), however, is the same as the life table function L_x (see Chapter 4). Thus

$$A_x = \text{number of births } x + \tfrac{1}{2} \text{ years ago} \times L_x.\tag{13.2}$$

We now need to find out how many births there were $x + \frac{1}{2}$ years ago. Suppose that in the current year the crude birth rate is b. Then the number of births in the current year is bP, where P is the current total population size.

Now, in a stable population, since the ASFRs are constant over time, and the proportion of the population at each age is constant over time, then the crude birth rate is constant over time. This must be so, because, if the population at each age x last birthday is denoted by the symbol P_x, and the total population by the symbol P, then the crude birth rate is defined as

$$b = \frac{\sum_x f_x P_x}{P}.$$

If we denote the proportion of the population which is aged x last birthday by the symbol c_x, then P_x is equal to $c_x P$. Thus

$$b = \frac{\sum_x f_x c_x P}{P} = \sum_x f_x c_x.$$

Since we have already said that in a stable population the ASFRs are constant over time, and the proportion of the population at each age is constant over time, then $\sum_x f_x c_x$ must be constant over time, and so the crude birth rate is constant over time.

Thus the number of births $x + \frac{1}{2}$ years ago is equal to the crude birth rate, b, multiplied by the total population $x + \frac{1}{2}$ years ago.

In Section 12.4 we showed that, in a population which is growing at a constant annual rate, the total population in t years' time, P_t, is given by the equation

$$P_t = P e^{rt},$$

where r is the constant annual rate of growth, and P is the current population (here we use the symbol P for the current population, rather than P_0, to avoid confusing subscripts which denote age last birthday with subscripts which denote calendar time). Thus the total population $x + \frac{1}{2}$ years ago, which we can denote by the symbol $P_{-(x+\frac{1}{2})}$ since it may be regarded as the population in $-(x + \frac{1}{2})$ years' time, must be related to the current population by the equation

$$P_{-(x+\frac{1}{2})} = P e^{-r(x+\frac{1}{2})}. \tag{13.3}$$

Since

$$\text{number of births } x + \tfrac{1}{2} \text{ years ago} = b P_{-(x+\frac{1}{2})},$$

then, substituting from equation (13.3) for $P_{-(x+\frac{1}{2})}$, we obtain

$$\text{number of births } x + \tfrac{1}{2} \text{ years ago} = b P e^{-r(x+\frac{1}{2})}. \tag{13.4}$$

Thus, substituting from equation (13.4) into equation (13.2), we find that the number of people now alive aged x last birthday is given by the equation

$$A_x = b P e^{-r(x+\frac{1}{2})} L_x.$$

Finally, the proportion of people alive aged x last birthday, c_x, is equal to A_x/P. Thus

$$c_x = b e^{-r(x+\frac{1}{2})} L_x. \tag{13.5}$$

Equation (13.5) is the required expression relating the age structure of a stable population to its fertility, its mortality and its rate of growth.

If the population is stationary, then the rate of growth, r, is zero. This means that in equation (13.5), $e^{-r(x+\frac{1}{2})}$ is equal to 1, and the equation reduces to

$$c_x = b L_x. \tag{13.6}$$

Equation (13.6) tells us that in a stationary population (which is, as we have seen, characterized by a constant set of ASDRs), the L_x column of the life table derived from that constant set of ASDRs (with l_0 set equal to 1) is proportional to the age structure of the population. This fact was deduced by a different route in Chapter 4.

CONTINUOUS AND DISCRETE FORMS OF THE STABLE POPULATION MODEL

The above derivation of the relationship between the age-specific fertility and death rates, the rate of growth and the age structure in a stable population uses a number of approximations. For example, we assumed that the people who were, at some point in time, aged between exactly x and $x + 1$ years were, on average, aged exactly $x + \frac{1}{2}$ years. Whereas these approximations are usually quite adequate for practical work, it is worth noting that in growing stable populations the average age of such people will be rather less than this, since there will usually be more people alive aged between exactly x years and exactly $x + \frac{1}{2}$ years than there will be aged between exactly $x + \frac{1}{2}$ years and $x + 1$ years.

We can avoid making assumptions of this kind by treating age and calendar time as continuous, and using calculus. This produces nice theoretical relationships between the various quantities. In the remainder of this chapter, we shall treat age and calendar time as continuous in this way.

We first sketch a proof of the theorem that a population which experiences a constant set of ASFRs and ASDRs will eventually develop a constant age structure (expressed as the proportion of the population at each age). We then prove that such a population will grow at a constant rate.

DISCRETE FORM OF THE MODEL

In practical work we usually need to resort to approximations, since the nice theoretical relationships tend to involve evaluating integrals, and this can usually only be carried out numerically. The continuous form of the model is fine for demonstrating the properties of stable populations, but for practical work we need tractable calculations. Thus, in practical work, we tend to treat age and calendar time as being divided up into discrete chunks, such as single years of age or five-year age groups. The discrete form of the theory is an approximation to the continuous form. The smaller the discrete units of age we use, the better the approximation.

13.7   Outline of a proof that a stable population has a constant age structure

We have said that a population which experiences a constant set of age-specific fertility and death rates will eventually develop a constant age structure (expressed as the percentage of the population at each age). This means that, *whatever its initial age structure*, a population with a constant set of ASFRs and ASDRs will eventually develop an age structure which does not change over time, and is completely determined by the constant ASFRs and ASDRs.

Thus, two populations with very different initial age structures will, if they experience the same ASFRs and ASDRs for long enough, come to have the same age structure. This will happen no matter what the initial age structure is (except in some 'special' cases such as an initial population with nobody younger than 60 years – this population will die out, regardless of its age-specific fertility rates).

This phenomenon is known as *ergodicity*, and the theorem that a population in which there is no migration, and which is exposed to a constant set of ASFRs and ASDRs, will arrive at a constant age structure which depends only on the ASFRs and ASDRs is known as the *strong ergodic theorem*. It was first proved by Alfred Lotka (see Lotka and Sharpe, 1911) and a number of other proofs have been suggested since then.

In fact, it is also true that even populations in which the ASFRs and ASDRs are not constant over time will ultimately have age structures which are determined only by these ASFRs and ASDRs. In other words, all populations eventually 'forget' their past age structures. This phenomenon is known as *weak ergodicity*.

There are several proofs of the strong ergodic theorem available. The one described in this section is based closely on that outlined by Coale (1987), which is in turn derived from Lopez (1967).

Consider a closed population (one not subject to migration). For simplicity, we shall assume that it is a single-sex population, and that it is female. In this population, we shall denote the number of people alive aged exactly x at time t by the symbol $N(x, t)$. (Strictly speaking, $N(x, t)$ is the number of people at time t aged between exact age x and some exact age $x + dx$ which is very slightly older than exact age x.)

The number of people alive at time t aged x is the product of the number of people born $t - x$ years ago, multiplied by the chance of surviving to age x. If $B(t)$ is the number of births at time t (strictly speaking, it is the number of births in a very small time interval between t and $t + dt$), and $l(x)$ is the chance of surviving to exact age x, then we can write

$$N(x, t) = B(t - x)l(x).$$
(13.7)

The number of births at time t to women aged x years (strictly speaking, aged between x and $x + dx$ years), $B(x, t)$, is the product of the number of people alive at each age x, multiplied by the ASFR at age x, $f(x)$. In symbols,

$$B(x, t) = N(x, t)f(x).$$

The total number of births at time t is simply the sum of the number of births at each age over all the fertile ages. Because we are treating age as continuous, rather than being divided into discrete chunks, we can write

$$B(t) = \int_\alpha^\beta B(x, t)\, dx = \int_\alpha^\beta N(x, t)f(x)\, dx,$$
(13.8)

where, in equation (13.8) and henceforth, α is the youngest age at which childbearing takes place, and β is the oldest age at which childbearing takes place. (In fact, we could have 0 and ω as the limits of the integral, as $f(x) = 0$ at ages where no childbearing takes place.) Substituting for $N(x, t)$ in equation (13.8) from equation (13.7), we obtain

$$B(t) = \int_\alpha^\beta B(t - x)l(x)f(x)\, dx.$$
(13.9)

Now, suppose we have two populations, population 1 and population 2. These populations have different age structures, so they have different numbers of people alive at each age. Let:

- the number of people alive aged x at time t in population 1 be $N_1(x, t)$;
- the number of people alive aged x at time t in population 2 be $N_2(x, t)$;
- the total number of births at time t in population 1 be $B_1(t)$; and
- the total number of births at time t in population 2 be $B_2(t)$.

Assume now that populations 1 and 2 experience the same constant age-specific fertility and death rates. If this is the case, then we can apply the reasoning behind equation (13.9) above to write

$$B_1(t) = \int_\alpha^\beta B_1(t-x)l(x)f(x)\,\mathrm{d}x \qquad (13.10)$$

and

$$B_2(t) = \int_\alpha^\beta B_2(t-x)l(x)f(x)\,\mathrm{d}x.$$

Now suppose that

$$B_1(t-x) = \gamma(t-x)B_2(t-x). \qquad (13.11)$$

This is an algebraic device: whatever the values of $B_1(t-x)$ and $B_2(t-x)$, we can always find some $\gamma(t-x)$ so that equation (13.11) holds.

Using equation (13.11), we can substitute for $B_1(t-x)$ in equation (13.10) to give

$$B_1(t) = \int_\alpha^\beta \gamma(t-x)B_2(t-x)l(x)f(x)\,\mathrm{d}x. \qquad (13.12)$$

Now, equation (13.11) implies that, if x is zero,

$$B_1(t) = \gamma(t)B_2(t).$$

Thus

$$\gamma(t) = \frac{B_1(t)}{B_2(t)}. \qquad (13.13)$$

Substituting from equation (13.12) into equation (13.13) produces

$$\gamma(t) = \frac{\int_\alpha^\beta \gamma(t-x)B_2(t-x)l(x)f(x)\,\mathrm{d}x}{B_2(t)}$$

$$= \int_\alpha^\beta \gamma(t-x)\left(\frac{B_2(t-x)l(x)f(x)\,\mathrm{d}x}{B_2(t)}\right).$$

The term

$$\left(\frac{B_2(t-x)l(x)f(x)\,\mathrm{d}x}{B_2(t)}\right)$$

is just the frequency distribution of the number of births $B_2(t)$ by the age of the mother. To see this, we note that $B_2(t-x)l(x)$ is just the number of women who survive to be aged x in year t out of those born $t-x$ years ago, and $f(x)$ is their age-specific fertility rate. Thus

$$B_2(t-x)l(x)f(x) = \text{number of women alive aged } x \times \text{ASFR for women aged } x$$

$$= \text{number of births to women aged } x.$$

Summing this over all ages, we obtain the total number of births in population 2 at time t. Since this is just $B_2(t)$, we can write

$$\int_\alpha^\beta \frac{B_2(t-x)l(x)f(x)\,\mathrm{d}x}{B_2(t)} = 1.0.$$

If we put

$$\left(\frac{B_2(t-x)l(x)f(x)\,\mathrm{d}x}{B_2(t)} \right) = \phi(x,t)\,\mathrm{d}x,$$

then we can write

$$\gamma(t) = \int_\alpha^\beta \gamma(t-x)\phi(x,t)\,\mathrm{d}x. \tag{13.14}$$

Substituting from equations (13.11) and (13.13) into equation (13.14) produces

$$\frac{B_1(t)}{B_2(t)} = \int_\alpha^\beta \frac{B_1(t-x)}{B_2(t-x)}\phi(x,t)\,\mathrm{d}x. \tag{13.15}$$

Equation (13.15) says that the ratio between the number of births in the two populations at time t is a *weighted average* of the corresponding ratios in the period between α and β years prior to time t (remember that the weights, $\phi(x,t)$, sum to 1.0). In other words, $B_1(t)/B_2(t)$ is a weighted average of the ratios $B_1(t-x)/B_2(t-x)$ over values of x ranging from α to β. But, each of the ratios $B_1(t-x)/B_2(t-x)$ is itself a weighted average of the corresponding ratios in a preceding period. Thus $B_1(t)/B_2(t)$ is a weighted average of a weighted average. Repeatedly applying equation (13.15) over many generations will produce a value of $B_1(t)/B_2(t)$ which is a weighted average of a weighted average of a weighted average....

It can be proved that this means that eventually $B_1(t)/B_2(t)$ will become constant. When $B_1(t)/B_2(t)$ has been constant for at least ω years (where ω is the limiting age – the oldest age to which anyone lives), then $N_1(x,t)/N_2(x,t)$ will be the same at all ages (remember mortality is the same in the two populations). Once that point has been reached, populations 1 and 2 will have the same proportion of their populations at each age x, although their absolute sizes will probably differ.

13.8   The rate of growth

We now prove that a population with constant age-specific fertility and death rates will eventually have a constant rate of growth.

If the proportion of the population between ages x and $x + \mathrm{d}x$ is $c(x)$, then

$$c(x) = \frac{N(x,t)}{\int_0^\omega N(x,t)\,\mathrm{d}x},$$

where $N(x,t)$ is the number alive aged x to $x + \mathrm{d}x$ at time t. Remember that $c(x)$ will be the same for all t in a stable population. Recall also that ω is the limiting age (the oldest age to which anyone survives).

The birth rate, b, in any female population is equal to the total number of births divided by the total population. Thus we have

$$b = \frac{\int_\alpha^\beta f(x)N(x,t)\,\mathrm{d}x}{\int_0^\omega N(x,t)\,\mathrm{d}x} = \int_\alpha^\beta f(x)c(x)\,\mathrm{d}x. \tag{13.16}$$

Moreover, the death rate, d, is given by

$$d = \frac{\int_0^\omega m(x)N(x,t)\,dx}{\int_0^\omega N(x,t)\,dx} = \int_0^\omega m(x)c(x)\,dx, \tag{13.17}$$

where $m(x)$ is the ASDR at age x.

Since we are told that $f(x)$ and $m(x)$ are constant over time, and we have sketched a proof in the previous section of the fact that, if this is so, then $c(x)$ will eventually be constant over time, then equations (13.16) and (13.17) tell us that this eventually will mean constant birth and death rates. Since the rate of increase is simply the difference between the birth rate and the death rate, the rate of increase is the difference between two constant terms, which must itself be constant.

Thus we have proved that a population which experiences, for a long period, constant age-specific fertility and death rates will, eventually, increase in size at a constant rate. Let this rate be r. Since a stable population is increasing in size at a constant rate r, and the birth rate is also constant, then the number of births is also increasing at a rate r. Thus the number of births x to $x + dx$ years ago is equal to $B_0\,e^{-rx}\,dx$, where B_0 is the number of births in the current year. The number of women now alive aged x to $x + dx$ is equal to the number of survivors out of the births x to $x + dx$ years ago. If the chance of surviving to age x is equal to $l(x)$, then the number of women now alive aged x to $x + dx$ years, $N(x)\,dx$, is given by the equation

$$N(x)\,dx = B_0\,e^{-rx}l(x)\,dx.$$

Thus the number of births in the current year to women aged x to $x + dx$ years is equal to $B_0\,e^{-rx}l(x)f(x)\,dx$. Integrating this term over all the childbearing ages gives us the equation

$$\text{total number of births to women at all ages in the current year} = \int_\alpha^\beta B_0\,e^{-rx}l(x)f(x)\,dx.$$

But we have already defined the total number of births in the current year to be B_0. Thus we have

$$B_0 = \int_\alpha^\beta B_0\,e^{-rx}l(x)f(x)\,dx, \tag{13.18}$$

and, dividing both sides of equation (13.18) by B_0, we obtain

$$1 = \int_\alpha^\beta e^{-rx}l(x)f(x)\,dx. \tag{13.19}$$

Equation (13.19) is called the *characteristic equation of a stable population*.

13.9 The length of a generation

In Section 12.3, we defined the net reproduction rate (NRR) using the equation

$$\text{NRR} = \sum_{x=15}^{49} f_x^d L_x, \tag{13.20}$$

where f_x^d is the ASFR for daughters only for women aged x last birthday, and L_x is the number of woman-years lived between exact ages x and $x + 1$ by a birth cohort of one woman.

Treating age as continuous, rather than as being made up of discrete chunks of one year, we can write a continuous version of equation (13.20) as follows:

$$\text{NRR} = \int_\alpha^\beta f^{\mathrm{d}}(x)l(x)\,\mathrm{d}x, \tag{13.21}$$

in which α and β are, respectively, the youngest and oldest ages at which childbearing takes place, $f_x^{\mathrm{d}}(x)$ is the ASFR (for daughters only) for women aged x (or, strictly speaking, between exact ages x and $x + \mathrm{d}x$), and $l(x)$ is the proportion of those born who survive to exact age x.

Now, in a stable population, the annual rate of growth, r, is constant. Thus, as we saw in Section 12.5, we can write

$$\text{NRR} = \mathrm{e}^{rg}, \tag{13.22}$$

where g is the *length of a generation*. Dividing equation (13.22) by e^{rg}, we obtain

$$\frac{\text{NRR}}{\mathrm{e}^{rg}} = 1,$$

or

$$\text{NRR} \times \mathrm{e}^{-rg} = 1. \tag{13.23}$$

Substituting for the NRR from equation (13.21) into equation (13.23), we obtain

$$\mathrm{e}^{-rg} \int_\alpha^\beta f(x)l(x)\,\mathrm{d}x = 1.$$

(Here we have omitted the superscript d from the symbol $f^{\mathrm{d}}(x)$ because we normally consider a stable population to be single-sex, and female. Thus in the stable population $f(x)$ and $f^{\mathrm{d}}(x)$ can be regarded as the same.)

Finally, since the characteristic equation of a stable population (see equation (13.19)) tells us that

$$\int_\alpha^\beta \mathrm{e}^{-rx} f(x)l(x)\,\mathrm{d}x = 1,$$

it is clear that

$$\mathrm{e}^{-rg} \int_\alpha^\beta f(x)l(x)\,\mathrm{d}x = \int_\alpha^\beta \mathrm{e}^{-rx} f(x)l(x)\,\mathrm{d}x. \tag{13.24}$$

The *length of a generation in a stable population* is that value of g which satisfies equation (13.24), given the values of $f(x)$ and $l(x)$. Note that the annual rate of growth, r, is itself determined by $f(x)$ and $l(x)$. It can be found by solving the *characteristic equation* for r, given the values of $f(x)$ and $l(x)$; see Coale (1972).

Unfortunately, equation (13.24) is hard to solve. The solution must be obtained iteratively, using numerical approximations. However, one fact about the solution is important to us: it can be shown that, in practice, the length of a generation is close to the *mean age at childbearing*. This, of course, seems intuitively reasonable. It also means that we can get a good idea of the value of g, the mean length of a generation, by calculating the mean age at childbearing.

How can we calculate the mean age at childbearing? Well, it is the average age of women at the birth of their children. Treating age as continuous, the appropriate formula is

$$\text{mean age at childbearing} = \frac{\int_\alpha^\beta xf(x)l(x)\,dx}{\int_\alpha^\beta f(x)l(x)\,dx}. \tag{13.25}$$

Equation (13.25) says that the mean age at childbearing is a weighted average of the childbearing ages, where the weights are the proportions of children born at each age in a population with given age-specific fertility and death rates.

In practice we cannot use equation (13.25) as it stands to work out the mean length of a generation. Instead, we use a discrete-form approximation, which is

$$\text{mean age at childbearing} = \frac{\sum_x xf_xL_x}{\sum_x f_xL_x},$$

where the summation is over all the childbearing ages x, and the symbols have their usual meanings.

In low-mortality populations, L_x is close to 1.0 throughout the childbearing age range (remember we are assuming that $l_0 = 1$). This means that a fairly close approximation to the mean age at childbearing in these populations can be obtained by using the simpler formula

$$\text{mean age at childbearing} \cong \frac{\sum_x xf_x}{\sum_x f_x}.$$

We have said that the mean age at childbearing is very similar to the length of a generation. In fact, if the population is stationary, then the mean age at childbearing is exactly the same as the length of a generation. In a stationary population, of course, the annual rate of growth, r, is equal to zero.

In a growing population, where $r > 0$, the mean age at childbearing is greater than the length of a generation (but not by much). In a declining population, where $r < 0$, then the mean age at childbearing is less than the length of a generation (but not by much).

Finally, it is worth noting that in real populations the mean age at childbearing does not vary very much. It is usually around 28 years. This fact is very useful, since it means that we can make the assumption that it is around 28 years in applied work, as shown in Chapter 14.

Further reading

The mathematics of stable and stationary populations is described more fully in Keyfitz (1977) and Coale (1972). Both these books make liberal use of calculus.

Exercises

13.1 Sketch age pyramids for the following populations:
   (a) the population of a developing country, with a life expectation at birth of about 50 years, and an annual growth rate of 2.5% per year;
   (b) the population of a developed country, with a life expectation at birth of about 75 years, and an annual growth rate close to zero;
   (c) a population which is declining at an annual rate of 1% per year.

Table 13E.1

Age x	Number surviving to age x out of 1000 births
0	1000
1	950
5	900
10	880
20	850
30	790
40	700
50	580
60	450
70	340
80	230

13.2 Table 13E.1 is the life table for the population of a developing country.
 (a) Calculate the percentage of the population in each age group (where the exact ages in the table above denote the boundaries between the age groups), assuming that the population is stationary. (*Note*: you may assume in your calculations that 1.5% of the stationary population is aged 80 years and over, and that the average age of people aged 80 years and over is 85 years.)
 (b) Calculate the percentage of the population in each age group, assuming that the population is growing exponentially at an annual rate of 2%.
 (c) Comment on the differences between the two age structures.

13.3 In a stable population which is declining in size, there are typically more people of middle age (aged, say, 40–60 years) than at younger or older ages. Explain why this is.

13.4 Suppose that each of the following changes were to occur to a stable population which is growing in size. In each case, sketch an age pyramid to illustrate the effect that the changes will have had on the age structure of the population after 25 years. Write brief notes beside the age pyramid to emphasize the main effects.
 (a) The elimination through a medical breakthrough of a major cause of mortality among males and females aged 45 years and over.
 (b) A gradual decline in age-specific fertility rates by 50% over a 20-year period to a new constant level which is then maintained for a long period.
 (c) A temporary one-year fall in the number of children born caused by a religious belief that children born in that year would be condemned to lives of misery and vice.
 (d) A rapid fall in infant mortality over a five-year period, to a new level. After five years, infant mortality remains constant at its new level.

13.5 In a stable population, you are told that the number of persons alive between exact ages 40 and $40\frac{1}{2}$ is the same as the number of persons alive between exact ages $40\frac{1}{2}$ and 41. Find an expression for the rate of growth, r, in terms of the constant set of l_xs representing the mortality of this population.

13.6 Table 13E.2 gives m-type age-specific death rates for females in England and Wales in 1992.

Table 13E.2

Age group	m-type death rate	Age group	m-type death rate
0	0.00571	45–49	0.00208
1–4	0.00029	50–54	0.00353
5–9	0.00014	55–59	0.00579
10–14	0.00013	60–64	0.01002
15–19	0.00029	65–69	0.01677
20–24	0.00032	70–74	0.02656
25–29	0.00035	75–79	0.04419
30–34	0.00051	80–84	0.07342
35–39	0.00082	85–89	0.12020
40–44	0.00132	90 and over	0.21060

Source: Office of Population Censuses and Surveys (1994, p. 6).

(a) Calculate the stationary population corresponding to this set of age-specific death rates, assuming there are 100 000 births per year. To calculate $_nL_{90}$ use the formula $_nL_{90} = {_nd_{90}}/{_nm_{90}}$.

(b) Calculate the percentage of the population in each age group in this stationary population.

(c) Calculate the percentage of the population in each age group in a stable population with the same set of age-specific death rates which is growing in size at the following rates: 1% per year; 2% per year; 3% per year.

(d) Calculate the percentage of the population in each age group in a stable population with the same set of age-specific death rates which is declining in size at a rate of 1% per year.

13.7 Explain why the length of a generation is less than the mean age at childbearing in a growing population, and greater than the mean age at childbearing in a declining population.

13.8 Using the data provided in Exercise 12.1 (Table 12E.1), calculate the mean age at childbearing in England and Wales in 1991.

14

Applications of Stable Population Theory

14.1 Introduction

Stable population theory has many applications. This chapter describes and illustrates two important sets of such applications. The first set concerns what might be called 'quasi-demographic entities'. These are collections of individual items, or people, about which, at first sight, demography might not seem to have much to say, but which, on closer inspection, are found to have similar dynamics to human populations. In other words, their form is fundamentally similar to that of the stable population, and therefore stable population theory may be used to analyse them. The second set of applications of stable population theory involves making estimates of the demographic characteristics of populations for which data are lacking, or of poor quality.

We begin in Section 14.2 by noting some general features of a stable population. In Section 14.3 we show how the model may be applied to the analysis of turnover and recruitment in a company's workforce, using a (rather stylized) example. Section 14.4 introduces the idea of demographic reconstruction, which describes the process by which stable population theory may be used to estimate certain demographic characteristics of populations in which data are limited or defective.

Demographic reconstruction makes use of what are called model life tables and model stable populations. These are described in Section 14.5. Section 14.6 describes the process of demographic reconstruction using data from two censuses. Finally, in Section 14.7 some miscellaneous approximations which are often found useful in practical work are described.

14.2 General features of a stable population

In order to understand the application of stable population theory to 'quasi-demographic entities', we need to form a general idea of what a stable population is. Stripped down to its bare essentials, a stable population has the following features:

1 It is a population of individuals. Individuals continually enter and leave the population.
2 At any one point in time, the individuals can be classified according to how long they have been in the population. (In human populations, this means that individuals can be classified by their age.)

3 We can measure, using some metric of time, the length of time for which any individual has been in the population.
4 The chance that an individual will leave the population varies with the length of time for which that individual has been in the population. (In human populations, this amounts to saying that death rates are age-specific.)
5 The rate at which individuals enter the population is constant over time, and the duration-specific rates of departure are also constant over time. (Birth and age-specific death rates are constant over time.)

There are many entities which fulfil (at least to a close approximation) these criteria. We can analyse any of these 'populations' using stable population theory. In the next section we consider one such entity: a company's workforce. However, many others can be imagined – for example, the set of personal computers owned and operated by a university for the use of its staff and students (see Exercise 14.1). At any one time the university will own a certain number of these computers: the 'population size' at some point in time. Assuming that the university purchases only new computers, the rate at which these computers either break down or become obsolete will probably depend on how long the university has owned them: in other words, 'death' rates are duration-specific. Since the time since a computer was purchased by the university is our metric of time, and this can presumably be measured, conditions 1 to 4 above are fulfilled. It only remains to posit that the university purchases each year a number of new computers in proportion to the number currently owned, and to assume that the duration-specific breakdown and obsolescence rates are constant to fulfil condition 5.

14.3 Example: a company's workforce

Suppose we are in charge of staffing and recruitment for a medium-sized company. The company is expanding at a good rate; indeed, so successful has it been that its workforce has doubled on average every ten years for the last few decades. It is planned to maintain this rate of growth in the future. The policy of the company is to recruit graduates aged exactly 21 years (and only graduates aged exactly 21 years). Company policy says that all employees should retire on their 61st birthday.

Of course, not all those who are recruited stay with the company until retirement: some die before their 61st birthday; others are fired; and others leave to join other firms. Studies made by the Personnel Department over the last few years have revealed that for every 100 graduates recruited, 80 are still with the company after ten years, 60 are still with the company after 20 years, 40 are still with the company after 30 years, and 20 stay with the company for 40 years (that is, until their 61st birthday).

As part of our job, we are required to make estimates about future staffing and recruitment needs. The first thing which we have been asked to do is to estimate the percentage by which the number of graduates recruited should be increased each year in order that the workforce will double every ten years.

How can we solve this problem? We first assume that this company's workforce is a stable population. From the description given above, it seems quite a reasonable assumption. Furthermore, if we assume that people are recruited throughout each year (that is, the new recruits in a given year do not all join on the same date), and that people leave the company throughout each year, the growth of the company's workforce will be exponential, and at a constant rate.

If W_t is the size of the workforce in year t, we have

$$W_{t+10} = W_t e^{10r}, \tag{14.1}$$

where r is the annual rate of growth. For the workforce to double in size in ten years, we require

$$W_{t+10} = 2W_t. \tag{14.2}$$

Substituting for W_{t+10} from equation (14.1) into equation (14.2) gives

$$2W_t = W_t e^{10r}, \qquad \Rightarrow 2 = e^{10r}$$

which may be solved for r to give $Ln\ 2 = 10\ r$

$$r = \frac{\ln 2}{10} = 0.0693.$$

Thus the company's workforce must increase at a rate of 0.0693 (or 6.93%) per year.

Now, recruitment is equivalent to birth in the stable population; and death, movement to another company and being sacked are together equivalent to death in the stable population. The problem now resolves itself into considering by what percentage the number of births in a stable population must rise each year in order that the size of the population must increase at a rate of 0.0693 per year.

Well, in a stable population the birth rate is constant. Thus, in the company, the recruitment rate must also be constant. The recruitment rate is simply the number of new recruits in a year divided by the total workforce in that year. Thus, if the number of new recruits in year t is denoted by the symbol R_t, and the recruitment rate is denoted by the symbol ν, then

$$\frac{R_t}{W_t} = \nu. \tag{14.3}$$

But we have already established that

$$W_{t+1} = W_t e^r.$$

Since, from equation (14.3),

$$R_t = W_t \nu,$$

we have

$$R_{t+1} = W_{t+1}\nu = W_t e^r \nu = \frac{R_t}{\nu} e^r \nu = R_t e^r.$$

Thus the number of employees recruited must grow each year at the same rate as the workforce, which is 0.0693.

The company places a very high priority on the training of its employees. Training is undertaken by all employees throughout their first ten years with the company. Once an employee has attained his/her 31st birthday, that employee is considered fully trained. The programme of training is carried out by senior staff within the company (who may be regarded as those aged 51–60 years last birthday). Company policy states that all senior staff must do their share of instruction. Studies by the Personnel Department have revealed that one senior employee cannot effectively train more than 15 junior colleagues at the same time. If there are insufficient senior employees available to do the training, additional instructors are hired from outside. We have been asked to find out whether, if the workforce is to double every ten years, it will be necessary to hire outside instructors, and, if so, what percentage of the 21–30-year-old employees will have to be trained by outside instructors at any one time.

The company's workforce can be regarded as a stable population with a rate of growth of 0.0693 and a life table such that l_0 is equal to 100, $l_{10} = 80$, $l_{20} = 60$, $l_{30} = 40$ and $l_{40} = 20$. In this life table, the subscripts denote duration of employment with the company. Assuming that the employees who leave the firm before they are 61 years of age do so evenly across the age groups (or duration groups), then we can estimate the constant proportions in each age group by first using the formula

$$_{10}L_x = \tfrac{10}{2}(l_x + l_{x+10}),$$

where x denotes duration with the company, to estimate $_{10}L_x$. Recall that $_{10}L_x$ measures the number of persons alive between ages x and $x + 10$ years in a stationary population in which the number of births each year is equal to l_0 (see Chapter 4). Now, in a stable or stationary population (where there is no migration), 'age' is just another way of saying 'duration since a person entered the population'. So, if we are considering using this company's workforce as a stable population, 'duration with the company' is the exact equivalent of 'age' in the conventional population. Therefore, $_{10}L_x$ in the context of a company's workforce may be defined as the number of employees who have been with the company between x and $x + 10$ years if the workforce is constant over time, and if the number of employees recruited each year is l_0. The same subscript, x, is used to refer to 'duration with the company' as is used to denote 'age' in the ordinary population to emphasize the equivalence between these two quantities.

Having estimated $_{10}L_x$, we can then estimate $_{10}A_x$ using the formula

$$_{10}A_x = {}_{10}L_x \cdot e^{-r(x+5)}, \tag{14.4}$$

where $_{10}A_x$ is the number of employees who have been with the company for between x and $x + 10$ years, assuming the workforce increases at a rate r per year, and the number of entrants in the current year is equal to l_0. The calculations are shown in Table 14.1. The value of r is, as calculated above, 0.0693.

If the company's workforce is a stable population, the proportions at each age are constant. Thus the ratio of employees in the age group 21–30 years to employees in the age group 51–60 years is equal to 636/26, which is 24.46 (see the $_{10}A_x$ column in Table 14.1). This is greater than 15, so it will be necessary to hire outside instructors.

The percentage of 21–30-year-old employees who will have to be trained by outside instructors at any one time is given by the equation

$$\text{proportion needing to be trained by outside instructors} = \frac{636 - (15 \times 26)}{636} = 38.7\%.$$

Table 14.1 Calculation of age structure of company's employees

Duration with company x	l_x	$_{10}L_x$	$e^{-r(x+5)}$	$_{10}A_x$	Age of employees
0	100	900	0.707	636	21
10	80	700	0.354	248	31
20	60	500	0.177	89	41
30	40	300	0.088	26	51
40	20				61

Now suppose that the Board of Directors is concerned about the quality of the training provided by the outside instructors. The directors want to know what the maximum rate of growth of the workforce is which would allow all training to be done 'in house' by senior employees, while maintaining the rule that one senior employee should not be responsible for the training of more than 15 junior colleagues at the same time.

We require there to be exactly 15 times as many employees in the 21–30 age group as there are in the 51–60 age group. Using the notation given above, this means that we need

$$_{10}A_0 = 15 \cdot {}_{10}A_{30}.$$

Using equation (14.4), this may be written

$$_{10}L_0 \, e^{-5r} = 15 \cdot {}_{10}L_{30} \cdot e^{-35r},$$

and, substituting values for $_{10}L_0$ and $_{10}L_{30}$ from Table 14.1, this becomes

$$900e^{-5r} = (15 \times 300)e^{-35r}.$$

Dividing both sides of this equation by $900e^{-5r}$, we obtain

$$1 = 5e^{-30r},$$

which can be solved for r to give

$$r = -\frac{\ln 0.2}{30} = 0.0536,$$

so the company's workforce may grow at a maximum rate of 5.36% per year, which, using the 'rule of 70' (see Exercise 12.3), implies a doubling about every 13 years.

14.4   Demographic reconstruction

Stable population theory is widely used by demographers to make estimates of the demographic characteristics of populations for which good-quality data are absent. Many developing countries fall into this category. In the remainder of this chapter we look briefly at a few of the methods used.

THE NEED FOR ESTIMATES

As we saw in Section 1.6, there are three main sources of demographic data:

- *population censuses*, which provide information about the exposed-to-risk for the calculation of demographic rates;
- *vital registration*, which provides information about the events which are summed in the numerators of demographic rates;
- *surveys*, which may provide any kind of data we like, provided the right questions are asked.

Almost every country has had at least one census, and, even in developing countries, regular censuses are quite usual.

In developing countries, however, systems of vital registration are rare, and comprehensive systems even rarer. This means that estimates of demographic rates and other population characteristics for these countries have to be obtained either by carrying out special surveys or by using indirect methods of estimation.

Stable population theory provides one very simple, useful and efficient method of making such estimates (it is not the only method). The term *demographic reconstruction* (Woods, 1982) may be used to describe the use of stable population theory to make estimates of the characteristics of populations with limited or defective data, and, in particular, populations with census data but without vital registration data.

SOME GENERAL POINTS ABOUT ESTIMATING DEMOGRAPHIC CHARACTERISTICS

It is worth making a few general points about indirect methods of demographic estimation. All the methods are, for obvious reasons, approximations. They all involve assumptions. Thus in many applications it is not clear that there is a single 'best' approach. Indeed, since the only reason why we are having to make estimates is that we do not know the real situation, we cannot even properly test our estimates against reality.

In practice, therefore, debate about the merits of various approaches to making estimates usually focuses on one of the following: the sensibleness of the assumptions each approach requires, and the speed and simplicity of the approaches (quick, simple approaches are generally preferred to long, complicated ones).

14.5   Model life tables

Probably the most important tool which demographers use in demographic reconstruction is the *model life table*. The critical principle underlying model life tables is that, although mortality varies with age, the age pattern of mortality has some common features regardless of the level of mortality. In fact, it is observed that the age pattern of mortality is very similar within quite large regions of the world, although there are differences between these regions. Some of the regularities were described in Section 4.8 above.

As a result, it has proved possible to draw up model life tables which represent average age patterns of mortality. These model life tables have been drawn up for the whole range of mortality levels commonly found in human populations.

There are two types of model life table in common use: empirical model life tables and relational model life tables.

EMPIRICAL MODEL LIFE TABLES

Empirical model life tables are based on averages of observed mortality experiences. There are several sets of such life tables in existence, and in this subsection we briefly describe the two most important.

The United Nations model life tables (United Nations, 1982) are available for different regions of the developing world. For example, there is a 'Far Eastern' set, which has been estimated using typical age patterns of mortality in Far Eastern countries. They are presented for a wide range of levels of mortality.

The Coale and Demeny model life tables were first constructed in 1966, but were revised and extended in 1983 (Coale and Demeny, 1983). There are four sets, called 'North', 'South', 'East' and 'West'. Each of these sets has a slightly different age pattern of mortality. Each set is presented for 25 levels of mortality, which have life expectations at birth ranging

Table 14.2 Extract from Coale and Demeny's model 'South' life table for females, level 12

Age x	$_nq_x$	l_x	$_nL_x$	T_x	e_x
0	0.14067	100 000	90 856	4 750 000	47.50
1	0.11496	85 933	316 455	4 659 144	54.22
5	0.02601	76 054	374 830	4 342 688	57.10
10	0.01511	74 076	367 693	3 967 859	53.57
15	0.02151	72 957	361 018	3 600 165	49.35
20	0.02728	71 388	352 264	3 239 148	45.37
25	0.03007	69 440	342 189	2 886 884	41.57
30	0.03161	67 352	331 650	2 544 695	37.78
35	0.03442	65 223	320 726	2 213 045	33.93
40	0.03762	62 978	309 203	1 892 319	30.05
45	0.04254	60 609	296 856	1 583 117	26.12
50	0.05715	58 031	282 194	1 286 260	22.17
55	0.07921	54 714	263 169	1 004 067	18.35
60	0.12661	50 360	236 591	740 898	14.71
65	0.19597	44 001	199 312	504 307	11.46
70	0.30970	35 378	150 595	304 995	8.62
75	0.46184	24 422	93 911	154 399	6.32
80	0.61290	13 143	43 942	60 489	4.60
85	0.77403	5 088	13 965	16 547	3.25
90	0.90280	1 150	2 412	2 582	2.25
95	0.97409	112	167	170	1.52
100	1.00000	3	3	3	1.01

Source: Coale and Demeny (1983, p. 389).

from 20 years to almost 80 years. Separate model life tables for males and females have been constructed for each set, and for each tabulated value of the life expectation at birth. An example is shown in Table 14.2.

RELATIONAL MODEL LIFE TABLES

Relational model life tables are based on finding a simple formula which relates the values of one or more life table functions to the values of the corresponding function in a standard life table. Suppose that the life table function in question is G_x, where x represents age. Then, for any life table in the set of relational model life tables, we can write

zeta

$$G_x = f(G_x^s, \zeta),$$

where ζ is a set of parameters. Because the age pattern of human mortality tends to be similar, even when the level of mortality differs, the number of parameters required to describe the range of human mortality experience is small.

In the most widely used set of relational model life tables, those derived by the British demographer William Brass, G_x is defined as $0.5 \ln((1 - l_x)/l_x)$, where l_x is the standard life table function defined in Chapter 4. Brass's formula is

$$0.5 \ln((1 - l_x)/l_x) = \alpha + 0.5\beta \ln((1 - l_x^s)/l_x^s).$$

In this formula, the parameter α represents the *level* of mortality, and the parameter β varies with the *age pattern* of mortality. The l_x^ss represent a *baseline life table*. In Brass's system of relational model life tables, therefore, model life tables are constructed by varying the level and/or the age pattern of mortality in some baseline life table according to a simple formula. Using different combinations of values of α and β, a whole family of life tables may be constructed from any baseline table; see Newell (1988) for further details.

Brass's system of relational model life tables is a two-parameter system. Coale and Demeny's empirical model life tables are also based on two parameters: life expectation at birth (measuring the level) and the four sets ('North', 'South', 'East' and 'West') forming four different age patterns. The United Nations' set is similar to Coale and Demeny's set, save that instead of four rather abstract 'regional' age patterns, the different age patterns are derived from the mortality experience of defined regions of the world. It is possible to draw up sets of model life tables with more than two parameters. However, two parameters are sufficient to obtain an approximation to most human mortality experience which is close enough to be useful in demographic reconstruction.

14.6   Demographic reconstruction with two censuses

Since almost all developing countries have censuses, some of the most widely used methods of estimation are designed to make use only of census data. In this section, we describe a method by which it is possible to obtain estimates of almost all the demographic characteristics of a population, provided that *data from at least two censuses are available*. (It is possible to do a certain amount of estimation with data from only one census, but it is a rather more risky business: see Keyfitz (1985) for details.) Drawing on the description in Woods (1982, pp. 63–81), we present below a simple approach to demographic reconstruction with two censuses.

CHECK THE QUALITY OF THE CENSUS DATA

First, the quality of the census data must be checked, to see if there are any particular problems. Two problems are typically encountered. One is *age heaping*: people often round their exact ages up or down, usually to the nearest five or ten years. Thus the raw census data suggest that there are large numbers of people aged 20, 25, 30, ... years last birthday, but relatively few people aged 19, 21, 24, 26, 29, 31, ... years last birthday.

The other problem is under-reporting and over-reporting at certain ages. This can affect any age group, although teenagers are particularly often affected. Usually, under-reporting in one age group is balanced by over-reporting in an adjacent age group.

These problems, it should be stressed, do not usually involve a systematic under-counting of the population. In other words, the total population is usually approximately correct – people just give the wrong ages. If age-heaping or under-reporting is suspected, adjustments to the age structure reported in the census can be made.

CALCULATE SURVIVORSHIP RATIOS

Once the age data have been checked, the two censuses can be used to calculate *survivorship ratios*. Suppose the censuses are ten years apart. Then, assuming that there is no migration (remember that this is not such a bad assumption for national populations) the number of people alive at a particular age x last birthday in the second census must be equal to the survivors out of those aged $x - 10$ last birthday in the previous census (Figure 14.1).

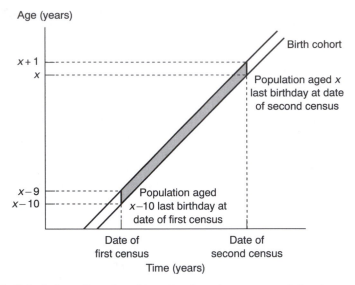

Figure 14.1 Calculation of survivorship ratios from two censuses taken ten years apart

Now, since we are dealing with the history of a specific birth cohort, we can write

$$\frac{\text{population alive aged } x \text{ last birthday in second census}}{\text{population alive aged } x - 10 \text{ last birthday in first census}} = \frac{L_x}{L_{x-10}},$$

where L_x refers to the life table function in the life table which represents the mortality experience of this birth cohort during the period between the two censuses. (Remember that L_x in a stationary population is proportional to the age structure.)

Now using single years of age is sometimes problematic, as random fluctuations or vagaries of the data can affect the number of people reported in censuses as being alive aged x last birthday. The problem can be tackled by using a coarser age classification – for example, ten-year age groups. This produces the equation

$$\frac{\text{population alive aged between } x \text{ and } x + 10 \text{ in second census}}{\text{population alive aged between } x - 10 \text{ and } x \text{ in first census}} = \frac{{}_{10}L_x}{{}_{10}L_{x-10}}.$$

Even better, and less prone to fluctuations caused by transient factors, is the use of the following equation:

$$\frac{\text{population alive at ages over } x \text{ in second census}}{\text{population alive at ages over } x - 10 \text{ in first census}} = \frac{T_x}{T_{x-10}},$$

where T_x refers to the life table function in the life table which best represents the mortality experience of all birth cohorts born more than x years before the second census.

The ratios ${}_{10}L_x/{}_{10}L_{x-10}$ and T_x/T_{x-10} are the survivorship ratios. They are calculated for a number of values of x representing as much of the age range as possible.

CHOOSE A MODEL LIFE TABLE TO REPRESENT THE MORTALITY EXPERIENCE OF THE POPULATION BETWEEN THE TWO CENSUSES

The survivorship ratios are then used to find out which of the various model life tables available fits the mortality experience of the country best.

Choosing a model life table is not straightforward, because, in practice, the survivorship ratios for different ages x will not all be closest to the same model life table. There are, in fact, two distinct issues. First, which age pattern of mortality fits best? Second, given the best-fitting age pattern, which level of mortality most closely represents the observed survivorship ratios?

There is, inevitably, a measure of subjectivity about the choice of model life table, but, if empirical model life tables are used, one procedure which can be adopted to resolve these issues is to work out, for each age x, the level of mortality which best fits the observed survivorship ratios. This is done separately for a range of age patterns. Within each age pattern, the best-fitting levels of mortality will vary with age. The distribution of these best-fitting levels is then examined for each age pattern, and the age pattern for which this distribution has the *smallest variance* is then regarded as being the 'best' one. Given this age pattern, the 'best' level may be worked out by taking the mean of the best-fitting levels for all ages. Once this mean is evaluated, then either the closest published model life table in United Nations (1982) or Coale and Demeny (1983) can be taken as applying to the population under investigation, or a model life table calculated by interpolation between two actual published tables.

Once a model life table has been chosen, all other features of the mortality experience (for example, the life expectation at birth) can be read from it.

CALCULATE THE INTERCENSAL GROWTH RATE

The intercensal annual growth rate, r, can be estimated by assuming the population is growing exponentially between the two censuses, and using the equation

$$\frac{\text{total population in second census}}{\text{total population in first census}} = e^{10r}.$$

CHOOSE A MODEL STABLE POPULATION

The growth rate can then be used to select a suitable *model stable population* from a range of model stable populations published in United Nations (1982) or Coale and Demeny (1983).

What is a model stable population? For each different model life table, both the United Nations and Coale and Demeny have calculated the age structure (in percentage terms) of various stable populations. For each life table, these stable populations have the set of age-specific death rates represented by the life table, but they have different constant annual rates of growth. The formula used to calculate the percentage of the population at each age last birthday is similar to the one described in Section 13.6. The whole procedure is very similar to that needed for the solution to Exercise 13.6. The resulting populations are termed model stable populations. Some examples are shown in Table 14.3.

READ OFF THE ESTIMATES OF THE DEMOGRAPHIC CHARACTERISTICS

Once the model stable population has been chosen, various demographic characteristics of a population with the particular mortality experience and growth rates which we have used to choose the model can be read from published lists like those in the bottom panel of Table 14.3.

Table 14.3 Some model stable populations based on Coale and Demeny's model 'South' life table for females, level 12

	Annual rate of growth				
	−0.01	0.00	0.01	0.02	0.03
Percentage of population in age groups					
0	1.33	1.91	2.62	3.44	4.35
1–4	4.75	6.66	8.91	11.40	14.04
5–9	5.88	7.89	10.09	12.35	14.53
10–14	6.07	7.74	9.42	10.96	12.27
15–19	6.26	7.60	8.79	9.74	10.37
20–24	6.42	7.42	8.16	8.60	8.71
25–29	6.56	7.20	7.54	7.56	7.28
30–34	6.68	6.98	6.95	6.63	6.07
35–39	6.79	6.75	6.40	5.80	5.06
40–44	6.89	6.51	5.87	5.06	4.19
45–49	6.95	6.25	5.36	4.39	3.47
50–54	6.95	5.94	4.84	3.78	2.84
55–59	6.81	5.54	4.30	3.19	2.28
60–64	6.44	4.98	3.67	2.59	1.76
65–69	5.70	4.20	2.94	1.98	1.28
70–74	4.53	3.17	2.12	1.35	0.83
75–79	2.97	1.98	1.26	0.76	0.45
80–84	1.45	0.93	0.56	0.33	0.18
85–89	0.48	0.29	0.17	0.09	0.05
90–94	0.09	0.05	0.03	0.01	0.01
95 and over	0.01	0.00	0.00	0.00	0.00
Demographic parameters					
Birth rate	14.56	21.05	29.02	38.27	48.55
Death rate	24.56	21.05	19.02	18.27	18.55
GRR	1.106	1.477	1.963	2.599	3.425
Average age	39.34	34.40	29.75	25.58	21.97

Source: Coale and Demeny (1983, p. 408).

14.7   Some useful approximations

THE GROSS REPRODUCTION RATE

You will recall that the gross reproduction rate (GRR) is calculated in the same way as the total fertility rate (TFR), but it uses daughters only. Since the average value of the sex ratio of births in human populations is about 105 boys for every 100 girls, and the sex ratio of births does not vary very much with the age of the mother, we can write

$$\text{GRR} \cong \text{TFR} \times \frac{100}{205}. \tag{14.5}$$

THE NET REPRODUCTION RATE

The net reproduction rate (NRR) is the gross reproduction rate adjusted for mortality. The formula which is used in practice to estimate it is

$$\text{NRR} = \sum_{x=15}^{49} f_x^{\text{d}} L_x, \tag{14.6}$$

where f_x^d is the age-specific fertility rate for daughters only for women aged x last birthday, and L_x is the number of woman-years lived between exact ages x and $x + 1$ by a birth cohort of one woman.

The calculations which this formula requires are rather tedious. Thus it would be convenient to have an approximation. Since mortality in the childbearing age range is rather low in most populations, and since deaths within this age group are not far from evenly distributed by age, the following formula is quite a good approximation:

$$\text{NRR} \cong \text{GRR} \times l_{\bar{m}},$$

where $l_{\bar{m}}$ is the probability that a baby girl will survive to the *mean age at childbearing*, $\bar{m}$. In a life table based on l_0 births, the formula thus becomes

$$\text{NRR} \cong \text{GRR} \times \frac{l_{\bar{m}}}{l_0}. \tag{14.7}$$

Of course, this formula begs the question of what the mean age at childbearing is. In Section 13.9, two formulae were given for calculating the mean age at childbearing. Unfortunately, these are about as tedious to evaluate as equation (14.6). However, we may recall that the mean age at childbearing is close to 28 years in many populations. Therefore, we might just assume this to be the case, and calculate the NRR using the formula

$$\text{NRR} \cong \text{GRR} \times \frac{l_{28}}{l_0}.$$

Substituting for the GRR in equation (14.7) from equation (14.5) produces a very useful equation linking the net reproduction rate and the total fertility rate:

$$\text{NRR} = \text{TFR} \times \frac{100}{205} \times \frac{l_{\bar{m}}}{l_0}.$$

Further reading

Those interested in empirical model life tables and stable populations should consult United Nations (1982) and Coale and Demeny (1983). Demographic reconstruction is described in Woods (1982, pp. 63–81). The selection of both empirical and relational model life tables to represent the mortality of real populations is discussed at length in Newell (1988, pp. 106–113 and 130–150).

Exercises

14.1 A university owns and operates a population of P microcomputers for its staff, which it buys new (the microcomputers, not the staff!). Studies by the Department of Computer Services have shown that the chance, λ, that any one of these computers will break down irreparably at any time, given that it is still working at that time, does not depend on the age of the computer. It is the university's policy to sell off to students at a cheap rate all computers that are still in service on the fifth anniversary of their date of purchase.

(a) Find an expression, in terms of P and λ, for the number of new computers the university must buy each year in order to maintain a constant number in service.

(b) The Department of Computer Services decides to upgrade the university's standard spreadsheet package, which is installed on all its microcomputers.

Unfortunately, it turns out that the upgraded version of the package will only run on the operating system fitted to computers bought during the last three years. What proportion of the computers in service will be unable to run the new package?

14.2 A certain university runs a degree in mathematics which lasts four years. Examinations are held at the end of each year, and students who fail at the end of each of the first three years are not permitted to continue their studies (that is, they must leave the university). For many years now, an average of 10% of the students taking each examination have failed.

(a) By what percentage must the number of students admitted to the degree course increase each year in order that the total number of students on the degree course doubles every 20 years?

(b) Suppose that, after many years during which admissions have increased at the rate calculated in the answer to part (a), the university decides to freeze the total number of students on the course for one year. Because of government pressure to maintain the increase in admissions, it is suggested that this is done by increasing the failure rate by the same amount in all examinations. By how much must the failure rate be increased in order to achieve a freeze in the total numbers, while maintaining the annual increase in admissions?

14.3 Suppose you find that censuses in a developing country exhibit age-heaping on ages ending in the digits 0 and 5. You wish to use the census data for demographic reconstruction. Suggest ways of either correcting for the age misreporting implied by the age heaping, or of taking it into account in the analysis.

14.4 Consider the developing country whose mortality experience is represented by the life table presented in Exercise 13.2 (Table 13E.1).

(a) Assuming that the population of this country is growing at an annual rate of 2%, estimate its total fertility rate. State any assumptions you make.

(b) If the country were to have a total fertility rate of 6.0 (and the same mortality rates), estimate the annual rate of population growth it would experience.

14.5 A population of small furry animals lives on a remote island. These animals have a maximum life-span of five years. Studies by naturalists of the mortality of these animals have shown that mortality has been constant for many years, is the same for males and females of the species, and is represented by the life table shown in Table 14E.1.

(a) Each year, the naturalists take a census of these animals. The evidence from a long run of censuses is that their number has been increasing by 2% per year for many years. Estimate the age structure of the animal population (in percentage terms).

Table 14E.1

Age x	Number of animals surviving to exact age x
0	1000
1	800
2	700
3	500
4	200

(b) Female small furry animals bear young between exact ages one and three years. Their age-specific fertility rate at age 1 last birthday is twice that at age 2 last birthday. Assuming that equal numbers of males and females are born, estimate their age-specific fertility rates at ages 1 and 2 last birthday.

15

The Analysis of Migration

15.1 Introduction

Migration is the third component of demographic change and, in principle, should have the same amount of attention paid to its measurement as the other two components, fertility and mortality. However, in practice, migration has been very much the 'poor relation' of the other two components. Demographers have paid far less attention to its analysis than they have paid to the analysis of fertility and mortality.

In fact, the analysis of migration poses more conceptual and methodological challenges than does the analysis of either fertility or mortality. For a start, migration necessarily involves more than one population: a person moves from one *origin* population to another *destination* population. Second, practically all migratory transitions may occur in both directions.

There are also practical difficulties in measuring migration. Whereas people only die once, and the number of births which one woman may have is quite severely limited, the number of times a person may migrate in his/her lifetime is much less strictly limited. Probably because of this, the data which are normally available on migration are much less complete as records of all moves than are the vital registration data usually available on mortality, fertility and nuptiality.

Section 15.2 deals with some preliminary issues (mainly concerned with definition, but partly concerned with measurement). Section 15.3 looks at the kinds of data which are typically available on migration. Section 15.4 deals with the calculation of migration rates. Unfortunately, in many cases migration rates cannot be calculated directly because the required data are not available. Accordingly, indirect methods must be used to estimate migration. Section 15.5 describes one commonly used approach to indirect estimation. Sections 15.4 and 15.5 are, essentially, concerned with measuring the effect of migration to and from a single population: we are not concerned in these sections with whence the migrants have come or whither they are going. However, in Section 15.6 we look at migration streams, which are flows of migrants between specific places.

It is stressed that this chapter serves only as an introduction to the analysis of migration, and does not pretend to give a comprehensive treatment of the topic. Readers who want to learn more are referred to the suggestions for further reading listed at the end of this chapter.

15.2 Some preliminary issues

Migration is generally defined as comprising a change in a person's permanent or 'usual' place of residence. This may seem simple enough, but it immediately raises questions. First, what is to be done about people with no 'usual' place of residence (such as members of nomadic tribes)? Second, some people live in different places in different seasons, and make regular moves between two or more 'usual' places of residence. The general practice is to exclude moves of these kinds from the definition.

Changes in a person's permanent place of residence can involve moves over a wide range of geographical distances, from a few yards to thousands of miles. Normally, very short moves are excluded from the definition of migration by adding a rider that migration should involve the 'taking up of life in a new or different place' (United Nations, 1970, p. 1). In practical work, very short moves are normally excluded because data on migration relate to moves between defined political or administrative areas. The smallest distances over which moves can be identified are therefore largely determined by the size of the smallest areal subdivisions available.

INTERNATIONAL AND INTERNAL MIGRATION

In the analysis of migration, it is conventional to distinguish between *international migration*, involving a move from one country to another, and *internal migration*, involving a move within a country. The vast majority of actual changes of residence are of the latter kind. Indeed, in general, the shorter the distance moved, the more common moves are.

People who move away from a population are called *emigrants* if their move involves crossing an international boundary, and *out-migrants* if not. People who move into a population are called *immigrants* if they have crossed an international boundary, and *in-migrants* if not. Although the distinctions between emigration and out-migration and between immigration and in-migration are substantively important, for the purposes of describing the construction of migration rates, they can often be ignored.

GROSS AND NET MIGRATION

The flow of migrants between any two places, i and j, is called a *migration stream*. Unlike ordinary streams, however, migration streams usually flow in both directions: people move from i to j, and from j to i. Suppose the number of persons moving from place i to place j in a given period is M_{ij} and the number moving from place j to place i is M_{ji}. Then we can define two additional quantities:

1 The total amount of migration between the two places. This is $M_{ij} + M_{ji}$ and is called the *gross migration* between places i and j.
2 The net effect of the two opposing flows, which is $M_{ij} - M_{ji}$. This is called the *net migration* between places i and j. Clearly, if $M_{ij} - M_{ji} < 0$, then place i, on balance, is losing population to place j; if $M_{ij} - M_{ji} > 0$, then place i is gaining at the expense of place j. If $M_{ij} = M_{ji}$ then net migration is zero.

Gross migration and net migration are both informative, yet it is important to be aware of what they do not reveal. Gross migration tells us how much movement is going on, but says nothing about its direction, and therefore about its effect on either of the two places. Net migration tells us a lot about the effect of the migratory flows on the two places, but nothing about how much movement, overall, is going on. As has been often remarked, there are in-migrants and out-migrants, but people never describe themselves as 'net migrants' or even as 'gross migrants'.

15.3   Data for the analysis of migration: moves and transitions

Unlike fertility and mortality, data on migration are not routinely collected by most vital registration systems. The exceptions are systems of continuous registration such as those operated in Scandinavia and certain other countries. In most countries, data on migration must be obtained either from population censuses or from various kinds of survey.

CENSUSES

Most population censuses contain questions about migration. These normally take one of three forms.

1   A question about each person's place of birth may be asked. This allows the measurement of what is called *lifetime migration*, or whether or not a person has ever changed his/her permanent place of residence. Clearly, lifetime migration is often a very poor measure of the number of moves actually made.
2   A question about how long a person has lived in his/her present place of residence, possibly coupled with a question about where that person resided prior to his/her latest move, provides information about the most recent move.
3   It is common for a question to be asked about each person's place of residence at a fixed prior date. In censuses of England and Wales, for example, a question is asked about place of residence one year prior to the census date (Flowerdew and Green, 1993).

Notice that, of these three types of question, only **2** gives information about the process of moving – that is, the date, origin and destination of the actual move. Questions **1** and **3** provide information about *transitions*. We learn from the answers to these questions only that a person was living in place X at a certain time t_1, and in place Y at a (later) time t_2. It is inferred that the person moved from place X to place Y between these two times. However, this inference is far from certain. The person might have moved twice between t_1 and t_2: once from place X to place Z, and then from place Z to place Y. Using transition data, therefore, can lead to both an underestimation of the number of moves which take place in a population, and an inaccurate picture of the origins and destinations of these moves. Clearly, these limitations are especially severe with respect to lifetime migration data.

SURVEY DATA

The survey data used to measure migration are of several types. First, there are surveys which incorporate the same kinds of question as censuses. An example is the Ghana Living Standards Survey of 1993, which asked questions about how long a person had lived at his/her current place of residence, and in which region of the country he/she had lived before.

   Second, it is possible to make use of survey data which are collected for some other purpose to measure migration. A good example is the use of the National Health Service

Central Register (NHSCR) in Britain to identify moves. The NHSCR records every instance of a person ceasing to be registered with one doctor and joining the register of another doctor, provided that the two doctors are in different Family Health Services Authorities (FHSAs). The FHSAs are geographically defined (usually being coterminous with counties in rural areas and with smaller districts in urban areas). Because people normally wish to be registered with a doctor close to their permanent place of residence, the moves recorded in the NHSCR provide a good indication of migration between different FHSAs. In England and Wales, the Office for National Statistics Longitudinal Study of a 1% sample of the population incorporates information from the NHSCR, along with data from censuses and vital registration.

Third, there are survey data which collect information about people travelling from place to place. For example, the International Passenger Survey collects information about people making journeys to and from the United Kingdom. From this survey, data about international migration can be extracted.

15.4   Migration rates

A change of residence can be treated as an event in a person's life, just like getting married, giving birth or dying, and so changes of residence may be used as the numerators in rates. A migration rate may, therefore, be defined in general as

$$\text{migration rate} = \frac{\text{number of persons moving in a given period}}{\text{population at risk of moving in the period}}.$$

So far, migration rates are defined in the same way as other demographic rates. However, it is immediately clear that things are rather more complicated than they are with, say, fertility or mortality. The difficulty is that migration may take place in two directions: people may move both into a population and out of a population. Every move, therefore, involves both an origin population and a destination population.

Consider, first, out-migration or emigration rates. The population at risk of making a move away from a population during a period (call it population i) are clearly those who are living in that population at the beginning of the period. Therefore, an out-migration rate can be defined as

$$\text{out-migration rate} = \frac{\text{number of out-migrants from area } i \text{ in a given period}}{\text{population of area } i \text{ at the beginning of the period}}.$$

However, there is a further problem, in that published data tend to report the numbers of migrants in, say, a given year, and the population on 30 June (that is, in the middle of the period). In practice, therefore, we often use this in the denominator. Thus, if the number of out-migrants from area i in a year is O_i, and the mid-year population of area i is P_i, we have

$$\text{out-migration rate} = \frac{O_i}{P_i},$$

or, if preferred,

$$\text{out-migration rate} = \frac{O_i}{P_i} \times 1000.$$

For in-migration or immigration rates the population at risk is, in theory, everyone who is *not* living in area i at the beginning of the period. This clearly poses practical difficulties, so it is usual to use the same denominator as for out-migration rates. Thus, if the number of in-migrants to area i in a year is I_i,

$$\text{in-migration rate} = \frac{I_i}{P_i}(\times 1000).$$

An advantage of using the same denominator for the out-migration and the in-migration rates is that it is then possible to define two further useful measures:

1   The *gross migration rate* in population i. This is simply

$$\text{gross migration rate} = \frac{I_i + O_i}{P_i}(\times 1000).$$

2   The *net migration rate* in population i, which is given by the formula

$$\text{net migration rate} = \frac{I_i - O_i}{P_i}(\times 1000).$$

AGE-SPECIFIC MIGRATION RATES

All the rates so far mentioned are crude rates. It is well known, however, that a person's propensity to migrate varies greatly with age. It is relatively low in early childhood, rises rapidly to reach a peak at around ages 18–23 years, and then falls away quite rapidly. Because of this, age-specific migration rates are of considerable value. They are calculated using the same principle as age-specific fertility and death rates.

15.5   Indirect estimation of net migration

Because many countries lack direct data on migration, but do have efficient vital registration of births and deaths, as well as regular censuses, it is common to use the basic demographic equation (see Section 1.2) to estimate net migration as a residual. Suppose two censuses are ten years apart, that the populations of an area (which might be either a country or some subdivision of a country) at the first census and the second census are P_t and P_{t+10}, respectively. Suppose, further, that the numbers of births and deaths in the intercensal period are B and D, respectively. Then, using equation (1.1) it is clear that

$$\text{net migration} = P_{t+10} - P_t - B + D.$$

For example, in the small English parish of Berwick St James in Wiltshire, the population in 1851 was 284, and in 1861 it was 248 (Hinde, 1987). During the intervening period there were 82 births and 50 deaths. Thus, during the period, there was a net migration of $(248 - 284 - 82 + 50)$ persons, or -68 persons. In other words, 68 more people left the parish to live elsewhere than came from elsewhere to live in the parish.

This procedure is very simple and has obvious attractions, but it is important to be aware of a number of limitations. These mainly stem from the fact that, since the net migration is estimated as a residual, the estimate will reflect any of the errors and inconsistencies in the census and vital registration data, as well as the true net migration itself (Bogue *et al.*, 1982). Such errors and inconsistencies include the following:

1   The boundaries of the area in question may change between the two censuses. This is a very common problem.
2   The areal units for which births and deaths are reported might not match up with the census-based units.

3 There may be a misalignment between the reference periods used to report births and deaths and the dates of the censuses. For example, data on births and deaths may be available on an annual basis, but the population censuses may take place, say, on 1 April. In such a case the intercensal births and deaths should be calculated by summing the reported totals of births and deaths in years t to $t + 10$ inclusive and then subtracting from the totals the births and deaths occurring between 1 January and 31 March in year t, and between 1 April and 31 December in year $t + 10$. It may be necessary to make the assumption that births and deaths are evenly spaced through years t and $t + 10$.

The second and third of these can in principle be overcome by estimating the number of births and deaths in the area for which the census populations are available during the intercensal period using some suitably representative crude birth and death rates. This procedure sounds very attractive in principle, but is subject to several uncertainties in practice. Since, typically, a ten-year period is being considered, it is likely that fertility and mortality will change during the period. Moreover, even if it is thought possible to use a single crude birth and death rate for the whole period, in order to estimate the numbers of intercensal births and deaths some estimate of the 'average' population at risk is required. The latter may be obtained by using the arithmetic average of the populations of the area at the two censuses. This amounts to assuming a linear growth of the population between the two censuses. Using the geometric mean of the two census populations is a slightly more attractive procedure in the light of what we know about how populations grow (see Chapter 12).

AGE-SPECIFIC NET MIGRATION

In Section 14.6, it was shown how data from two censuses taken ten years apart may be used to estimate survivorship probabilities in a life table. For example, it is clear that

$$\frac{\text{population alive aged } x + 10 \text{ last birthday in second census}}{\text{population alive aged } x \text{ last birthday in first census}} = \frac{L_{x+10}}{L_x},$$

where L_x and L_{x+10} are standard life table quantities. This equation assumed that there was *no migration*. If we relax the assumption that there is no migration, we can write

$$\frac{\text{population alive aged } x + 10 \text{ last birthday in second census}}{\text{population alive aged } x \text{ last birthday in first census}} = \frac{L_{x+10}}{L_x} + M,$$

where M depends upon the net migration of persons who are aged $x + 10$ years last birthday in the second census. Denoting the population aged x last birthday in year t by the symbol $P_{x,t}$, and the net migration of persons aged x in year t by $M_{x,t}$, we can then estimate $M_{x+10,t+10}$ as follows:

$$M_{x+10,t+10} = P_{x+10,t+10} - \frac{L_{x+10}}{L_x} \cdot P_{x,t}. \tag{15.1}$$

In equation (15.1), the term L_{x+10}/L_x is a *survivorship probability*. It represents the chance that a person aged x years last birthday at the first census will survive to be aged $x + 10$ years last birthday at the second census. All that is required in order to calculate the survivorship probabilities for a given area is a life table which may reasonably be expected to represent the mortality of the population in question.

Equation (15.1) may be used to estimate age-specific net migration. Since the resulting estimates are only approximate, in order to minimize the risk of large errors, it is usual to make the estimates by age groups (often quite wide age groups). For censuses taken ten years apart, it is convenient (though not necessary) to use ten-year age groups. Denoting the population aged between x and $x + 10$ years exactly in year t by the symbol $_{10}P_{x,t}$, and the net migration of persons aged x to $x + 10$ exactly in year t as $_{10}M_{x,t}$, we can write

$$_{10}M_{x+10,t+10} = {}_{10}P_{x+10,t+10} - \frac{_{10}L_{x+10}}{_{10}L_x} \cdot {}_{10}P_{x,t}. \tag{15.2}$$

The only age groups for which equation (15.2) cannot be used to estimate net intercensal migration are those which include persons at the second census who are too young to have been alive at the first census. For these age groups it is necessary to know the number of births during the intercensal period. If this is known, then the following formula may be used:

$$_{10}M_{0,t+10} = {}_{10}P_{0,t+10} - \frac{_{10}L_0}{10l_0} \cdot B, \tag{15.3}$$

where B is the intercensal number of births, and $_{10}L_0/10l_0$ is the chance that a baby born during the intercensal period will survive until the date of the second census.

15.6   Migration streams

So far in this chapter, the analysis has concentrated on measuring migration to and from a single place. In other words, we have been concerned to measure the impact of migration on the development of a single population. However, each change of residence necessarily involves an origin and a destination. Often interest centres on the magnitude and intensity of flows between pairs of places.

The flow of persons between an origin place i and a destination place j is known as a *migration stream*. The intensity of a migration stream may be measured by computing a rate of the following form:

$$\text{migration rate from place } i \text{ to place } j = \frac{M_{ij}}{P_i} (\times 1000). \tag{15.4}$$

In this formula, M_{ij} is the number of persons moving from place i to place j, and P_i is some measure of the population at risk of moving away from place i.

Unfortunately, because of the way in which migration data are usually collected, the definition of the appropriate exposed-to-risk, P_i, is not straightforward. For example, suppose data have been collected in a census on the basis of place of residence at a fixed prior date, say n years before the census. During the intervening period the population at risk in place i is changing. It is changing because of births and deaths, and also because of in-migration and out-migration. Various possible ways of defining the exposed-to-risk under these circumstances are possible, but perhaps the most useful is to restrict the denominator of equation (15.4) to refer to persons living in place i, n years before the census, who survived to the census date. We can estimate this quantity by noting that

$$\begin{matrix} \text{persons living in place } i, \\ n \text{ years before census date} \end{matrix} = \begin{matrix} \text{persons living in place } i \\ \text{at time of census} \end{matrix} - \begin{matrix} \text{in-migrants to place} \\ i \text{ during } n \text{ years} \\ \text{before census date} \end{matrix}$$

$$+ \begin{matrix} \text{out-migrants from place } i \text{ during} \\ n \text{ years before census date} \end{matrix}.$$

If the census is assumed to take place in year t, this equation can be written

$$P_{i,t-n} = P_{i,t} - \sum_j M_{ji} + \sum_j M_{ij},$$

where $\sum_j M_{ji}$ and $\sum_j M_{ij}$ relate to moves in the n years before the census date.

Using the right-hand side of this equation as the denominator of equation (15.4) therefore produces

$$\text{migration rate from place } i \text{ to place } j = \frac{M_{ij}}{P_{i,t} - \sum_j M_{ji} + \sum_j M_{ij}} (\times 1000).$$

A migration rate calculated in this way measures the probability that a person living in place i at time $t - n$ who survives until time t will be living in place j at that time.

Further reading

Migration data sources are reviewed briefly in Bulusu (1991). Boden *et al.* (1992) is a very useful evaluation of the quality of the National Health Service Central Register data in England and Wales, comparing them with data from the census question on migration. The census data themselves are described in Flowerdew and Green (1993).

A good collection of studies dealing with contemporary migration patterns, especially in the United Kingdom, is Champion and Fielding (1992).

More sophisticated modern analysis of the impact of migration on the size and structure of regional populations uses extensions of the life table to the multi-regional case. These extensions are considered in a number of excellent textbooks, notably Rees and Wilson (1977) and Rogers (1975). These books are not recommended for those who dislike mathematics: in particular, familiarity with matrix algebra is essential for those wishing to study multi-regional methods.

16

Introducing Population Projection

16.1 Introduction

The final three chapters of this book are about population projection. This is an important practical aspect of the work of demographers, many of whom are employed in various aspects of population forecasting. It is also, however, a subject which requires a knowledge of mortality, fertility and migration, and the ways in which they act together to determine population change and growth. Population projection, therefore, provides a real test of the demographer's understanding of the major components of population change.

This chapter introduces the topic, and describes one simple method by which population projection might be carried out. Section 16.2 lists some reasons why population forecasts are required. Section 16.3 describes the two main methods employed by demographers to make these forecasts. These are known as the mathematical (or formula) method and the component method. In Section 16.4, we show that these two methods are but applications of a general procedure, which we describe. Section 16.5 describes the mathematical method of population projection in detail, and Section 16.6 summarizes its advantages and disadvantages. The component method is the subject of Chapter 17.

A NOTE ON TERMINOLOGY

Population projection is all about trying to determine what the future population will be like. It involves trying to forecast what will happen, but, as will be seen, it involves rather more than crystal-ball gazing. Occasionally there is confusion over the terminology used to describe the looking into the future which demographers do. *Population projection* is a term used to refer to the whole of the fairly complex exercise we shall describe in this chapter and the next. The term *forecast* is used to indicate the actual predictions about which demographers feel reasonably confident (these may be what are called 'principal projections' – see Section 16.4 below – or projections which only go a few years into the future). Sometimes the term 'projection' is used to describe a predicted population. However, it has always seemed to me better to restrict the term 'projection' to describe the method or the technique, and to use some other term to describe the results (perhaps *predicted population* or *projected population*).

16.2 The need for population forecasts

Population forecasts are required for many purposes. Some examples are listed below.

1 Local education authorities need forecasts of the number of children for whom school places will be required in five or ten years' time. In the United Kingdom, such forecasts will typically be required for counties (and perhaps local authority districts).
2 National forecasts are needed to enable the future demand for food, power, transport and other services to be estimated.
3 Forecasts of the future age structure are required for national insurance purposes, since contribution rates are linked to the ratio of non-working to working people.
4 Forecasts of the numbers of people classified by marital condition may be required to estimate the future number of households, the potential membership of the workforce, and the need for social security benefits.
5 Forecasts are made of the number of students in higher education, which help governments to decide what proportion of the available public funds should be allocated to universities.

There are, of course, many other purposes for which forecasts are required.

16.3 Approaches to population projection

The process by which demographers make their forecasts about the future population is known as *population projection*. There are two methods of carrying out population projection in common use: the *mathematical method*, in which the evolution of the population is assumed to be described by some fairly simple mathematical formula (because of this, it is sometimes called the *formula method*); and the *component method*, in which the components of population change are taken into account explicitly.

The mathematical method is quick, simple, and requires little in the way of data. It is the approach of choice for many projections of the whole populations of countries.

The component method is much more cumbersome than the mathematical method, and has heavy data requirements. It is more time-consuming than the mathematical method, although the advent of computers has made it a great deal quicker than it used to be. It has the great advantage over the mathematical method that detailed aspects of the population structure can be forecast.

Which is the most convenient approach to use in any specific application depends on the degree of detail required in the projection. The latter is usually indicated by the purpose. For some purposes only the total population size is needed; for others the population structure in five-year age and sex groups may be necessary; occasionally very detailed projections (for example, of the population structure by single years of age and marital status) may be desired. These requirements dictate the degree of disaggregation required in the model for population change.

Often it is not possible to obtain the desired detail. This might be because the necessary data are not available or are unreliable; it might be because the resources required to carry out the projection are out of all proportion to its value; it might also be because the time taken to produce the projection would render it useless by the time it was finished. The choice of method and the degree of detail with which the chosen method is applied are influenced by all these considerations.

16.4   The general procedure

Population projection makes use of simple models of population change. It involves, first, choosing a model of population change; second, specifying the parameters of the chosen model (this may be done by estimating them using existing data, or by making intelligent assumptions); and, third, applying the model to current data to extrapolate into the future. The third of these stages is completely mechanical. In other words, once the model has been chosen and the parameters specified, the future population's characteristics are determined. Thus all the important effort has to go into the first two stages.

CHOOSING A MODEL

Although in theory there is a great variety of models, in practice the choice is rather limited. In the mathematical method, there are really only two commonly used models: the exponential growth model (and its discrete form, the geometric growth model) and the logistic growth model. These are described in Section 16.5 below. In the component method, there is considerably more choice of models: however, the range of models which can be applied tends to be determined by data availability.

SPECIFYING THE PARAMETERS

This is the hardest bit of the whole process. Once a model has been chosen, it needs numbers. The numbers are of two types: the *base population data*, which form a starting point, and the *parameters* which will determine how that population will evolve in the future.

Specifying the parameters invariably involves some degree of crystal-ball gazing. Because we can never know for sure what will happen in the future, population projection is necessarily an uncertain exercise. But, given a particular model, most of the uncertainty relates to the values of the parameters which will determine the future evolution of the population. The base population data are usually fairly accurate.

It is not easy to deal with this uncertainty. In practice, demographers tend to do three things.

1   They spend a great deal of time trying to make their specifications of the parameters as plausible as possible.
2   They conduct *sensitivity analyses* to see by how much their forecasts of the future population's characteristics change if their specifications of particular parameters change. It may be, for example, that uncertainty about the value of a particular parameter is great, but that the forecasts are not very sensitive to changes in the value of that parameter. In such a case, the great deal of uncertainty need not be too much of a worry.
3   They usually present, in their reports of population projection exercises, several different scenarios, each using different combinations of parameter values. It may be that one of these scenarios is the one for which they will go to the barricades if they have to. This forecast is often called a *principal projection*. The other scenarios are called *analytic projections*, and are included in the reports to give some idea of the sensitivity of the future population to changes in the parameter values.

This description of population projection emphasizes that it is a method of discovering what the future population will be like if the parameters governing population change in the chosen model take certain particular values. Therefore, its results should be presented in an 'if X then Y' format. It is not a kind of foretelling of the future. It does not produce results which can be described in a declaratory 'the population in year *t* will be *P*' format. Projected populations are always conditional on the actual specification of the parameters of the model used.

16.5   The mathematical method

In the mathematical approach to population projection, no attempt is made to model the components of population change explicitly. Normally the method is applied to the total population, but it is also possible to use it to project subgroups within the population (for example, the number of married women aged 20–24 years last birthday).

The method is very simple. A mathematical model of population change is chosen, and a formula expressing this model is written down. The parameters of this formula are then estimated by fitting it to past data on the population being projected. Predictions of the future numbers are obtained by extrapolating the fitted curves. If several subgroups within the same population are being projected simultaneously, they are usually projected independently of one another.

There are two commonly used models, the exponential model and the logistic model.

THE EXPONENTIAL MODEL

Exponential growth was described in detail in Chapter 12, so we shall only give it a brief examination here. The exponential growth formula is

$$P_t = P_0 e^{rt}, \tag{16.1}$$

where P_t is the population (either the total population, or that in the relevant subgroup) in some future year t, P_0 is the population in some base year (usually the latest year for which we have data), and r is the constant annual rate of population growth. Differentiating equation (16.1) with respect to t produces

$$\frac{dP_t}{dt} = P_0 r e^{rt}. \tag{16.2}$$

Solving equation (16.2) for r gives

$$r = \frac{dP_t}{dt} \cdot \frac{1}{P_0 e^{rt}}, \tag{16.3}$$

and, using equation (16.1), equation (16.3) can be simplified to read

$$r = \frac{dP_t}{dt} \cdot \frac{1}{P_t}. \tag{16.4}$$

THE LOGISTIC MODEL

A problem with the exponential growth model is that if population growth continues indefinitely at a constant annual rate, the population size will either increase without limit (which is generally thought to be an unrealistic implication) or approach zero (which is also thought to be unrealistic). One way of modifying the exponential model to avoid these implications is to allow the annual rate of growth in any year t to be determined in part by the population size in year t, P_t. This can be achieved by modifying equation (16.4) above as follows:

$$\frac{dP_t}{dt} \cdot \frac{1}{P_t} = r + f(P_t), \tag{16.5}$$

where $f(P_t)$ is some (as yet unspecified) function of P_t. If we suppose that $f(P_t)$ is equal to $-kP_t$ (where k is a parameter to be estimated), equation (16.5) becomes

$$\frac{dP_t}{dt} \cdot \frac{1}{P_t} = r - kP_t.$$

This differential equation may be solved (it is fairly straightforward, but we shall not go through the algebra here), to produce the following equation for P_t:

$$P_t = \frac{r}{Cre^{-rt} + k}, \tag{16.6}$$

where C is a constant associated with an initial population size (C is not the same as P_0). Equation (16.6) describes an S-shaped curve for P_t with an upper asymptote of r/k (Figure 16.1). Such a curve is known as a *logistic curve*. The logistic curve implies that, after a time, the rate of population growth slows down, and that the population size will never rise beyond a certain level – the upper asymptote shown in Figure 16.1.

Equation (16.6) can be divided by k to produce

$$P_t = \frac{r/k}{\left(\dfrac{Cre^{-rt}}{k} + 1\right)}.$$

This can be written

$$P_t = \frac{K}{1 + e^{(\alpha + \beta t)}}, \tag{16.7}$$

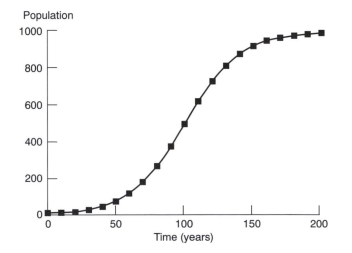

Figure 16.1 A logistic curve, drawn using equation (16.6) with $r = 0.05$, $k = 0.00005$ and $C = 0.148$

where

$$K = \frac{r}{k}, \qquad \beta = -r \qquad \text{and} \qquad \alpha = \ln(Cr/k).$$

Therefore, equations (16.6) and (16.7) are equivalent. They are both used in the literature to describe a logistic curve.

APPLYING THE MATHEMATICAL METHOD

The procedure which we adopt to apply the mathematical method of population projection is extremely simple in principle. The data required are simply the population size P_t for a number of values of t less than zero (that is, values of P_t in the past – the recent past is preferable). Using these data, the parameters of the chosen model are estimated.

The exponential model has a single parameter, the rate of growth r. This may be estimated by linear regression using the following equation:

$$\ln P_t = \ln P_0 + rt. \tag{16.8}$$

Equation (16.8) is obtained by taking natural logarithms of equation (16.1). Performing the regression based on equation (16.8) will also indicate how well the exponential growth model fits the past experience. This can also be ascertained by plotting past values of $\ln P_t$ against t. If the exponential model fits well, the result should be close to a straight line.

With the logistic model, things are a little more complicated. It is easiest to work with equation (16.7). This has three parameters: K, α and β. Of these, K is the upper asymptote (the maximum population size), which is normally specified in advance by the investigator. Assuming that K is specified in advance, then it is possible to evaluate the quantity P_t/K for all known past values of P_t. Dividing equation (16.7) by K produces

$$\frac{P_t}{K} = \frac{1}{1 + e^{(\alpha + \beta t)}}. \tag{16.9}$$

The right-hand side of equation (16.9) may be rewritten as $e^{(-\alpha - \beta t)}/(1 + e^{(-\alpha - \beta t)})$ (by multiplying numerator and denominator by $e^{(-\alpha - \beta t)}$). Therefore, we have

$$\frac{P_t}{K} = \frac{e^{(-\alpha - \beta t)}}{1 + e^{(-\alpha - \beta t)}},$$

and this equation can be rearranged to read

$$\ln\left(\frac{P_t/K}{1 - P_t/K}\right) = -\alpha - \beta t. \tag{16.10}$$

Equation (16.10) may then be used to estimate the parameters α and β by linear regression.

Whatever the model used, the minimum number of past time points for which data are required is equal to one more than the number of parameters to be estimated. However, the more time points we have, the more confident we are likely to be about our model.

16.6   The limitations of the mathematical method

The mathematical method of population projection has the decided advantage that its data requirements are nominal and that it is quick. These advantages should not be under-estimated.

However, it has a number of serious limitations.

1   It involves implicit assumptions about the continuity of population change, notably the assumption that the same model (exponential or logistic) will continue to apply throughout the period of the projection.
2   It takes no account of current levels and trends in the components of population change. In other words, population change is seen as an abstract phenomenon, without the investigator having to think about how and why it happens.
3   Empirically, it has been found in recent years that most mathematical formulae tend to overestimate future population sizes, mainly because of decreasing fertility levels which are not taken into account by simply extrapolating past trends.
4   Finally, when applied to projecting the size of several subgroups within the same population the mathematical method ignores logical interdependencies among these subgroups. (This point is considered more fully in Chapter 17.)

Because the increasing availability of computers has reduced the advantages of the mathematical method, while its disadvantages have remained roughly constant, it is less widely used than it used to be.

Exercise

16.1   Table 16E.1 gives the populations of Algeria and Jordan in various years between 1950 and 1985.
(a)   Plot the population (on the vertical axis) against time (on the horizontal axis) for both countries.
(b)   Plot the natural logarithms of the population against time. Is the exponential model a good description of the past growth of these two countries' populations?
(c)   Project the populations of these two countries in the years 1995 and 2005 using the mathematical formula method.

Table 16E.1

Year	Population (thousands)	
	Algeria	Jordan
1950	8 753	1 237
1955	9 715	1 447
1960	10 800	1 695
1965	11 923	1 962
1970	13 746	2 299
1975	16 018	2 600
1980	18 740	2 923
1985	21 788	3 407

Source: United Nations (1991, pp. 306, 448).

17

The Component Method of Population Projection

17.1 Introduction

This chapter describes the component method of population projection, using the United Kingdom's national population projections as an example.

In order to understand how the component method works, and why it is now the preferred method in most situations, we need to recall what was said in Chapter 1 about population structure. The *structure* of a population is the way in which the total population is distributed among different ages, marital statuses, employment statuses, and so on.

For many purposes, it is less the future total population size which is of interest than the future population structure. For example, education planners are typically interested in the future population of children aged, say, between 5 years and 16 years last birthday, classified by single years of age. For the purposes of planning the housing stock, it is important to have forecasts of the population classified by marital status, because the required housing stock depends on the number of households there will be, and the number of households is greatly influenced by the proportion of the population which is single (that is, never married), married, divorced and widowed.

Even for forecasting the number of births in future years, it may be considered necessary to have forecasts of the number of women of childbearing age classified by marital status, because the fertility of married women is different from that of single, divorced and widowed women. Other things being equal, changes in the distribution of the female population by marital status will affect the number of births. The need, therefore, is for forecasts of the future population structure. How are these to be made?

We could use the mathematical method, which was described in Section 16.5. To apply this method, the current population would be classified according to the relevant variables (age, sex, marital status and so on), and then a mathematical formula would be used to project each resulting cell. So, for example, we might obtain some data on how many married females are alive aged 25–29 years, fit a suitable mathematical formula to the past trends in the number of such women, and use this formula to project the future numbers of women in this category. Similar formulae (with different parameters) could then be used to project different categories (for example, married females aged 30–34 years).

This approach has serious drawbacks. For a start, it is rather cumbersome, since if there are a great many categories a lot of different sets of parameters will need to be estimated. There is also a logical weakness, in that there are clear interdependencies between the numbers in certain categories, yet using this method the projection of each category is done in isolation. For example, the number of married women aged 30–34 years last birthday in ten years' time is clearly dependent to a great extent on the number of married women aged 25–29 years last birthday in five years' time (the two categories both involve many of the same women). The mathematical method takes no account of such interdependencies.

Because of these drawbacks, projections of the population structure are rarely made using the mathematical method. Instead, the much more powerful component method is used, which explicitly takes into account the interdependencies.

The component approach allows us to project the whole population structure over a given future period. Of course, it also allows the projection of the total population size, since by summing the projected numbers in all of the categories the total population is obtained.

Applications of the component method of population projection tend to be somewhat complicated. However, its fundamental principles are rather simple. Section 17.2 describes these principles. The details of the method are described in Sections 17.3 and 17.4, in the context of a particular example, the United Kingdom national population projections. It will become clear as these sections unfold that the level of detail which can be achieved in component projection is often limited by the availability of data. Section 17.5 looks in detail at the data required for component projections like the UK national ones. Section 17.6 discusses more complex projection exercises, and Section 17.7 briefly considers projections for subnational units.

17.2   Principles of the method

The key to understanding how the population structure changes over time is to remember that changes in the population size and structure can only occur because of a relatively small, countable, number of different events. For example, the number of people living in England and Wales will only change when one of the following three events takes place:

- a birth in England and Wales;
- the death of someone living in England and Wales;
- a person migrating into or out of England and Wales.

These events are called the *components of population change*. Their intensity is measured by observable rates: the fertility rate, the mortality rate and the migration rates, respectively.

More complicated lists of components may be necessary when one is considering the numbers of people in specific categories of the population structure. For example, the number of currently married women aged 20–24 years last birthday living in England and Wales will change when any one of the following events occurs (Figure 17.1):

1   the 20th birthday of a currently married woman living in England and Wales;
2   the marriage of a woman aged 20–24 years last birthday who is living in England and Wales;
3   the migration into England and Wales of a married woman who is aged 20–24 years last birthday;
4   the death of a married woman aged 20–24 years last birthday who was living in England and Wales when she died;
5   the divorce of a woman aged 20–24 years last birthday who is living in England and Wales;

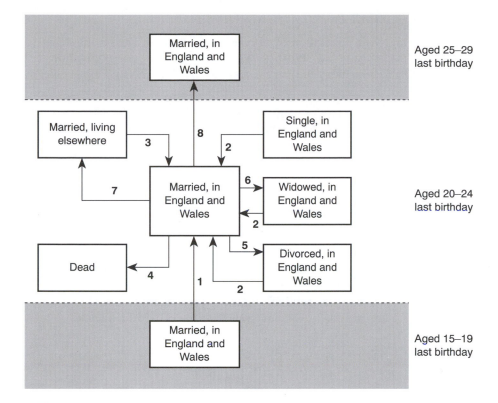

Figure 17.1 Multiple-state representation of changes in the number of married women aged 20–24 years last birthday in England and Wales (the numbers beside each transition refer to the list of components in the text)

6 the death of the husband of a woman aged 20–24 years last birthday who is living in England and Wales;
7 the migration out of England and Wales of a married woman who is aged 20–24 years last birthday;
8 the 25th birthday of a married woman living in England and Wales.

There are eight components of change in this example. The first three of these events will increase the number of women in this particular category; the remaining five will decrease that number. Even in this quite complicated example, the number of components of change is still fairly small.

17.3 The details of the method

In this section, the component projection method is described in detail using the example of the United Kingdom's official national population projections, which are produced every two years by the Government Actuary's Department, in association with the Office for National Statistics.

The UK national population projection is a projection of the population by sex and single years of age last birthday. The stages involved in this projection are as follows.

OBTAIN A BASE YEAR POPULATION STRUCTURE

Estimates of the mid-year population by sex and single years of age last birthday are produced by the Office for National Statistics. We denote the mid-year populations of males and females aged x last birthday in the year t by the symbols $P_{x,t}^{m}$ and $P_{x,t}^{f}$, respectively.

PROJECT THE POPULATION ONE YEAR ON FROM THE BASE YEAR AT ALL AGES LAST BIRTHDAY EXCEPT AGE 0

All the people alive aged at least 1 year last birthday at mid-year $t + 1$ were already born at mid-year t. Thus for these people the only components of change which we need to consider are mortality and migration. In the UK national population projections the following formula is used for males (Daykin, 1986) (here it is written in a general form, applying to a projection from any year t to the next year $t + 1$):

$$P_{x+1,t+1}^{m} = P_{x,t}^{m}(1 - q_{x+\frac{1}{2},t}^{m}) + M_{x+1,t+1}^{m}. \tag{17.1}$$

In equation (17.1), $q_{x+\frac{1}{2},t}^{m}$ is the probability that a live male aged exactly $x + \frac{1}{2}$ years at mid-year t will die before mid-year $t + 1$; and $M_{x+1,t+1}^{m}$ is the net number of male migrants into the population between mid-year t and mid-year $t + 1$ who survive to mid-year $t + 1$, when they are aged $x + 1$ last birthday. An analogous formula is used for females.

PROJECT THE NUMBER OF PEOPLE WHO WILL BE ALIVE ONE YEAR ON FROM THE BASE YEAR AGED 0 LAST BIRTHDAY

These people will all be born during the period between the middle of the base year and the middle of the next year (remember, we are using mid-year populations for each year). Thus all three components of population change – fertility, mortality and migration – must be included in the equation.

The equation which is used for males may be written as follows (Daykin, 1986):

$$P_{0,t+1}^{m} = B_{t}^{m}(1 - \tfrac{1}{2}q_{0,t}^{m}) + M_{0,t+1}^{m}. \tag{17.2}$$

In this equation, B_{t}^{m} is the number of male babies born between mid-year t and mid-year $t + 1$; $\tfrac{1}{2}q_{0,t}^{m}$ is the probability that a male baby born between mid-year t and mid-year $t + 1$ will die before mid-year $t + 1$; and the remaining terms have already been defined. A similar equation is used for females.

How do we determine the number of male births between mid-year t and mid-year $t + 1$, B_{t}^{m}? This is done by working out the total number of births (of both sexes combined) during this period, and than splitting them up into males and females using the sex ratio of births.

The number of births in the whole of the *calendar* year t, which we will denote by the symbol $B_{t-\frac{1}{2}}$ (since it begins half a year before the middle of year t) is given by the equation

$$B_{t-\frac{1}{2}} = \sum_{x} f_{x,t-\frac{1}{2}} P_{x,t}^{f}, \tag{17.3}$$

where $f_{x,t-\frac{1}{2}}$ is the age-specific fertility rate (AFSR) at age x last birthday in calendar year t. The number of births of both sexes combined in the whole of the calendar year $t + 1$, which we will denote by the symbol $B_{t+\frac{1}{2}}$, is given by the equation

$$B_{t+\frac{1}{2}} = \sum_{x} f_{x,t+\frac{1}{2}} P_{x,t+1}^{f}. \tag{17.4}$$

Assuming that births are evenly distributed across each calendar year, the number of births in the year between mid-year t and mid-year $t+1$, B_t, is therefore $\frac{1}{2}(B_{t-\frac{1}{2}} + B_{t+\frac{1}{2}})$. Thus, substituting for $B_{t-\frac{1}{2}}$ from equation (17.3) and for $B_{t+\frac{1}{2}}$ from equation (17.4), we obtain

$$B_t = \frac{1}{2}\left(\sum_x f_{x,t-\frac{1}{2}}P^f_{x,t} + \sum_x f_{x,t+\frac{1}{2}}P^f_{x,t+1}\right). \tag{17.5}$$

If we assume that the ASFRs in the two calendar years are the same, then $f_{x,t-\frac{1}{2}}$ is equal to $f_{x,t+\frac{1}{2}}$ and equation (17.5) reduces to

$$B_t = \frac{1}{2}\left(\sum_x f_{x,t}P^f_{x,t} + \sum_x f_{x,t}P^f_{x,t+1}\right), \tag{17.6}$$

where $f_{x,t}$ is the constant ASFR in the year between mid-year t and mid-year $t+1$. Equation (17.6) may be rearranged to read

$$B_t = \sum_x f_{x,t}[\tfrac{1}{2}(P^f_{x,t} + P^f_{x,t+1})]. \tag{17.7}$$

In other words, the number of births between mid-year t and mid-year $t+1$ is equal to the sum (over all ages x) of the ASFRs multiplied by the average of the female population aged x last birthday at the beginning and the end of the period. Since the $P^f_{x,t}$ and $P^f_{x,t+1}$ values are already known as a result of previous stages of the projection, we can calculate B_t using equation (17.7).

The number of male and female births can then be calculated by multiplying B_t by $\frac{105}{205}$ and $\frac{100}{205}$ respectively, assuming that the sex ratio of births is 105 males for every 100 females. In general, if the proportion of births which are males is s, then the number of males born between mid-year t and mid-year $t+1$ is $B_t s$, and the number of females is $B_t(1-s)$.

A number of points about equations (17.1), (17.2) and (17.7) are worth considering in a little more detail.

1   Why is the quantity $(1 - q_{x+\frac{1}{2}})$ used in equation (17.1) to denote the proportions surviving (we ignore the subscript denoting calendar year for the moment)? The normal representation of this *survivorship probability* is L_{x+1}/L_x (see Section 15.5). The rationale for using $(1 - q_{x+\frac{1}{2}})$ is that those aged x years last birthday are, on average, aged $x + \frac{1}{2}$ years exactly. The quantity $(1 - q_{x+\frac{1}{2}})$ measures the probability that someone aged $x + \frac{1}{2}$ years exactly will survive to age $x + 1\frac{1}{2}$ years exactly. It is certainly true that

$$1 - q_{x+\frac{1}{2}} \cong \frac{L_{x+1}}{L_x}$$

to a good approximation. In fact, the quantity L_{x+1}/L_x is used in many projections.

2   Similarly, it might be wondered why the quantity $(1 - \frac{1}{2}q_{0,t})$ is used in equation (17.2) to denote the proportions of children born between mid-year t and mid-year $t+1$ who survive to mid-year $t+1$. The rationale for using $(1 - \frac{1}{2}q_0)$ (suppressing the subscripts denoting year for the moment) is that those aged under 1 year at the middle of year $t+1$ are, on average, aged $\frac{1}{2}$ *year exactly*. The term $(1 - \frac{1}{2}q_0)$ is the probability that someone will survive to exact age $\frac{1}{2}$ year. However, the usual representation of the probability that someone born during a year will survive until the end of that year is L_0/l_0, and this is often used in population projections instead of $(1 - \frac{1}{2}q_0)$.

3   In equation (17.7), it is often reasonable to assume that the female population aged x last birthday at mid-year t is the same as the female population aged x last birthday at

Table 17.1 Projecting the population structure from mid-year t to mid-year $t + 1$

Age last birthday x	Base year population	Survivorship factor	Population in year $t + 1$
0	$P_{0,t}$	$1 - q_{\frac{1}{2},t}$	$P_{0,t+1}$
1	$P_{1,t}$	$1 - q_{1\frac{1}{2},t}$	$P_{1,t+1}$
2	$P_{2,t}$	$1 - q_{2\frac{1}{2},t}$	$P_{2,t+1}$
$\vdots$	$\vdots$	$\vdots$	$\vdots$
ω	$P_{\omega,t}$	0	$P_{\omega,t+1}$

mid-year $t + 1$ (this is reasonable because mortality in the childbearing age range is low, and in national populations net migration rates are usually small). If we make this assumption, then equation (17.7) takes the simpler form

$$B_t = \sum_x f_{x,t} P^f_{x,t}.$$

4   Migration is incorporated after mortality and fertility have been included. In a way, the term $M^m_{x+1,t+1}$ in equation (17.1) and the term $M^m_{0,t+1}$ in equation (17.2) can be seen as adjustment factors. Of course, one could derive equations which explicitly account for the mortality of migrants. In practice, however, since migration is not very important in national projections, the gain from this is small, and the additional complexity is considerable. In projections for *subnational units*, however, migration is much more important, and the mortality of migrants should be taken into account explicitly (see Section 17.7).

The similarity of equations (17.1) and (17.2) to the equations used to estimate age-specific net migration rates from data in two censuses together with information about survivorship ratios (equations (15.1) and (15.3)) should be noted.

REPEAT THE WHOLE PROCEDURE FOR SUBSEQUENT YEARS

Equations (17.1), (17.2) and (17.7) are then applied again for the period between mid-year $t + 1$ and mid-year $t + 2$. The procedure is iterated for as many years as are needed to obtain the population forecasts required.

   The whole projection procedure can be programmed in a series of spreadsheets on a computer. Two tables are required, one for projecting the population structure, and a subsidiary one for calculating the number of births. These are illustrated in Tables 17.1 and 17.2. For simplicity, we have ignored the adjustments for migration in these tables.

17.4   The use of broader age groups

In Section 17.3, the component projection method was outlined for the case when the population was classified by single years of age. There is no reason why the component method cannot be used to project the numbers of people in different five-year or ten-year age groups (or, indeed, age groups of any width), and the method is often used with coarser groupings of age than single years.

   When using the component method to project the population in age groups broader than single years, however, two points should be borne in mind:

Table 17.2 Working out the number of births between mid-year t and mid-year $t+1$

Age last birthday x	Female population aged x last birthday in year t	Female population aged x last birthday in year $t+1$	Age-specific fertility rate	Births to women aged x last birthday
15	$P^f_{15,t}$	$P^f_{15,t+1}$	$f_{15,t}$	$f_{15,t}[\frac{1}{2}(P^f_{15,t}+P^f_{15,t+1})]$
16	$P^f_{16,t}$	$P^f_{16,t+1}$	$f_{16,t}$	$f_{16,t}[\frac{1}{2}(P^f_{16,t}+P^f_{16,t+1})]$
17	$P^f_{17,t}$	$P^f_{17,t+1}$	$f_{17,t}$	$f_{17,t}[\frac{1}{2}(P^f_{17,t}+P^f_{17,t+1})]$
⋮	⋮	⋮	⋮	⋮
49	$P^f_{49,t}$	$P^f_{49,t+1}$	$f_{49,t}$	$f_{49,t}[\frac{1}{2}(P^f_{49,t}+P^f_{49,t+1})]$

1 The age groups should all be of the same width (except for the oldest one).
2 If the width of the age groups is n years, then it is usual to project the population forward by n years at a time.

These two criteria are not necessary but, if they are not met, the projection becomes very awkward. If the two criteria are met, then the projection of broader age groups follows exactly the same procedure as outlined in Section 17.3. The three important formulae (equations (17.1), (17.2) and (17.7)) become

$$_nP^m_{x+n,t+n} = {}_nP^m_{x,t}\left(\frac{_nL_{x+n,t}}{_nL_{x,t}}\right) + {}_nM^m_{x+n,t+n}, \tag{17.8}$$

$$_nP^m_{0,t+n} = {}_nB^m_t\left(\frac{_nL_{0,t}}{_nl_0}\right) + {}_nM^m_{0,t+n}, \tag{17.9}$$

$$_nB_t = n\sum_x {}_nf_{x,t}[\tfrac{1}{2}({}_nP^f_{x,t} + {}_nP^f_{x,t+n})]. \tag{17.10}$$

Note that in equations (17.8) and (17.9) we have used the alternative estimate of the survivorship probabilities which is expressed in terms of L_x rather than q_x. In equations (17.8) and (17.9), n is the width of the age groups used (and also the number of years for which the population is projected at each iteration). The values of $_nL_{x,t}$ and $_nL_{x+n,t}$ are taken from a life table which describes the mortality between years t and $t+n$; $_nP^m_{x,t}$ is the male population aged between x and $x+n$ years at mid-year t; and $_nM^m_{x,t+n}$ is the net number of migrants between mid-year t and mid-year $t+n$ who survive to mid-year $t+n$ when they are aged between x and $x+n$ years. In equation (17.10), $_nB_t$ is the number of births between mid-year t and mid-year $t+n$, and $_nf_{x,t}$ is the age-specific fertility rate in the age group x to $x+n$ years between mid-year t and mid-year $t+n$.

17.5 Data requirements

The data requirements for population projections made using the component method are quite onerous. To consider the UK national population projections, it is clear that four sets of data are required:

1 a base year population subdivided by age and sex;
2 sex-specific life tables for the projection period;
3 age-specific fertility rates for the projection period;
4 age- and sex-specific net migration rates for the projection period.

All these data except the base population must themselves be forecasts. In other words, the value of the forecast population structure depends completely on the value of the forecasts made of fertility, mortality and migration. The degree to which the predicted future population structures reflect what will actually occur is entirely dependent upon the degree to which the assumptions about future fertility, mortality and migration reflect what will come to pass. The projection equations (equations (17.1), (17.2) and (17.7) or equations (17.8), (17.9) and (17.10)) describe purely mechanical operations. If the assumptions about fertility, mortality and migration are silly, the results of the projection will also be silly.

The four data requirements will now be considered in turn, using the United Kingdom national population projections as an example.

THE BASE YEAR POPULATION

This is the least troublesome of the requirements. Since almost every country in the world has had at least one census, it is not usually too hard to find some kind of population structure for a recent year. Clearly, since the base year population is the one solid item in the list of data, it is important that it is as accurate as possible. In certain cases, adjustments to the age distribution may need to be made to remove the effects of age heaping or systematic misstatement of ages.

Mid-year estimates of the population of the United Kingdom and its constituent countries (England, Wales, Scotland and Northern Ireland) are made each year by the Office for National Statistics for England and Wales and by the General Register Offices of Scotland and Northern Ireland for those two countries. These estimates are based on the most recent population census, and on the numbers of births and deaths registered since the most recent census.

MORTALITY

The mortality assumptions usually begin with the latest available life table. In countries without vital registration systems, this life table will typically be a model life table selected from one of the published sets (usually either the United Nations set or Coale and Demeny's set; see Section 14.5). Because these model life tables are available in abridged form, interpolation might be necessary to convert an abridged life table into a full life table (by single years of age). More usually, however, the population of countries without vital registration systems is projected using five-year age groups.

In the case of developed countries and other countries with vital registration systems, the latest published life table is the obvious starting point. In the United Kingdom, therefore, English Life Table 15 (which is based on data for the years 1990–92) might be used (Office for National Statistics, 1997a).

We now have to consider how mortality will evolve over the period of the projection. (The *period of the projection* is the period between the present and the latest year for which we are trying to forecast the population.) Generally speaking, the assumption is made of a slow decline in mortality. Is this reasonable?

In the United Kingdom, two main approaches have been adopted to try to forecast mortality rates. One approach is the analysis of trends by age. During the past few decades, the rate of decline of age-specific death rates has varied greatly with age. It has been greatest for infants, and least for the very elderly. Certain other age groups (notably males aged 18–25 years) have experienced very slow improvements (in the case of males aged 18–25 years, the increase in the death rate from road accidents is responsible). Assumptions about future developments can, therefore, be made on an age-specific basis. For example, the 1983-based UK national projections assumed that mortality would decrease over the period 1983–2023 by 10% at the younger working ages and 25% at ages over 65 years.

The other approach is the analysis of trends by cause of death. One way to proceed is to calculate what would happen to mortality rates if certain causes of death were eliminated (through medical advances, for example). This analysis uses cause-specific death rates to construct multiple-decrement life tables, and then work out the death rates assuming that various causes of death have been eliminated (Exercise 5.1 is an example of the principle involved).

An example of both kinds of analysis is provided by a study by Benjamin and Overton (1981). They projected mortality using three assumptions. The first was a *pessimistic assumption*, in which mortality improvements in the future would be confined to ages under 15 years. The second was a *neutral assumption* of an exponential decrease in sex-specific death rates. The third was a (very) *optimistic assumption*, in which deaths from congenital malformations would be eliminated; 90% of lung cancer deaths and about one-third of deaths from ischaemic heart disease below the age of 65 years would be eliminated (these are, broadly-speaking, those deaths thought to be caused by cigarette smoking); all remaining deaths from heath disease would be postponed by ten years; all deaths from bronchitis and emphysema would be prevented (again, these deaths were thought mainly to be due to cigarette smoking, although air pollution is now thought to play more of a part); all cancer deaths except those from lung cancer would be avoided; deaths from unspecified causes would be deferred for ten years, except deaths to infants, which would be prevented altogether (this presumably meant, for example, that 'cot deaths' would cease to occur).

Using these three assumptions, Benjamin and Overton projected the life expectation at birth over the period between the late 1970s (which was the period to which their base data referred) and 2017. Their results are shown in Table 17.3. One of the striking features of their analysis is that the differences between the life expectations at birth resulting from the three assumptions were not very great. The optimistic assumption resulted in a life expectation at birth just under seven years higher than the neutral assumption.

It seems, then, that in attempting to forecast future mortality levels, we shall not do too badly by forecasting a slow decline. For anything other than this to happen, some quite remarkable developments would be needed. In the 1991-based UK national population projections it was assumed that, in England and Wales, life expectation at birth, e_0, would rise gradually from 79 years for females and about 73.5 years for males in 1991 to just over 82 years for females and 77.5 years for males in 2026. For Scotland and Northern

Table 17.3 Benjamin and Overton's results

Assumption	Life expectation at birth in 2017	
	Males	Females
Pessimistic	71.3	77.3
Neutral	74.4	80.4
Optimistic	81.3	87.1

Source: Benjamin and Overton (1981, pp. 24–26).

Ireland a similar trend was assumed, but the values of e_0 were about two years and one year, respectively, less than those assumed for England and Wales (Shaw, 1993).

FERTILITY

The greatest area of uncertainty in national-level population projections concerns fertility rates. This is because fertility behaviour is mainly the result of human choice, and is therefore subject to influences from many factors (economic events, social changes and cultural factors). Unfortunately for those wishing to do population projections, fertility rates, through their effect on the number of births, will have a greater influence on future population numbers, and the future age distribution, than will mortality rates.

In the United Kingdom, the Government Actuary's Department and the Office for National Statistics try to reduce their uncertainty about future fertility trends by looking at the childbearing of recent birth cohorts. The method is described in detail in Daykin (1986), and is summarized briefly below.

The argument that is used is that there is little reason to suspect that the cohort total fertility rate (TFR) will change very much. Past experience suggests that cohort TFRs change only slowly, and that, once they have reached between about 1.8 and 2.1 in developed countries, they do not rise much above that level or fall much below it. (Of course, period TFRs may fluctuate quite a lot from year to year.)

If we make this assumption, then we can work out, for the cohorts which are still in the childbearing ages (that is, the ones that have not completed their childbearing yet), how many more children they must have in order to reach the assumed cohort TFR. By doing this, we can calculate the age-specific fertility rates which they must experience between their current age and exact age 50 years in order to reach the assumed TFR. By doing this for all the cohorts currently in their childbearing years, valuable information can be gained which will help us estimate future ASFRs for different periods in the short-term future. ASFRs for the longer-term future can then be estimated by extrapolation.

The process is illustrated in Figure 17.2, which shows the results of an exercise carried out in the mid-1980s (Daykin, 1986). Fertility data were observed for years up to 1985. The solid lines denote fertility which had already been observed in 1985, the dotted lines denote projected fertility. The projected fertility is worked out by assuming an average completed family size for each birth cohort (this was 2.2 children for the 1945 birth cohort and about 2.0 for subsequent ones), and drawing the cumulated fertility curves in so that the assumed completed fertility is achieved. From the projected fertility curves, a set of ASFRs for each birth cohort can then be worked out. A Lexis chart can then be used to work out which of these cohort ASFRs should be combined together to get a set of ASFRs for each time period in the future.

Of course, this procedure depends on the assumed cohort TFR being a good reflection of what will actually prove to be the case. Evidence to support this assumption is taken in the United Kingdom from various social surveys which ask people about the number of children they want. Such social surveys which have been undertaken in recent years suggest that desired family size is not changing a great deal. However, it does seem that the mean achieved family size of recent birth cohorts has a tendency to fall below their mean desired family size (as reported in surveys conducted when they were in their late teens or early twenties).

The 1991-based UK national population projections assumed that the TFRs in England, Wales and Scotland would rise from their 1991 values (just under 1.9 in Wales, 1.82 in

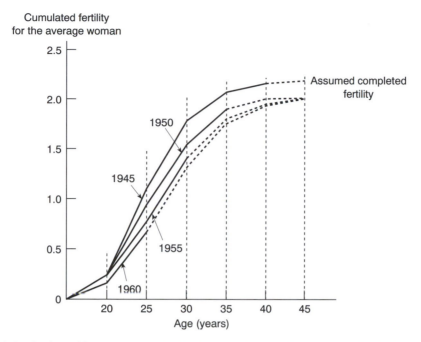

Figure 17.2 Projection of future age-specific fertility rates for four birth cohorts: 1945, 1950, 1955 and 1960. Source: Based on data given in Daykin (1986, p. 31)

England and about 1.7 in Scotland) to reach 1.9 by the end of the century, and that they would remain constant at that value thereafter. For Northern Ireland, the assumption was made that the 1991 TFR of just over 2.1 would fall to 2.05 by the end of the century, and thereafter remain constant (Shaw, 1993).

MIGRATION

The uncertainty about future migration trends is even greater than that about future fertility trends. However, for national-level population projections the importance of migration in determining the future population size and structure is usually not so great. Nevertheless, since its effects tend to be age-specific (especially affecting young adults), some analysis is necessary.

Past trends in net migration rates form the core of this analysis. During the 1980s, the United Kingdom was gaining population through net migration, whereas the opposite was the case during the 1970s. The 1991-based UK national population projections assumed that England will make an annual gain through net migration of about 50 000 persons per year between 1996 and about 2010, but that this net gain will then fall to zero by the year 2016. A much smaller annual gain was forecast for Wales, and small annual losses for Scotland and Northern Ireland (Shaw, 1993).

17.6   More complex component projections

Component projections are not restricted just to projecting the age structure. It is possible to project any aspect of the population structure. The principle remains the same, but projections which involve other aspects of the population structure clearly require additional data. In general, sets of age- and sex-specific transition rates are required for every one of the relevant components of change (except those which involve people celebrating birthdays, since these may be derived as a residual – people get older at a constant rate). As we saw in Section 17.2, a component projection of the population by age, sex and marital status has eight components of change to deal with. Thus we need the following data:

1   a base year population classified by age, sex and marital status;
2   marriage rates specific to age and sex, as well as to marital status prior to marriage;
3   net migration rates specific to age, sex and marital status;
4   death rates specific to age, sex and marital status;
5   divorce rates specific to age and sex;
6   widowhood rates specific to age and sex (although these would probably be derived from the death rates specific to marital status).

These data requirements are very onerous. Indeed, some of these rates may not be available even in developed countries. Of course, assumptions can be made, but this increases the uncertainty in the projections. Therefore, although it may seem that more complex projections will be better than simpler ones, the lack of suitable data for complex projections may mean that a lot of assumptions must be made. If the assumptions are silly, the results will be silly. So, although a very complex projection may look sophisticated and impressive, the results it produces will be worth very little if the assumptions which have been incorporated into the equations are not sensible.

One way to try to reduce the possibility of silly results is to check the plausibility of combinations of assumptions. For example, in a projection of the United Kingdom population by age, sex and marital status, it is reasonable to suppose that the total number of married males living in the United Kingdom at some future time *t* will be very close to the total number of married females living in the United Kingdom in the same year (it need not be *exactly* the same, for some husbands do live in different countries from their wives). If the assumptions about the marriage rates that we have made in a specific application suggest that there will be a large difference between the number of married males and the number of married females at some future time *t*, we should look at them again.

17.7   Population projections for subnational units

Information about the future population of subnational units (such as counties, local authority districts, or towns) is often required. The planning of educational and housing provision, for example, is in the United Kingdom the responsibility of local authorities.

The projection of the population of, say, a county may be achieved either by the mathematical method or using the component method. The component method is more useful, however, since it is often the population in specific subgroups which is of interest (such as school-age children in the case of educational provision).

The main additional complication with the component method, compared with national population projections, is that migration is much more important in determining the future of local area populations than it is in determining the future of national populations. Thus we usually treat migration more seriously in local than we do in national projections. We need to incorporate it into the analysis explicitly, and to look at the mortality of migrants. This involves additional terms in the component projection equations.

There is, of course, a great deal of uncertainty about future migration into and out of local authority areas. Indeed, in the United Kingdom there is even great uncertainty about past and current migration within the nation, so even the existing data are not very good (recall that migration within the United Kingdom need not be registered in any way). Information about recent migration patterns is collected in the Office for National Statistics Longitudinal Study (a 1% sample of the population), and is based on people *changing their doctors*, since change of doctor is registered in the National Health Service Central Register (see Section 15.3). Of course, this is an imperfect measure: no one is forced to change doctor when moving from one place to another, and many people do not register with a doctor until a long time after they move to a place – they wait until they fall ill. The situation is made worse by the fact that the people who are most likely to move from place to place (young adults) are also the least likely to fall ill and need to visit a doctor, and so their migration is the least well recorded of any age group within the population.

Further reading

The United Kingdom national population projections are described in Daykin (1986) and Shaw (1993). For mortality projections, see Benjamin and Overton (1981).

Exercises

17.1   Table 17E.1 shows the age structure of the female population of England and Wales in 1841 and 1861. Prepare a projection to 1881 of the female population of England and Wales. Assume that childbearing is restricted to the 20–39 year age group. State any other assumptions you make.

17.2   You are asked to project the number of men aged 65 years and over in Great Britain by single years of age, for each of the next 20 years. State exactly what data you would require.

17.3   You have been asked to prepare a projection of the number of children aged 3 and 4 years last birthday in England and Wales on 30 June 2000, 30 June 2001 and 30 June 2002. This is to facilitate the planning of nursery schools. Describe in detail:
(a)   the data you would require to make your projection;
(b)   the approach you would adopt.

17.4   (a)   Explain briefly what factors might cause the projection you made in Exercise 17.3 to differ from the actual outcome.
(b)   Assess the likely importance of each factor in accounting for the difference.

Table 17E.1

Age group	Female population (thousands)	
	1841	1861
0–19	3667	4537
20–39	2551	3164
40+	1919	2589

Sources: Census of England and Wales, 1841 (1843); Census of England and Wales, 1861 (1863).

Table 17E.2

Age group	Population on 30 June 1995
45–49	1500
50–54	1468
55–59	1370
60–64	980
65–69	940
70–74	800
75–79	730

17.5 List some circumstances under which you might wish to project the population classified by marital status.

17.6 You are given the data in Table 17E.2, which refer to the population of a small seaside town in the south of England. You are also given the data in Table 17E.3, which come from a life table which is believed to reflect the current mortality in that region of England.

(a) Using these data, estimate the number of inhabitants there will be in this seaside town in each five-year age group in the age interval from 50 to 79 years on 30 June 2000. State any assumptions you make.

(b) Someone now informs you that, each year for the foreseeable future, there is expected to be a net migration into this town of 100 recently retired persons, aged 65 nearest birthday on the day they move in. Describe in detail how you might incorporate this net migration into the population projection you made in part (a) above.

Table 17E.3

Exact age x	l_x ($l_0 = 100\,000$)	Exact age x	l_x ($l_0 = 100\,000$)
45	96 573	63	86 695
46	96 361	64	85 582
47	96 125	65	84 384
48	95 862	66	83 095
49	95 569	67	81 708
50	95 244	68	80 214
51	94 884	69	78 603
52	94 486	70	76 864
53	94 047	71	74 987
54	93 564	72	72 959
55	93 034	73	70 772
56	92 453	74	68 416
57	91 819	75	65 886
58	91 129	76	63 178
59	90 379	77	60 294
60	89 564	78	57 236
61	88 681	79	54 010
62	87 726	80	50 300

Source: Office of Population Censuses and Surveys (1987a, p. 8).

18

Population Projection and Population Dynamics

18.1 Introduction

This chapter concludes the study of population projection by looking at some examples. These examples are instructive not only from the point of view of understanding population projection, for they also provide excellent illustrations of the ways in which fertility and mortality act together to influence population growth. They therefore provide illustrations of some of the ideas described in Chapter 13.

Section 18.2 looks at some results from recent United Kingdom national population projections. Section 18.3 examines further the impact of uncertainty about future fertility, which was revealed in Section 17.5 as the most important reason why national-level population projections might not reflect the ensuing reality. This section introduces another important set of population projections: those made by the United Nations. The discussion in Section 18.3 leads naturally to a consideration of population momentum, which is the subject of Section 18.4. Finally, Section 18.5 looks at one past projection exercise: a (now notorious) set of projections made in the 1940s of the UK population.

18.2 The United Kingdom national population projections

As the United Kingdom national population projections was the main example used in Chapter 17, it is perhaps useful to begin by looking at some of the results of recent efforts made by the Government Actuary in this regard.

Figure 18.1 shows the predicted population of the United Kingdom and its constituent countries up to 2031, using the 1991-based principal projection. As can be seen, a modest rise in the population was projected, although this rise will cease in the 2020s, at which point the population of the United Kingdom will be about 62 million, compared with about 58 million at the time of writing (1997).

As we saw in Chapter 17, the component projection method used also allows the numbers of people in various age groups to be forecast. The main trends to note are (Shaw, 1993):

- the number of people aged 75 years and over is expected to rise quickly between 2021 and 2041;
- the number of people aged 60–74 years will rise quickly between the first and third decades of the twenty-first century;
- the numbers of people in most of the younger age groups will gradually decline.

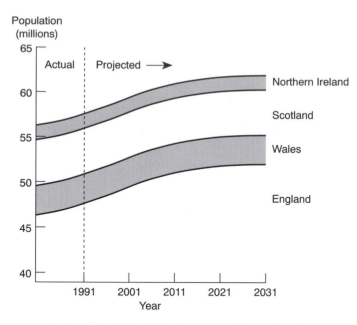

Figure 18.1 Future population of the United Kingdom according to the principal 1991-based projection. Source: Shaw (1993, p. 47)

18.3   The impact of uncertainty about fertility

In Section 17.5, the point was made that uncertainty about future fertility is the major source of concern about the accuracy of national population projections. We can illustrate this using two examples: the difference between the principal and analytic projections of fertility used in the 1983-based United Kingdom national population projection; and an example taken from the United Nations projections of the populations of developing countries.

THE UNITED KINGDOM NATIONAL PROJECTIONS OF FERTILITY

We could use any set of projections to illustrate the difficulty with fertility projections. Here we choose the 1983-based population projections, since the impact of different fertility assumptions has been analysed in detail by Daykin (1986).

The 1983-based projections used three sets of assumptions about fertility: a *principal projection*, which assumed that the period total fertility rate (TFR) would be constant at a level of 2.1 children per woman (on average); a *higher variant*, which assumed a constant TFR of 2.3 children per woman (on average); and a *lower variant*, which assumed a constant TFR of 1.8 children per woman (on average). The numbers of births per year which these three assumptions imply are shown in Figure 18.2.

It is clear that even the quite small differences in the TFR between these three assumptions result in large differences in the forecast numbers of children born. Once they are born, children remain in the population, and the iterative nature of the component projection method means that the projected population in any future year t will depend on the projected number of births in all the preceding years $t-1, t-2, t-3, \ldots$. This means that the three assumptions will result in forecasts of the total population size which diverge quite

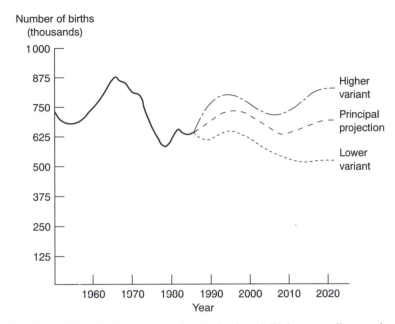

Figure 18.2 Number of live births per year in England and Wales according to the 1983-based national population projections. Source: Daykin (1986, p. 32)

markedly, with the magnitude of the divergence increasing as we move further into the future.

In fact, the actual period TFR during the remainder of the 1980s and the early 1990s followed the *lower variant* very closely. As a result of this, the assumptions used in the United Kingdom national population projections were changed in about 1990 so that the principal 1991-based projection assumed that the TFR would be constant at 1.9 births per woman (at least from 1996 onwards). It is most important that the assumptions used in national population projections are kept under review, so that, if they prove to be at odds with actual developments, more appropriate assumptions can be made in future projections.

THE UNITED NATIONS NATIONAL PROJECTIONS FOR DEVELOPING COUNTRIES

The United Nations regularly publishes population projections for the different countries of the world, and, in particular, for developing countries (see, for example, United Nations, 1995a). These projections usually have to rely on model life tables for their mortality data (since high-quality vital registration systems are rare in developing countries). Nevertheless, because future mortality in most countries of the world is likely to show a slow decline, the major uncertainty with these projections does not lie with the assumptions about mortality. The major problem is with fertility. To illustrate the magnitude of the uncertainty which surrounds the impact of future fertility trends on the future population size, we can consider the United Nations projections of the population of the north African country of Tunisia. The population was projected up to the year 2025 using base population data for the years 1985–89.

United Nations projections take the form of four separate forecasts of the population, based on different assumptions about fertility: a 'high' variant; a 'medium' variant; a 'low' variant; and a constant fertility variant. In the case of Tunisia, the details of the four variants were as follows.

1 With the 'high' variant assumption, the period TFR falls from its 1985–89 level of 4.1 to reach 2.56 by 2015. Thereafter, the period TFR remains constant until 2025.
2 With the 'medium' variant assumption, the period TFR falls from its 1985–89 level of 4.1 to reach 2.10 by 2015. Thereafter, it remains constant until 2025. The value 2.1 is close to *replacement level* – the level at which the population is just replacing itself, so that it will not increase or decrease (this point is discussed further in Section 18.4).
3 With the 'low' variant assumption, the period TFR falls from its 1985–89 level of 4.1 to reach 1.54 by 2015. Thereafter, the period TFR remains constant until 2025. This variant is designed to cater for the possibility that Tunisia will emulate the western and southern European countries in seeing the period TFR fall below replacement level.
4 With the constant fertility variant assumption, fertility remains at its 1985–89 level (a period TFR of 4.1) until the end of the projection period.

The results of the three variants are shown in Figure 18.3 and Table 18.1. The relative difference between the predicted population totals becomes greater as we move further into the future.

An important point to note about these projections is that, with the 'medium' and the 'low' variants, the total population continues to increase even after the period TFR has

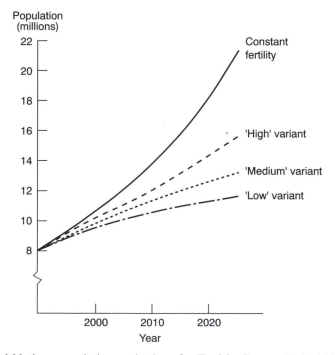

Figure 18.3 United Nations population projections for Tunisia. Source: United Nations (1991, pp. 354–355)

Table 18.1 Forecast population of Tunisia as a percentage of the medium variant forecast, 1995–2025

Year	Forecast population		
	Constant fertility variant	'High' variant	'Low' variant
1995	105	102	98
2000	111	104	97
2005	118	107	95
2010	125	109	93
2015	135	112	91
2020	145	114	89
2025	158	116	87

Source: United Nations (1991, pp. 354–355).

fallen to replacement level (or below it, in the case of the 'low' variant). This phenomenon is known as population momentum. In order to understand it, we need to understand what is meant by the term 'replacement level'. This is the subject of the next section.

18.4 Replacement level and population momentum

REPLACEMENT LEVEL

For a birth cohort, *replacement level* means a net reproduction rate (NRR) of 1.0. In other words, it means that fertility is at such a level that, given the prevailing mortality rates, the rate of population growth is zero.

Now we can write

$$\text{NRR} \cong \text{TFR} \times \frac{100}{205} \times \frac{l_{28}}{l_0}$$

(this approximation was derived in Section 14.7). Using this equation, and assuming that l_{28}/l_0 is about 0.98 (this value has been taken from English Life Table 15 (Office for National Statistics, 1997a) and is representative of most populations in which fertility has fallen close to replacement level), we can estimate that the NRR for that birth cohort will reach 1.0 when the cohort TFR reaches 2.09. Thus a TFR of about 2.1 will imply replacement-level fertility.

POPULATION MOMENTUM

Now consider a population in which the period TFR falls rapidly to replacement level. When the period TFR reaches replacement level, the cohorts who are of childbearing age in that period will have had many of their children in previous years (for example, those women who are aged 35–39 years last birthday when the period TFR reaches replacement level will have been bearing children for the past 20–25 years). In those previous years, the period TFR was above replacement level (possibly considerably above), and so the cohort TFRs will reflect to various degrees the previous higher fertility. Thus the cohort TFRs will be above replacement level, meaning that they will produce a generation which is more numerous than they are themselves. The clear implication of this is that the population will continue to grow. Only when the period TFR has been at replacement level for long enough for a cohort to spend its entire childbearing age range experiencing the age-specific fertility rates which lead to a period TFR of replacement level will that cohort produce a generation which is the same size as the cohort itself. This phenomenon, by which a population will continue to grow for several decades after its period fertility falls to replacement level, is known as *population momentum*.

Even when the whole population has an NRR of 1.0, there will still be population momentum remaining. This is because, at the point when the NRR reaches 1.0, the people born during the previous period, when the NRR was greater than 1.0, will still be alive. During this previous period, when fertility was above replacement level, the birth rate was larger than is consistent with an NRR of 1.0, resulting in a larger number of children being born (relative to the total population size). Even after the NRR has reached 1.0, the large number of children born during the previous years of higher fertility cannot be unborn: they must be allowed to grow older gradually, and, in particular, to go through the childbearing age range and have their children.

The source of the population momentum in a population with an NRR of 1.0 is, therefore, its age distribution. The number of children born during a particular period is determined not just by the fertility rates but also by the number of women of childbearing age. For some time after the NRR has reached 1.0, there will be a relatively large proportion of the population in the childbearing age range (as a result of the previous high fertility). This will inflate the number of children born relative to the total population, thus raising the birth rate and delaying the arrival of *zero population growth*.

This phenomenon is an application of the mathematics of stable population theory (Sections 13.6 and 13.7), in which we saw that it will take a considerable period of time before a population with constant fertility and mortality rates achieves the constant rate of growth determined purely by those fertility and mortality rates.

18.5   A cautionary tale of population projection

Finally, by way of finishing this chapter, and thereby this book, we shall look at a notorious population projection exercise, in which the forecast population proved not to reflect the future experience. Of course, population projections are necessarily uncertain, and we should not expect them to predict the future accurately. Nevertheless, in this example, the departure of the projected population from the eventual experience is quite spectacular.

Figure 18.4 shows five projections of the population of England and Wales made in the 1940s. In the event, the population rose to a value of about 50 million in 1991. What went wrong? Even projection IV, which was thought at the time to be about the highest plausible forecast, has fallen far below the actual experience.

These projections were based on period fertility data from the 1930s and, in particular, calculations of the net reproduction rate. The net reproduction rate in England and Wales had been falling for many years before the 1930s, and during the first few years of that decade had reached values between 0.7 and 0.8, suggesting a decline in the population of 20–30% per generation. It is no surprise, therefore, that population projections based on such a net reproduction rate predicted quite substantial falls in the future population.

However, the net reproduction rates used in the projections were based on data for a short period of time (one or two calendar years). We know that any measures of fertility calculated from figures relating to one or two calendar years tell us almost nothing about the long-run growth rates of a population (this was demonstrated in Section 8.7 using the

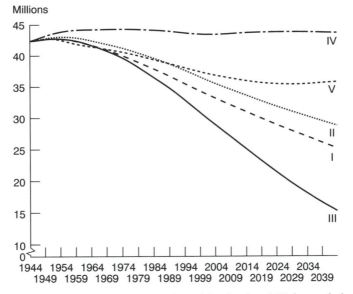

Figure 18.4 Results of projections of the population of England and Wales made in the mid-1940s: projection I assumed that fertility and mortality would remain constant at their 1938–39 level; projection II assumed that fertility would remain constant at its 1938–39 level and that mortality would fall slowly; projection III assumed that both fertility and mortality would fall slowly; projection IV assumed that mortality would fall slowly, but that fertility would remain constant at replacement level; and projection V assumed that mortality would fall slowly, and that fertility would first fall and then rise to replacement level by 1974. Source: Political and Economic Planning (1948, p. 56)

example of Japanese fertility in 1966). In the late 1930s, period fertility in England and Wales was very low, well below replacement level. The net reproduction rates for those years reflected this. However, they proved to be a very poor guide to the future growth of the population.

After the Second World War, period fertility rose substantially, to reach levels well above replacement level during the 1960s (for example, the period TFR in 1964 was 2.9). Consequently the population continued to grow. This example demonstrates superbly how unwise it is to base conclusions about the long-run population growth on data about fertility and mortality for a single calendar year, or even a few calendar years.

Further reading

Readers interested in the United Nations population projections are advised to look at United Nations (1995a). The accuracy of the UK national population projections is discussed in Shaw (1994). Details of the projections of the population of Great Britain made using data from the 1930s are included in Political and Economic Planning (1948, pp. 54–64).

Appendix

The website **http://www.arnoldpublishers.com/download/hinde.htm** contains the data for all the numerical exercises. It also contains other data sets.

Exercises

The data for the numerical exercises are contained in notebook files designed to be read by either of the widely available spreadsheet packages Corel® Quattro® Pro v. 6.0 and Microsoft® Excel v. 7.0. There is one file for each chapter. Within each notebook file, the data for each numerical exercise are on a separate sheet. Sheets are distinguished by the number of the exercise.

In order to use the data, you will need to have available either Quattro® Pro v. 6.0 or higher, running under the Windows® operating system, or Microsoft® Excel (v. 7.0 or Office 97 version). Readers unfamiliar with either of these two spreadsheet packages should consult a relevant manual or guide, of which there are many on the market.

For ease of downloading, all notebook files for Microsoft® Excel are compressed together in a single file titled `hinde-excel.zip`, and all files for Quattro® Pro in `hinde-quattro.zip`. Once downloaded, these files should first be decompressed using the winzip program; instructions for downloading this are included in the website.

Once the .zip files have been decompressed, in order to read a notebook file, enter either spreadsheet package and click on File|Open. What happens next will depend on which spreadsheet you are using.

1 In Quattro® Pro the Open File window will appear. Select the directory containing the files in the box marked Directories, and a list of notebook files will appear in the box on the left. These will be named Chap2ex.wb2, Chap3ex.wb2, ... Double clicking on any of these will open the notebook.
2 In Microsoft® Excel the Open window will appear. Click on the Look in box to find the directory containing the downloaded files. A list of notebook files will appear in the large box below. These will be named Chap2ex, Chap3ex, ... Double clicking on any of these will open the notebook.

Other data

For those who wish to have further practice in analysing read demographic data, a number of additional data sets have been put on the website. These are drawn from a variety of sources, including official statistics from the United Kingdom and Demographic and Health Survey data from some developing countries.

These data are also contained in notebook files for Quattro® Pro v. 6.0 and Microsoft® Excel v. 7.0, which are compressed into the files `hinde-quattro2.zip` and `hinde-excel2.zip`, respectively. They may be opened in the same way as the files containing data for the exercises.

Details of these other data files are included in the `Readme.txt` file which is also on the website. This file is a text file, and may be opened by any widely available word-processing package (for example, Corel® WordPerfect® and Microsoft® Word).

Solutions to Exercises

Chapter 1

1.1 See Figure 7.1.

1.2 See Figure S.1.

Chapter 2

2.1 The crude death rate is obtained by dividing the number of deaths by the mid-year population. Thus, for Argentina, 1990, we have

$$\text{crude death rate} = \frac{295\,796}{32\,322\,000} = 0.00915,$$

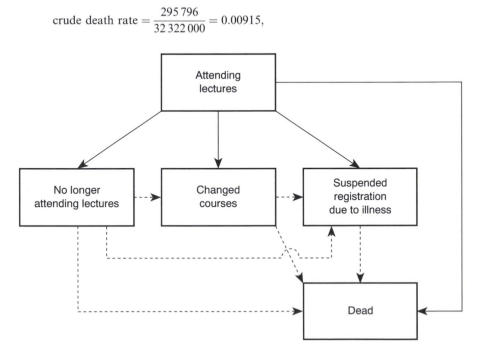

Figure S.1 Solution to Exercise 1.2 (the solid lines denote transitions which directly reduce the number of students attending the lectures; the dotted lines denote other possible transitions)

Table S.1

Age group	Males	Females
1–4	1.15	0.96
5–14	0.45	0.31
15–24	1.16	0.62
25–44	2.36	1.48
45–64	13.28	6.73

or

$$\text{crude death rate} = \frac{295\,796}{32\,322\,000} \times 1000 = 9.15 \text{ per thousand.}$$

The figures for the remaining countries are:

Brazil 1989	0.00790 (7.90 per thousand),
Colombia 1990	0.00610 (6.10 per thousand),
Costa Rica 1991	0.00406 (4.06 per thousand),
Mexico 1991	0.00570 (5.70 per thousand).

2.2 The age-specific death rate for a particular age group is obtained by dividing the number of deaths of persons in that age group by the mid-year population in that age group. Thus, for males aged 1–4 years,

$$\text{age-specific death rate} = \frac{1637}{1\,422\,000} = 0.00115,$$

or

$$\text{age-specific death rate} = \frac{1637}{1\,422\,000} \times 1000 = 1.15 \text{ per thousand.}$$

Age-specific death rates per thousand for all age groups for males and females are shown in Table S.1.

2.3 See Table S.2. Male mortality is greater than female mortality at all ages. The relative differential is greatest in the age group 15–24 years, and least in the age group 1–4 years.

2.4 (a) The infant mortality rate for a particular year is obtained by dividing the number of deaths of infants in that year by the number of births in that year. Thus, for males in 1971,

$$\text{infant mortality rate} = \frac{7970}{402\,500} = 0.0198,$$

Table S.2

Age group	Males	Females
1–4	0.00029	0.00025
5–14	0.00018	0.00013
15–24	0.00073	0.00029
25–34	0.00096	0.00045
35–44	0.00166	0.00105
45–64	0.00785	0.00471
65–74	0.03585	0.02128
75–84	0.08876	0.05664
85 and over	0.19417	0.15184

Table S.3

Year	Males	Females	Both sexes combined
1971	0.0198	0.0151	0.0175
1976	0.0162	0.0122	0.0143
1981	0.0126	0.0094	0.0110
1986	0.0110	0.0080	0.0095
1991	0.0083	0.0064	0.0074
1993	0.0070	0.0056	0.0063
1994	0.0069	0.0054	0.0062
1995	0.0069	0.0053	0.0061

or

$$\text{infant mortality rate} = \frac{7970}{402\,500} \times 1000 = 19.8 \text{ per thousand.}$$

The infant mortality rates for males and females, and for both sexes combined, for all the years in question are shown in Table S.3.

(b) The rates for both sexes combined are obtained by dividing the sum of the male and female deaths in a year by the sum of the male and female births in that year.

(c) Infant mortality declined by more than 50% for both sexes between 1971 and 1991. The rate for males is about one-third higher than that for females throughout the period. During the 1990s, the rate of decline has slowed for both sexes.

2.5 See Figure S2. (a) AB, (b) CD, (c) EF, (d) FG.

2.6 See Figure S3. (a) ABCD, (b) CFGH, (c) ADFC, (d) DEF.

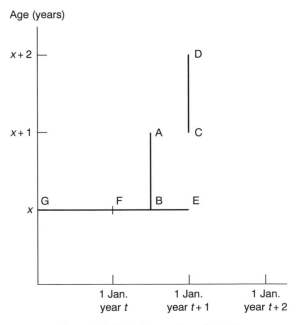

Figure S.2 Solution to Exercise 2.5

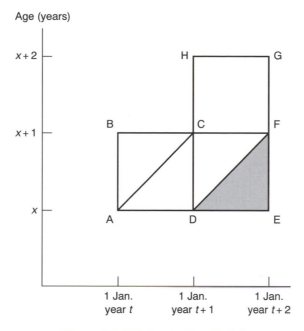

Figure S.3 Solution to Exercise 2.6

2.7 Assume that deaths are evenly distributed within the age group between exact ages x and $x + n$, and that mortality does not vary with calendar time.

Consider the Lexis chart in Figure S.4. The q-type mortality rate between ages x and $x + n$, $_nq_x$, can be obtained by dividing the deaths represented by the parallelogram ABCD by the

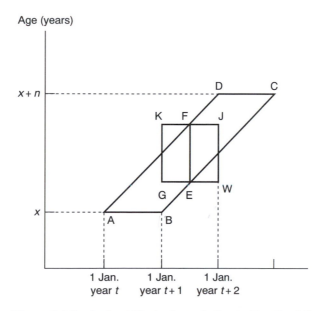

Figure S.4 Lexis chart illustrating solution to Exercise 2.7

population attaining exact age x in year t, P_x (represented by the line AB on the chart):

$$_nq_x = \frac{\text{deaths in parallelogram ABCD}}{P_x}.$$

How many deaths are there in parallelogram ABCD?

Suppose that there are θ_x deaths to people aged x last birthday when they die. Parallelogram ABCD includes all deaths of people aged $x, x+1, x+2, \ldots, x+n-1$ last birthday when they die. Given our assumption about deaths being evenly distributed within each age group, it is clear that the total number of such deaths is equal to $n\theta_x$. Thus, we have

$$_nq_x = \frac{n\theta_x}{P_x}. \tag{1}$$

What about the m-type death rate? We assume that the m-type death rate is the same throughout the age group. Using our two assumptions above, the m-type rate between ages x and $x+n$, $_nm_x$, can be represented on the Lexis chart as

$$_nm_x = \frac{\text{deaths in rectangle GWJK}}{\text{population EF}}.$$

Now, the deaths in rectangle GWJK are equal to θ_x, and population EF is equal to $P_x - \frac{1}{2}n\theta_x$ (this should be clear by noting that the line EF divides the parallelogram ABCD into two halves. Thus we have

$$_nm_x = \frac{\theta_x}{P_x - \frac{1}{2}n\theta_x}. \tag{2}$$

Rearranging equation (2) produces

$$\theta_x = \left(P_x - \tfrac{1}{2}n\theta_x\right) \cdot {}_nm_x = P_x \cdot {}_nm_x - \tfrac{1}{2}n\theta_x \cdot {}_nm_x,$$

whence it is straightforward to see that

$$P_x = \frac{\theta_x}{{}_nm_x} + \tfrac{1}{2}n\theta_x. \tag{3}$$

Substituting for P_x from equation (3) into equation (1) produces

$$_nq_x = \frac{n\theta_x}{(\theta_x/{}_nm_x) + \frac{1}{2}n\theta_x} = \frac{n\theta_x \cdot {}_nm_x}{\theta_x + \frac{1}{2}n\theta_x \cdot {}_nm_x} = \frac{n \cdot {}_nm_x}{1 + \frac{1}{2}n \cdot {}_nm_x}.$$

Thus the required expression is

$$_nq_x = \frac{2n \cdot {}_nm_x}{2 + n \cdot {}_nm_x}.$$

2.8 The solution is shown in Table S.4.

Table S.4

Year	Percentage of infant deaths which were neonatal deaths	Infant mortality rate	Neonatal mortality rate
1971	66.7	0.0180	0.0120
1976	68.2	0.0145	0.0099
1981	60.4	0.0112	0.0067
1986	55.7	0.0095	0.0053
1991	59.5	0.0073	0.0044
1995	67.9	0.0062	0.0042

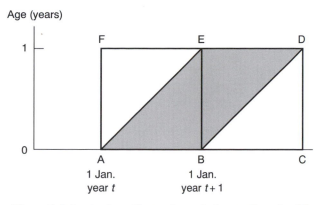

Figure S.5 Lexis chart illustrating solution to Exercise 2.9

(a) Consider Figure S.5. The births in year t are represented by the line AB, and the deaths occurring at ages under 1 year for those births are represented by the parallelogram ABDE. Provided that deaths are evenly spread across the year of age, then the deaths in the triangle ABE will be half of the total number of deaths occurring in year t, and the deaths in the triangle BED will be half of the total number of deaths occurring in year $t + 1$. Therefore

$$\text{deaths ABDE} = \tfrac{1}{2}(\text{deaths in year } t) + \tfrac{1}{2}(\text{deaths in year } t + 1), \qquad (1)$$

which is just the average of the number of deaths occurring in each of these years. Notice that this procedure does not require the assumption that mortality is constant over calendar time. Mortality can vary between year t and year $t + 1$.

(b) Unfortunately, deaths during the first year of life are not evenly spaced through the year of age. Most occur during the first few weeks of life. This means that many more than half of the deaths in the diamond ABDE occur in the triangle ABE.

(c) One solution to this problem is to modify equation (1) to read

$$\text{deaths ABDE} = \alpha(\text{deaths in year } t) + (1 - \alpha)(\text{deaths in year } t + 1),$$

where $\tfrac{1}{2} < \alpha < 1$. The exact value of α will vary from population to population.

Chapter 3

3.1 Mortality in Burnley was 23% higher than that of England and Wales taken as a whole, after controlling for differences in the age structure of the population of Burnley, and that of England and Wales as a whole.

3.2 (a) Argentina: 8.69 per thousand,
Colombia: 4.90 per thousand,
Panama: 4.49 per thousand.

(b) Table S.5 shows the necessary intermediate calculations for the standardized death rates. Summing the two right-hand columns in this table gives the total expected number of deaths in Argentina's population if the age-specific death rates (ASDRs) of Colombia and Panama applied. Dividing these by 15 244 000 (Argentina's total population) gives the standardized death rates, which are, for Colombia, 7.66 per thousand, and for Panama, 6.18 per thousand.

(c) The crude death rates show that mortality in Argentina is 77% higher than that in Colombia, and 94% higher than that in Panama. However, the standardized death rates

Table S.5

Age group	ASDRs (per thousand)		Expected deaths in Argentina with ASDRs of	
	Colombia	Panama	Colombia	Panama
0–4	2.79	5.73	4 930	10 125
5–14	0.68	0.46	2 082	1 409
15–24	2.13	1.33	5 176	3 231
25–44	3.41	2.09	13 984	8 571
45–64	9.73	6.90	26 806	19 010
65 and over	56.56	45.94	63 856	51 866
			$\sum$ 116 835	94 212

reveal that mortality in Argentina is only 13% higher than that in Colombia, and 41% higher than that in Panama. Because the age structure of Argentina is older than that of Colombia or Panama, the crude rates overstate the true mortality differential.

3.3   Table S.6 shows the intermediate calculations necessary for the standardized mortality ratios (SMRs). Summing the two right-hand columns of this table gives the total number of deaths which would be expected in Colombia and Panama if the age-specific death rates of Argentina applied. These are the denominators of the respective SMRs. The numerators are the total number of deaths actually recorded in Colombia and Panama (69 302 and 5196, respectively). Thus, we have

$$\text{SMR for Colombia} = \frac{69\,302}{77\,140} = 0.898,$$

$$\text{SMR for Panama} = \frac{5196}{7057} = 0.736.$$

The SMRs indicate that mortality in Argentina is 11% higher than that in Colombia, and 36% higher than that in Panama, broadly confirming the conclusions reached in Exercise 3.2 using standardized death rates.

3.4   (a)   England and Wales: 11.65 per thousand,
Scotland: 12.32 per thousand,
Northern Ireland: 10.63 per thousand.

(b)   Only SMRs can be calculated using the available data. The SMR for Scotland is 1.125, and that for Northern Ireland 1.120.

Table S.6

Age group	Age-specific death rate in Argentina (per thousand)	Expected deaths if Argentina's ASDRs applied	
		Colombia	Panama
0–4	6.70	12 442	1005
5–14	0.45	1 517	129
15–24	1.16	3 623	282
25–44	2.36	8 788	694
45–64	13.28	21 075	1780
65 and over	62.12	29 693	3168
		$\sum$ 77 140	7 057

(c)   The SMRs indicate that mortality in Scotland and Northern Ireland is about 12% higher overall than that in England and Wales. The crude death rates suggest that Scotland's mortality is about 6% higher than England's, and that Northern Ireland's mortality is lower than England's. Because the age structure of Scotland is (slightly) younger than that of England, the crude death rates understate the true mortality differential. Northern Ireland's age structure is substantially younger than that of England so the crude death rates suggest that its mortality is lower, whereas, in reality, it is higher.

3.5   The SMR is not an ideal index in this case, although it will provide some indication of the relative mortality of the two regions. The ratio between the age-specific death rate in age group x in region C, ^Cm_x, and the corresponding age-specific death rate in region D, ^Dm_x, is not the same for all age groups. In the 0–19 age group, for example, it is 0.5, whereas among those aged 80 years and over it is 2.0. Since the standardized mortality ratio is a weighted average of these ratios, the value which the SMR actually takes will depend crucially upon the weights used.

3.6   We shall make the assumption that all the age groups have equal width. This amounts to assuming that the oldest age group refers to ages 80–99 years.
    If we do this, the Yerushalmy index for region C using the whole country as the standard, YER_C, is given by

$$YER_C = \frac{\left(\frac{0.010}{0.015}\right) + \left(\frac{0.005}{0.005}\right) + \left(\frac{0.015}{0.012}\right) + \left(\frac{0.035}{0.030}\right) + \left(\frac{0.100}{0.075}\right)}{5} = 1.08.$$

The Yerushalmy index for region D using the whole country as the standard, YER_D, is given by

$$YER_D = \frac{\left(\frac{0.020}{0.015}\right) + \left(\frac{0.005}{0.005}\right) + \left(\frac{0.010}{0.012}\right) + \left(\frac{0.020}{0.030}\right) + \left(\frac{0.050}{0.075}\right)}{5} = 0.90.$$

    The interpretation is that values in excess of 1.00 mean that mortality in the population of interest is heavier than that in the standard population, while values below 1.000 mean that mortality in the population of interest is lighter than that in the standard population. Thus mortality in region C is 8% heavier than that in the country as a whole, and mortality in region D is 10% lighter than that in the country as a whole.
(b)   Yerushalmy's index has advantages over the standardized death rate when attention is focused on the relative mortality levels at all ages, and not just at those ages where there are most deaths. The standardized death rate will assign a weight to each age group in the comparison proportional to the number of deaths in that age group in the standard population. Thus it tends to be heavily influenced by the relative mortality levels in the older age groups, where there are more deaths.

3.7   The high SMR for male hairdressers and barbers probably arises from a lack of correspondence between the deaths and the exposed-to-risk.
    In this investigation, the information about deaths was taken from the death certificates for the years 1979–80 and 1982–83. Information about the exposed-to-risk was taken from the 1981 census. Since the census asked more detailed questions about a person's occupation, we may assume that it was possible accurately to distinguish between hairdressers and barbers on the one hand, and hairdressing supervisors on the other. On the death certificates, the information necessary to make this distinction may not have been available. As a result, some hairdressing supervisors have probably classified at death as ordinary hairdressers, inflating the SMR for the latter and deflating it for the former.
    In the case of females, the problem is probably not so great, since there are relatively few female hairdressing supervisors.

It is likely that there is a small real difference in the mortality of these two groups, since hairdressing supervisors are probably better paid than ordinary hairdressers. The real difference, though, is almost certainly much smaller than the apparent difference in the table.

The high SMR of fishermen probably results from the arduous and dangerous nature of their occupation (deaths from drowning are common), from their low and insecure incomes, and from the high levels of stress associated with the occupation. In addition, deaths from cancer of the lip and skin are common (Benjamin and Pollard, 1993).

In the case of travel stewards and attendants, hospital and hotel porters, there is a problem with heterogeneity. This group includes a wide range of social classes, from airline stewards (social class III, or even II at the top end) to hospital porters (social class V). Furthermore, the better-off people in this group are disproportionately female (airline stewardesses), and the worse-off disproportionately male (hospital and hotel porters, who experience particularly poor social conditions). This goes a long way towards explaining the different SMRs for males and females. In addition, for males, the high rate may be due to selection of chronically sick people into certain occupations within this category.

Teachers and local government officers tend to be relatively well educated, and have secure incomes. Therefore they are likely to seek advice early about potentially life-threatening health problems. This probably explains much of their low SMR.

3.8   (a)   The simplest single-figure index is the crude death rate (CDR). From the data, we have:

$$\text{CDR (whole country)} = \frac{20 + 30 + 200}{10\,000 + 7000 + 5000} = 0.0114,$$

$$\text{CDR (Graveside)} = \frac{0.15 + 0.24 + 1.00}{50 + 40 + 30} = 0.0116,$$

$$\text{CDR (Croakingham)} = \frac{0.12 + 0.25 + 3.00}{40 + 40 + 40} = 0.0281.$$

In order to calculate the standardized death rates for Graveside and Croakingham (using the whole country as the standard population), we need to calculate the m-type age-specific death rates (ASDRs) for the two towns. The m-type ASDR for a particular age group is obtained by dividing the number of deaths in that age group by the population in that age group. Thus, for the age group 0–29 years in Graveside, the ASDR is equal to 0.15/50. The complete set of ASDRs for the two towns, and for the country as a whole, is shown in Table S.7.

The standardized death rate for each town is then calculated by working out the expected number of deaths which the population of the whole country would have if it experienced the ASDRs of the relevant town, and dividing this by the whole population of the country. Thus we have:

$$\text{standardized death rate (Graveside)} = \frac{30 + 42 + 167}{22\,000} = 0.0109,$$

$$\text{standardized death rate (Croakingham)} = \frac{30 + 44 + 375}{22\,000} = 0.0204.$$

Table S.7

Age group	Age-specific death rate		
	Whole country	Graveside	Croakingham
0–29	0.0020	0.0030	0.0030
30–59	0.0043	0.0060	0.0063
60 and over	0.0400	0.0333	0.0750

Table S.8

Age group	ASDR (in institutions)		
	Whole country	Graveside	Croakingham
0–29	0.0050	0.0040	0.0040
30–59	0.0120	0.0117	0.0120
60 and over	0.1500	0.1533	0.1500

The standardized mortality ratio (SMR) for each town can be calculated by dividing the actual total number of deaths experienced in that town by the number of deaths that would be expected if people in that town experienced the ASDRs of the country as a whole. Using the ASDRs for the country as whole from the table above, we have

$$\text{SMR (Graveside)} = \frac{1.390}{0.100 + 0.172 + 1.200} = 0.944,$$

$$\text{SMR (Croakingham)} = \frac{3.370}{0.080 + 0.172 + 1.600} = 1.820.$$

Now, we are also given the population in institutions (by age) and the number of deaths to people in institutions (by age) in each town and in the whole country. Therefore, we can calculate CDRs, standardized death rates and SMRs for the population in institutions and the population not in institutions separately. To do this, we need to calculate:

- the ASDRs for the population in institutions,
- the population not in institutions, and the number of deaths to people not in institutions, by age group,
- the ASDRs for the population not in institutions.

These three sets of calculations are shown in Tables S.8–S.10. Using these data, we can calculate the single-figure indices shown in Table S.11.

Finally, we can calculate standardized death rates and SMRs which standardize simultaneously for the proportion of the population in institutions and the age structure. To calculate these standardized death rates, we work out the expected number of deaths in the whole country in each age group for the institutionalized and non-institutionalized populations separately, using the relevant ASDRs (specific to whether or not they apply

Table S.9

Age group	Population not in institutions (thousands)			Deaths to people not in institutions (thousands)		
	Whole country	Graveside	Croakingham	Whole country	Graveside	Croakingham
0–29	9950	49.75	39.75	19.75	0.149	0.119
30–59	6900	39.40	39.50	28.80	0.233	0.244
60 +	4500	27.00	28.00	125.00	0.550	1.200

Table S.10

Age group	ASDR (not in institutions)		
	Whole country	Graveside	Croakingham
0–29	0.0020	0.0030	0.0030
30–59	0.0042	0.0059	0.0062
60 and over	0.0278	0.0204	0.0429

Table S.11

Index	In institutions			Not in institutions		
	Whole country	Graveside	Croakingham	Whole country	Graveside	Croakingham
CDR	0.1176	0.1190	0.1417	0.0081	0.0080	0.0146
Standardized death rate	0.1176	0.1175	0.1175	0.0081	0.0076	0.0125
SMR	1.000	0.999	1.000	1.000	0.917	1.526

to the institutionalized or non-institutionalized population), sum these, and divide the result by the total population of the whole country. Thus we have, for Graveside,

$$\text{standardized death rate} = \frac{30 + 41 + 92 + 0.2 + 1.17 + 75}{22\,000} = 0.0109,$$

and, for Croakingham,

$$\text{standardized death rate} = \frac{30 + 43 + 193 + 0.2 + 1.2 + 75}{22\,000} = 0.0156.$$

The SMRs are calculated by dividing the observed number of deaths in each town by the expected number of deaths which would be obtained if the populations in institutions and not in institutions at each age had the ASDRs (specific to whether or not they apply to the institutionalized or non-institutionalized population) of the whole country. Thus, for Graveside, we have

$$\text{SMR} = \frac{1.39}{0.00125 + 0.0072 + 0.450 + 0.1 + 0.165 + 0.751} = 0.943,$$

and, for Croakingham, we have

$$\text{SMR} = \frac{3.37}{0.00125 + 0.006 + 1.8 + 0.08 + 0.166 + 0.778} = 1.190.$$

(b) The crude death rates suggest that mortality in Graveside is almost the same as in the country as a whole, but that mortality in Croakingham is much higher (more than twice as high).

Just under half the difference in the crude death rate between Croakingham and Graveside is accounted for by the older age structure of the population of Croakingham compared with the country as a whole, or with Graveside (the age structure of Graveside is very similar to that of the country as a whole).

The crude death rate for the population in institutions suggests that mortality in institutions is slightly lighter in Graveside than in the country as a whole, but that mortality in institutions in Croakingham is rather heavier than in the country as a whole. However, these differences are entirely accounted for by differences in the age structure of the institutionalized populations of the two towns (as is revealed by the standardized death rates and the SMRs for the institutionalized populations).

The crude death rate for the population not in institutions suggests that mortality is heavier for this group in Croakingham than in Graveside or the country as a whole. Some, but by no means all, of this difference is accounted for by differences in the age structure of the non-institutionalized populations, specifically that the non-institutionalized population of Croakingham is older than that of either Graveside or the country as a whole.

Standardizing for both age structure and the proportion of the population in institutions at each age reveals that mortality is heavier in Croakingham than in Graveside, but that the difference is by no means as great as the crude death rates for the whole population showed. Indeed, it is smaller than suggested by the standardized death rates and SMRs (which standardize only for age). Part of the heavier mortality in Croakingham is due to its having a higher proportion of its population in institutions than Graveside.

The heavier mortality in Croakingham, even after standardizing both for age and for the institutionalized population, is entirely due to higher mortality among those aged 60 years and over who are not in institutions.

Chapter 4

4.1 The life expectation at birth for males is 47.09 years. The calculations are shown in Table S.12. The following formulae were used in calculating this table:

$$l_{x+n} = l_x(1 - {}_nq_x), \qquad {}_nL_x = (n/2)(l_x + l_{x+n})$$

(except for the youngest age group, where the formula

$$L_0 = 0.2l_0 + 0.8l_1$$

was used) and

$$e_0 = \frac{\sum_{u=0}^{\infty} {}_nL_u}{l_0},$$

where the symbols have the usual life table meanings. The value of l_0 was taken to be 100 000, but any value could have been chosen.

4.2 Assume that $l_0 = 100\,000$ in each population. Since

$$T_0 = e_0 l_0,$$

we can immediately calculate T_0 from the data given.

a₀ = .2

Table S.12

Age (years) x	n	l_x	$_nL_x$	e_x
0	1	100 000	88 784	47.09
1	4	85 980	331 642	
5	5	79 841	395 073	
10	5	78 188	388 010	
15	5	77 015	380 842	
20	5	75 321	370 712	
25	5	72 964	358 561	
30	5	70 461	345 382	
35	5	67 692	330 573	
40	5	64 537	313 361	
45	5	60 807	293 166	
50	5	56 459	268 930	
55	5	51 113	239 591	
60	5	44 724	204 040	
65	5	36 893	162 216	
70	5	29 330	116 441	
75	5	22 255	71 779	
80	5	14 773	35 133	
85	5	7 905	12 060	
90	5	3 062	2 478	
95	5	702	229	

Now
$$l_1 = \frac{T_1}{e_1} = \frac{T_0 - L_0}{e_1}.$$

Then, assuming $a_0 = 0.2$, we have
$$l_1 = \frac{T_0 - (0.2l_0 + 0.8l_1)}{e_1}, \tag{1}$$

and equation (1) may be rearranged to give
$$l_1 = \frac{T_0 - 0.2l_0}{e_1 + 0.8} \tag{2}$$

In a similar way, assuming that deaths are evenly distributed across the remaining age groups, it may be shown that
$$l_{x+n} = \frac{2T_x - nl_x}{2e_{x+n} + n}. \tag{3}$$

Equations (2) and (3) allow the values of l_x to be calculated for all x, whence the values of $_nq_x$ may be calculated from the formula
$$_nq_x = \frac{l_x - l_{x+n}}{l_x}.$$

The results are shown in Table S.13.

4.3 This result follows from noting that the stationary population has l_0 births per year, and a constant total population size of T_0. The crude birth rate is, therefore, equal to l_0/T_0, which is $1/e_0$.

4.4 We have
$$e_0 = \frac{\int_0^\infty l_u\, du}{l_0}, \tag{1}$$

and
$$e_1 = \frac{\int_1^\infty l_u\, du}{l_1}. \tag{2}$$

It is also clear that
$$\frac{\int_0^\infty l_u\, du}{l_0} = \frac{\int_0^1 l_u\, du + \int_1^\infty l_u\, du}{l_0}. \tag{3}$$

We are told that e_0 is equal to e_1. Using equations (1) and (2) this means that
$$\frac{\int_0^\infty l_u\, du}{l_0} = \frac{\int_1^\infty l_u\, du}{l_1}. \tag{4}$$

Table S.13

Age x	n	l_x		$_nq_x$	
		Nicaragua, 1990–95	United States, 1989	Nicaragua, 1990–95	United States, 1989
0	1	100 000	100 000	0.04661	0.00885
1	4	95 389	99 115	0.02439	0.00131
5	5	93 014	98 985	0.00761	0.00139
10	5	92 306	98 847	0.00492	0.00149
15		91 852	98 701		

Substituting from equation (3) into equation (4) gives

$$\frac{\int_0^1 l_u\,du + \int_1^\infty l_u\,du}{l_0} = \frac{\int_1^\infty l_u\,du}{l_1}. \tag{5}$$

By rearranging equation (2), we obtain

$$\int_1^\infty l_u\,du = e_1 l_1 \tag{6}$$

so that, substituting from equation (6) into equation (5) produces

$$\frac{\int_0^1 l_u\,du + e_1 l_1}{l_0} = e_1. \tag{7}$$

Rearranging equation (7) gives

$$\int_0^1 l_u\,du = e_1(l_0 - l_1). \tag{8}$$

We are also told that

$$l_x = \left(\frac{l_1}{l_0}\right)^x l_0. \tag{9}$$

Substituting from equation (9) into equation (8) gives

$$\int_0^1 \left(\frac{l_1}{l_0}\right)^x l_0\,dx = e_1(l_0 - l_1).$$

Performing the integration gives

$$l_0\left[\frac{(l_1/l_0)^x}{\ln(l_1/l_0)}\right]_0^1 = e_1(l_0 - l_1),$$

or

$$l_0\left[\frac{(l_1/l_0) - 1}{\ln(l_1/l_0)}\right] = e_1(l_0 - l_1).$$

Simplifying the left-hand side of this equation produces

$$\frac{l_1 - l_0}{\ln(l_1/l_0)} = e_1(l_0 - l_1),$$

which may be rewritten

$$\frac{l_1 - l_0}{\ln(l_1/l_0)} = -e_1(l_1 - l_0).$$

Dividing both sides by $(l_1 - l_0)$ gives

$$\frac{1}{\ln(l_1/l_0)} = -e_1,$$

so

$$\ln(l_1/l_0) = -\left(\frac{1}{e_1}\right),$$

or

$$l_1/l_0 = \exp\left(\frac{-1}{e_1}\right).$$

But l_1/l_0 is just p_0, so we have shown that

$$p_0 = \exp\left(\frac{-1}{e_1}\right).$$

4.5 The force of mortality, μ_x, is given by

$$\mu_x = -\frac{d}{dx}(\ln l_x). \tag{1}$$

Integrating equation (1) produces

$$\int_0^x \mu_u \, du = \ln l_x,$$

whence

$$l_x = \exp\left(-\int_0^x \mu_u \, du\right). \tag{2}$$

Since, by the definition of q_x, we can write

$$q_x = \frac{l_x - l_{x+1}}{l_x} = 1 - \frac{l_{x+1}}{l_x}, \tag{3}$$

substituting from equation (2) into equation (3) we obtain

$$q_x = 1 - \frac{\exp(-\int_0^{x+1} \mu_u \, du)}{\exp(-\int_0^x \mu_u \, du)}$$

$$= 1 - \frac{\exp(-\int_0^x \mu_u \, du - \int_x^{x+1} \mu_u \, du)}{\exp(-\int_0^x \mu_u \, du)}$$

$$= 1 - \frac{\exp(-\int_0^x \mu_u \, du)\exp(-\int_x^{x+1} \mu_u \, du)}{\exp(-\int_0^x \mu_u \, du)}$$

$$= 1 - \exp\left(-\int_x^{x+1} \mu_u \, du\right).$$

4.6 The solution is shown in Table S.14. In working this table out, a_0 has been taken as 0.2, a_1 as 0.35, and $a_2, \ldots, a_{89}$ as 0.5. For the oldest age group, 90 years and over, $_nL_{90}$ has been calculated using the formula

$$_nL_{90} = \frac{_nd_{90}}{_nm_{90}}.$$

4.7 (a) The probability that a woman aged 20 years will survive to her 40th birthday is equal to l_{40}/l_{20}, which is $97\,346/98\,497$, or 0.988.

(b) The infant mortality rate is the proportion of persons dying before their first birthday. In this case for every 100 000 births, only 99 016 attain their first birthday, so

$$\text{infant mortality rate} = \frac{100\,000 - 99\,016}{100\,000} = 0.00984.$$

(c) The life expectation at any age x, e_x, is given by the formula

$$e_x = \frac{T_x}{l_x},$$

where the usual life table notation is used. Thus we have:

$$\text{life expectation at birth} = \frac{7\,700\,187}{100\,000} = 77.00 \text{ years};$$

Table S.14

Age x	q_x	l_x	L_x	T_x	e_x
0	0.00728	100 000	99 418	7 380 705	73.81
1	0.00053	99 272	99 238	7 281 287	73.35
2	0.00034	99 220	99 203	7 182 049	72.39
3	0.00026	99 186	99 173	7 082 846	71.41
4	0.00025	99 160	99 148	6 983 673	70.43
5	0.00021	99 135	99 125	6 884 525	69.45
6	0.00018	99 115	99 106	6 785 400	68.46
7	0.00020	99 097	99 087	6 686 295	67.47
8	0.00015	99 077	99 069	6 587 208	66.49
9	0.00016	99 062	99 054	6 488 138	65.50
10	0.00017	99 046	99 038	6 389 084	64.51
11	0.00016	99 029	99 021	6 290 047	63.52
12	0.00022	99 014	99 003	6 191 025	62.53
13	0.00019	98 992	98 982	6 092 023	61.54
14	0.00026	98 973	98 960	5 993 040	60.55
15	0.00033	98 947	98 931	5 894 080	59.57
16	0.00042	98 915	98 894	5 795 149	58.59
17	0.00064	98 873	98 841	5 696 256	57.61
18	0.00078	98 810	98 771	5 597 414	56.65
19	0.00080	98 733	98 693	5 498 643	55.69
20	0.00076	98 654	98 616	5 399 950	54.74
21	0.00084	98 579	98 537	5 301 333	53.78
22	0.00085	98 496	98 454	5 202 796	52.82
23	0.00078	98 412	98 374	5 104 342	51.87
24	0.00085	98 336	98 294	5 005 968	50.91
25	0.00075	98 252	98 215	4 907 674	49.95
26	0.00082	98 178	98 138	4 809 459	48.99
27	0.00083	98 098	98 057	4 711 321	48.03
28	0.00088	98 017	97 973	4 613 263	47.07
29	0.00087	97 930	97 888	4 515 290	46.11
30	0.00091	97 845	97 801	4 417 402	45.15
31	0.00098	97 756	97 708	4 319 602	44.19
32	0.00096	97 660	97 614	4 221 893	43.23
33	0.00102	97 567	97 517	4 124 280	42.27
34	0.00111	97 467	97 413	4 026 763	41.31
35	0.00121	97 359	97 300	3 929 350	40.36
36	0.00126	97 241	97 180	3 832 049	39.41
37	0.00131	97 119	97 055	3 734 869	38.46
38	0.00165	96 992	96 912	3 637 814	37.51
39	0.00164	96 832	96 753	3 540 902	36.57
40	0.00168	96 673	96 592	3 444 150	35.63
41	0.00188	96 511	96 420	3 347 557	34.69
42	0.00190	96 330	96 238	3 251 137	33.75
43	0.00215	96 147	96 044	3 154 899	32.81
44	0.00238	95 940	95 826	3 058 855	31.88
45	0.00241	95 712	95 597	2 963 029	30.96
46	0.00291	95 482	95 343	2 867 432	30.03
47	0.00333	95 204	95 046	2 772 089	29.12
48	0.00357	94 888	94 719	2 677 043	28.21
49	0.00396	94 550	94 363	2 582 324	27.31
50	0.00450	94 176	93 965	2 487 961	26.42
51	0.00504	93 753	93 518	2 393 996	25.54
52	0.00570	93 282	93 017	2 300 479	24.66
53	0.00635	92 752	92 458	2 207 462	23.80

Table S.14 Continued

Age x	q_x	l_x	L_x	T_x	e_x
54	0.00672	92 165	91 856	2 115 004	22.95
55	0.00785	91 547	91 189	2 023 148	22.10
56	0.00853	90 831	90 446	1 931 958	21.27
57	0.00953	90 060	89 633	1 841 513	20.45
58	0.01104	89 206	88 716	1 751 880	19.64
59	0.01213	88 226	87 694	1 663 164	18.85
60	0.01358	87 163	86 575	1 575 469	18.08
61	0.01514	85 987	85 341	1 488 894	17.32
62	0.01710	84 695	83 977	1 403 554	16.57
63	0.01921	83 259	82 467	1 319 577	15.85
64	0.02150	81 675	80 806	1 237 110	15.15
65	0.02344	79 937	79 011	1 156 304	14.47
66	0.02638	78 085	77 069	1 077 292	13.80
67	0.02974	76 052	74 938	1 000 224	13.15
68	0.03260	73 824	72 640	925 286	12.53
69	0.03625	71 456	70 183	852 646	11.93
70	0.03928	68 911	67 584	782 463	11.35
71	0.04390	66 257	64 834	714 879	10.79
72	0.04553	63 411	61 999	650 045	10.25
73	0.05374	60 588	59 002	588 046	9.71
74	0.05907	57 417	55 770	529 044	9.21
75	0.06098	54 123	52 521	473 274	8.74
76	0.06840	50 920	49 236	420 753	8.26
77	0.07625	47 552	45 806	371 517	7.81
78	0.08275	44 059	42 309	325 711	7.39
79	0.08829	40 558	38 844	283 402	6.99
80	0.09995	37 129	35 362	244 558	6.59
81	0.10618	33 594	31 901	209 197	6.23
82	0.11813	30 207	28 523	177 296	5.87
83	0.12796	26 838	25 224	148 773	5.54
84	0.13729	23 610	22 094	123 549	5.23
85	0.14903	20 577	19 150	101 456	4.93
86	0.16365	17 723	16 383	82 306	4.64
87	0.18104	15 042	13 793	65 923	4.38
88	0.19402	12 545	11 436	52 130	4.16
89	0.21259	10 326	9 334	40 694	3.94
90	1.00000	8 342	31 360	31 360	3.76

and

$$\text{life expectation at exact age 1 year} = \frac{7\,601\,014}{99\,016} = 76.77 \text{ years.}$$

(d)   The probability that a girl who survives until her first birthday will die between her 10th and 20th birthdays is equal to the number of girls dying between their 10th and 20th birthdays divided by the number who survive until their first birthday. In symbols, therefore, we have

$$\begin{array}{l}\text{probability that a girl who survives to age 1 year} \\ \text{will die between exact ages 10 and 20 years}\end{array} = \frac{l_{10} - l_{20}}{l_1}$$

$$= \frac{98\,746 - 98\,497}{99\,016} = 0.0025.$$

(e) The expected age at death of those who die when they are aged 20–30 years is given by the following formula:

$$\text{expected age at death} = 20 + \frac{\text{person-years lived between exact ages 20 and 30 years by those who die within that age range}}{\text{number of persons dying between exact ages 20 and 30 years}}$$

$$= 20 + \frac{T_{20} - T_{30} - 10 l_{30}}{l_{20} - l_{30}}$$

$$= 20 + \frac{5\,725\,004 - 4\,741\,877 - 10(98\,105)}{98\,497 - 98\,105}$$

143

$$= 20 + \frac{2077}{392}$$

$$= 20 + 5.30$$

$$= 25.30 \text{ years.}$$

(f) The expected age at death of those who die when they are aged under 1 year is given by the following formula:

$$\text{expected age at death} = \frac{\text{person-years lived at ages under exact age 1 year by those who die within that age span}}{\text{number of persons dying at ages under exact age 1 year}}$$

$$= \frac{T_0 - T_1 - l_1}{l_0 - l_1}$$

$$= \frac{7\,700\,187 - 7\,601\,014 - 99\,016}{100\,000 - 99\,016}$$

$$= \frac{157}{984} = 0.160 \text{ years.}$$

If the assumption that deaths are evenly distributed within each age group is valid, then the expected age at death of those who die within an age group should be equal to the mid-point of that age group. Thus the expected age at death of those who die between their 20th and 30th birthdays should be exactly 25 years, and the expected age at death of those who die aged under 1 year should be exactly 0.5 years. The results of (e) and (f) above demonstrate that for English females in 1980–82 this is approximately true for women aged 20–30 years, but is far from being the case for infants aged under 1 year. The assumption of an even distribution of deaths is thus very poor for infants, but quite good for young adults.

4.8 The average number of person-years lived between ages x and $x + 1$ by those who die within that age group is a_x. Therefore

$$L_x = l_{x+1} + a_x d_x = l_{x+1} + a_x q_x l_x = l_x(1 - q_x) + a_x q_x l_x. \tag{1}$$

But

$$m_x = \frac{d_x}{L_x} = \frac{q_x l_x}{L_x}, \tag{2}$$

and substituting from equation (1) in equation (2) produces

$$m_x = \frac{q_x l_x}{l_x(1 - q_x) + a_x q_x l_x}. \tag{3}$$

Cancelling the l_xs on the right-hand side of equation (3) produces

$$m_x = \frac{q_x}{(1 - q_x) + a_x q_x},$$

so

$$m_x(1 - q_x) + m_x a_x q_x = q_x,$$

whence

$$m_x - m_x q_x + m_x a_x q_x = q_x,$$

which may be re-arranged to give

$$q_x(1 + m_x - a_x m_x) = m_x,$$

and hence

$$q_x = \frac{m_x}{1 + (1 - a_x)m_x}.$$

Chapter 5

5.1 (a) Using the combined dependent rates of decrement $_n(aq)_x$ the calculations proceed as shown in Table S.15. Note that in this table a_0 has been taken to be 0.2, but that deaths have been assumed to be evenly distributed across all other age groups.

(b) Taking cerebrovascular disease as decrement α and all other causes as decrement β we have

$$_nq_x^\beta \cong \frac{_n(aq)_x^\beta}{[1 - {_n(aq)_x^\alpha}/2]}.$$

The revised life expectation at birth in the absence of deaths from cerebrovascular disease is then calculated from the table of $_nq_x^\beta$s (that is, using the associated single-decrement life table for all other causes of death), as shown in Table S.16.

Table S.15

Age x	$(al)_x$	$_n(aL)_x$	$(aT)_x$	e_x
0	100 000	99 577	7 966 906	79.67
1	99 472	397 688	7 867 329	79.09
5	99 372	496 715	7 469 641	75.17
10	99 314	496 401	6 972 926	70.21
15	99 247	495 912	6 476 525	65.26
20	99 118	495 225	5 980 613	60.34
25	98 972	494 394	5 485 388	55.42
30	98 786	493 272	4 990 993	50.52
35	98 523	491 645	4 497 741	45.65
40	98 135	489 053	4 006 076	40.82
45	97 486	484 875	3 517 023	36.08
50	96 464	478 159	3 032 149	31.43
55	94 800	467 559	2 553 990	26.94
60	92 224	450 773	2 086 431	22.62
65	88 085	423 549	1 653 658	18.57
70	81 335	381 047	1 212 109	14.90
75	71 084	320 897	831 062	11.69
80	57 275	241 803	510 165	8.91
85	39 446	151 661	268 361	6.80
90	21 218	116 700	116 700	5.50

Table S.16

Age x	$_nq^\beta_x$	l^β_x	$_nL^\beta_x$	T^β_x	e^β_x
0	0.00527	100 000	99 579	8 071 328	80.71
1	0.00098	99 473	397 698	7 971 749	80.14
5	0.00058	99 375	496 733	7 574 051	76.22
10	0.00066	99 318	496 425	7 077 319	71.26
15	0.00128	99 252	495 946	6 580 894	66.30
20	0.00145	99 126	495 270	6 084 948	61.39
25	0.00180	98 982	494 466	5 589 678	56.47
30	0.00252	98 804	493 399	5 095 212	51.57
35	0.00373	98 555	491 857	4 601 813	46.69
40	0.00618	98 188	489 420	4 109 956	41.86
45	0.00993	97 581	485 480	3 620 536	37.10
50	0.01623	96 612	479 138	3 135 055	32.45
55	0.02562	95 044	469 130	2 655 917	27.94
60	0.04208	92 608	453 299	2 186 786	23.61
65	0.07107	88 711	427 795	1 733 487	19.54
70	0.11417	82 407	388 512	1 305 692	15.84
75	0.17168	72 998	333 661	917 180	12.56
80	0.26957	60 466	261 581	583 519	9.65
85	0.40135	44 166	176 516	321 938	7.29
90	1.00000	26 440	145 422	145 422	5.50

Note that we have retained the assumption that $e_{90} = 5.5$ years, even though it is likely that the elimination of deaths from cardiovascular disease will lead to a slight increase in this figure.

The elimination of deaths from cardiovascular disease would, on this evidence, add about one year to the life expectation at birth of English women.

5.2 (a) The withdrawals can be treated as censored cases, so, using equation (5.16), the calculations are shown in Table S.17.

(b) Using the independent q-type death rates, and assuming a value for l_{60} of, say, 1000, the values of $l_{61}, \ldots, l_{70}$ can be calculated. It turns out that $l_{70} = 740$. These l_xs may then be used to estimate the L_xs (assuming an even distribution of deaths within each year of age). Since we know that $e_{70} = 12$, T_{70} is therefore equal to $740 \times 12 = 8880$. Using this value of T_{70}, the values of $T_{60}, \ldots, T_{70}$ may be calculated (Table S.18). The estimated life expectation at age 60 years is then obtained as $T_{60}/l_{60} = 17 733/1000 = 17.7$ years.

Table S.17

Exact age x	Independent death rate
60	$18/(1000 - (0.5 \times 30)) = 0.0183$
61	$19/(952 - (0.5 \times 20)) = 0.0202$
62	$18/(913 - (0.5 \times 10)) = 0.0198$
63	$22/(885 - (0.5 \times 30)) = 0.0253$
64	$24/(833 - (0.5 \times 20)) = 0.0292$
65	$20/(789 - (0.5 \times 50)) = 0.0262$
66	$25/(719 - (0.5 \times 30)) = 0.0355$
67	$25/(664 - (0.5 \times 10)) = 0.0379$
68	$25/(629 - (0.5 \times 80)) = 0.0424$
69	$21/(524 - (0.5 \times 50)) = 0.0421$

Table S.18

Exact age x	q_x	l_x	L_x	T_x
60	0.0183	1000	991	17733
61	0.0202	982	972	16742
62	0.0198	962	952	15770
63	0.0253	943	931	14817
64	0.0292	919	906	13887
65	0.0262	892	881	12981
66	0.0355	869	853	12100
67	0.0379	838	822	11247
68	0.0424	806	789	10425
69	0.0421	772	756	9636
70		740		8880

(c)   The average age at death of male policyholders dying between exact ages 60 and 70 years is obtained from the formula

$$\text{average age at death} = 60 + \frac{\text{person-years lived between exact ages 60 and 70 years by those who die within that age range}}{\text{number of persons dying between exact ages 60 and 70 years}}$$

$$= 60 + \frac{T_{60} - T_{70} - 10 l_{70}}{l_{60} - l_{70}}$$

$$= 60 + \frac{17733 - 8880 - 10(740)}{1000 - 740}$$

$$= 65.59 \text{ years.}$$

5.3 (a)   Let the 'natural' causes of death which operated before the pollution incident be decrement α, and let deaths due to the pollution be decrement β. Then before the pollution incident, mortality proceeded according to the *associated single-decrement life table* for decrement α.

The calculation of the life expectation is summarized in Table S.19, using the notation of the associated single-decrement life table for decrement α. We have

$$\text{life expectation at birth} = \frac{T_0^{\alpha}}{l_0^{\alpha}} = \frac{3214}{1618} = 1.986 \text{ years.}$$

(Note that, in this calculation, we have assumed that deaths are evenly distributed within each year of age.)

(b)   We need to estimate the new dependent rates of decrement $(aq)_x^{\alpha}$ and $(aq)_x^{\beta}$. We can do this using the standard formulae

$$(aq)_x^{\alpha} = q_x^{\alpha}(1 - \tfrac{1}{2} q_x^{\beta})$$

Table S.19

Age x	l_x^{α}	L_x^{α}	T_x^{α}
0	1618	1309.0	3214.0
1	1000	865.0	1905.0
2	730	622.5	1040.0
3	515	337.5	417.5
4	160	80.0	80.0
5	0		

Table S.20

Age x	l_x^α	q_x^α
0	1618	0.382
1	1000	0.270
2	730	0.295
3	515	0.689
4	160	1.000
5	0	

and

$$(aq)_x^\beta = q_x^\beta(1 - \tfrac{1}{2}q_x^\alpha),$$

once we know the independent rates of decrement, q_x^α and q_x^β. The q_x^α can be calculated from the l_x^α, as shown in Table S.20. We are told that $q_x^\beta = 0.100$, $x = 0, \dots, 4$.

The calculation of the new dependent rates is shown in Table S.21. To illustrate how the calculations proceed, we can take as an example the dependent rates for age 0 last birthday, $(aq)_x^\alpha$ and $(aq)_x^\beta$. Applying the formulae above, we have

$$(aq)_0^\alpha = q_0^\alpha(1 - \tfrac{1}{2}q_0^\beta) = 0.382[1 - \tfrac{1}{2}(0.1)] = 0.3629$$

and

$$(aq)_0^\beta = q_0^\beta(1 - \tfrac{1}{2}q_0^\alpha) = 0.100[1 - \tfrac{1}{2}(0.382)] = 0.0809.$$

Once we have calculated all the $(aq)_x^\alpha$s and $(aq)_x^\beta$s, we can calculate the combined dependent rate of decrement $(aq)_x$. This is simply the sum of the dependent rates from each individual decrement:

$$(aq)_x = (aq)_x^\alpha + (aq)_x^\beta.$$

Finally, we use the $(aq)_x$s to calculate the number of survivors at each exact age x years out of 1618 births. The results are shown in the $(al)_x$ column of Table S.22. The number of animals which survive to exact age 4 years, $(al)_4$, is 105. Thus, of the animals born in the year ended 30 June 1996, 105 will survive to exact age 4 years.

(c) The calculation of the new life expectation at birth (after the pollution incident) uses the data in Table S.22. We have

$$\text{new life expectation at birth} = \frac{(aT)_0}{(al)_0} = \frac{2780}{1618} = 1.718 \text{ years.}$$

(Note that we have again assumed that deaths are evenly distributed across each year of age.)

Thus the change in the life expectation at birth caused by the pollution incident is equal to $1.986 - 1.718$ years, or 0.268 years. The pollution incident has reduced the animals' life expectation at birth by 0.268 years.

Table S.21

Age x	q_x^α	q_x^β	$(aq)_x^\alpha$	$(aq)_x^\beta$	$(aq)_x$
0	0.382	0.100	0.3629	0.0809	0.4438
1	0.270	0.100	0.2565	0.0865	0.3430
2	0.295	0.100	0.2803	0.0853	0.3656
3	0.689	0.100	0.6546	0.0656	0.7202
4	1.000	0.100	0.9500	0.0500	1.0000

Table S.22

Age x	$(aq)_x$	$(al)_x$	$(aL)_x$	$(aT)_x$
0	0.4438	1618	1259.0	2780.0
1	0.3430	900	745.5	1521.0
2	0.3656	591	483.0	775.5
3	0.7202	375	240.0	292.5
4	1.0000	105	52.5	52.5
5		0	0.0	0.0

5.4   The stages in the analysis are as follows.

1   Obtain data on the number of deaths at each age from the major cause of death, and from all other causes. Call these numbers of deaths $(ad)_x^m$ and $(ad)_x^c$ respectively.

2   Using the initial life table estimates of l_x, the numbers of people still alive at exact age x, work out the dependent q-type death rates at each age for the major cause of death, $(aq)_x^m$, and all other causes, $(aq)_x^c$. The relevant formulae are

$$(aq)_x^m = \frac{(ad)_x^m}{l_x}$$

and

$$(aq)_x^c = \frac{(ad)_x^c}{l_x}.$$

3   Work out the independent rates of decrement from all other causes, q_x^c, using the formula

$$q_x^c \cong \frac{(aq)_x^c}{[1 - \frac{1}{2}(aq)_x^m]}.$$

4   Then use the q_x^cs to calculate the associated single-decrement life table for all other causes. This life table assumes that the death rate from the major cause of death is zero at all ages.

5   The value of e_0 from the associated single decrement life table for all other causes represents that life expectation which would obtain after elimination of the major cause of death. It may be compared with the life expectation at birth calculated using the initial life table to evaluate the required effect.

5.5   The critical point to understand is that the effect on the life expectation at birth is directly related to the number of additional person-years that would be lived by people after the elimination of the relevant deaths. The total additional number of person-years lived is given by the formula

$$\text{additional number of person-years lived} = \sum_{i=1}^{n}(d_i' - a_i),$$

where n people died from the eliminated cause prior to its elimination, a_i is the age at death of the ith person who would have died from the eliminated cause before elimination, and d_i' is the age at death of the ith person from whatever cause that person did die from after elimination.

In the light of this, consider the three causes. Remember that we are dealing with a developing country.

Eliminating maternal deaths. In this case, $d_i' - a_i$ is likely to be around 35 years on average (given the fact that the average age at giving birth is about 28 years). However, maternal deaths are not a great proportion of all deaths (perhaps 1–2%).

Eliminating infant deaths from infectious diseases. In this case, $a'_i - a_i$ might, on average, be very large, if we can assume that the infants who would have died from the infectious diseases will survive until old age. The total number of additional person-years lived depends on the number of infants affected (the proportion of all deaths which are infant deaths from infectious diseases), and the chance of the infants who would have died from infectious diseases dying in infancy or childhood from some other cause.

Eliminating deaths from cancer. In developing countries, cancer mortality is not as large a proportion of all deaths as it is in developed countries. Suppose it is 10%. The average age at death from cancer is perhaps 60 years. Thus $a'_i - a_i$ will not, on the average, be more than about 10 years.

An approximate indication of the relative magnitude of the effects can be obtained by multiplying the average value of $a'_i - a_i$ by the percentage of deaths from the eliminated cause. If we do this, we obtain:

- maternal deaths, $35 \times 2 = 70$;
- deaths from cancer, $10 \times 10 = 100$.

So, if, say, 5% of all deaths are infant deaths from infectious diseases, then, provided the average value of $a'_i - a_i$ for these infants is greater than 20, the elimination of infant deaths from infectious diseases would have the largest impact on the life expectation at birth.

Chapter 6

6.1 (a) See Figure S.6.

(b) The maximum likelihood estimate of the force of mortality is

$$\frac{\sum_{i=1}^{n} \delta_i}{\sum_{i=1}^{n} x_i},$$

where δ_i is an indicator variable taking the value 1 if person i died and 0 otherwise; x_i is the length of time for which person i was observed (that is, his/her age at death or censoring, minus 40 years); and n is, in this example, 20.

From the data given, we can work out δ_i and x_i for each of the 20 individuals in the investigation, as shown in Table S.23. The maximum likelihood estimate of the force of mortality is equal to 7/59, or 0.1186.

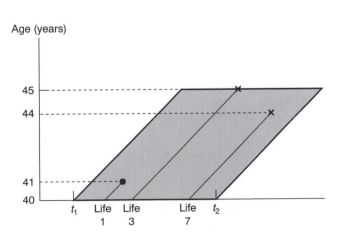

Figure S.6 Solution to Exercise 6.1

Table S.23

Person number, i	Last age at which person was observed (years)	Fate of person	δ_i	x_i
1	41.0	Died	1	1.0
2	42.0	Died	1	2.0
3	45.0	Survived	0	5.0
4	45.0	Survived	0	5.0
5	40.5	Withdrew	0	0.5
6	41.0	Withdrew	0	1.0
7	44.0	Withdrew	0	4.0
8	45.0	Survived	0	5.0
9	45.0	Survived	0	5.0
10	44.5	Died	1	4.5
11	44.0	Died	1	4.0
12	43.0	Died	1	3.0
13	42.0	Withdrew	0	2.0
14	45.0	Survived	0	5.0
15	42.5	Died	1	2.5
16	40.5	Died	1	0.5
17	41.0	Withdrew	0	1.0
18	42.0	Withdrew	0	2.0
19	43.0	Withdrew	0	3.0
20	43.0	Withdrew	0	3.0
			Σ 7	59.0

6.2 Suppose that the force of mortality at age between exact ages 0 and 1 is λ_0. Suppose also that the probability of survival to exact age x years is $S(x)$. Then the probability of survival to exact age 1 year, $S(1)$, is given by

$$S(1) = e^{-\lambda_0}.$$

From the data given, we know that $S(1)$ is equal to $1000/1618$, so

$$e^{-\lambda_0} = \frac{1000}{1618},$$

whence

$$\lambda_0 = -\ln\left(\frac{1000}{1618}\right) = 0.481.$$

The values of the force of mortality at ages 1, 2 and 3 years last birthday, λ_1, λ_2 and λ_3, are found using the formula

$$S(x+1) = S(x)\,e^{-\lambda_x},$$

from which it is clear that

$$\lambda_x = -\ln\left(\frac{S(x+1)}{S(x)}\right),$$

which (because $S(x)$ and l_x are equivalent), may be written

$$\lambda_x = -\ln\left(\frac{l_{x+1}}{l_x}\right).$$

Thus,

$$\lambda_1 = -\ln\left(\frac{730}{1000}\right) = 0.315,$$

$$\lambda_2 = -\ln\left(\frac{515}{730}\right) = 0.349$$

and

$$\lambda_3 = -\ln\left(\frac{160}{515}\right) = 1.169.$$

(Note that the last of these is an example of a force of mortality (or a hazard rate) greater than 1.0. The fact that the force of mortality (or hazard rate) may exceed 1.0 illustrates that it is not a true probability.)

6.3   Since

$$S(x) = \exp\left(-\int_0^x h(u)\,du\right),$$

then we have

$$S(x) = \exp\left(-\int_0^x (\alpha + \beta u)\,du\right) = \exp(-[\alpha u + \tfrac{1}{2}\beta u^2]_0^x) = \exp(-[\alpha x + \tfrac{1}{2}\beta x^2]).$$

Further, since we know that

$$f(x) = h(x)S(x)$$

then

$$f(x) = (\alpha + \beta x)\exp(-[\alpha x + \tfrac{1}{2}\beta x^2]).$$

6.4   If $S(x)$ is the probability that a person survives to at least age x, and $f(x)$ is the probability density function of the random variable X, then the likelihood of the data, L, is given by

$$L = \prod_{i=1}^n [f(x_i)]^{\delta_i}[S(x_i)]^{1-\delta_i}. \tag{1}$$

We know from the question that

$$h(x) = \beta\gamma^x, \tag{2}$$

and that

$$S(x) = \exp\left[-\int_0^x \beta\gamma^u\,du\right]. \tag{3}$$

We also know that

$$h(x) = \frac{1}{S(x)}f(x).$$

Rearranging this equation, we obtain

$$f(x) = h(x)S(x). \tag{4}$$

Substituting from equations (2) and (3) into equation (4), we obtain

$$f(x) = \beta\gamma^x \exp\left[-\int_0^x \beta\gamma^u\,du\right]. \tag{5}$$

We now use equations (3) and (5) to substitute into equation (1) expressions for $f(x_i)$ and $S(x_i)$ in terms of β and γ. This produces the following equation for the likelihood, L:

$$L = \prod_{i=1}^{n} \left(\beta\gamma^{x_i} \exp\left[-\int_0^{x_i} \beta\gamma^u\, du \right] \right)^{\delta_i} \left(\exp\left[-\int_0^{x_i} \beta\gamma^u\, du \right] \right)^{1-\delta_i}$$

$$= \prod_{i=1}^{n} [\beta\gamma^{x_i}]^{\delta_i} \exp\left[-\int_0^{x_i} \beta\gamma^u\, du \right].$$

Partially differentiating L with respect to β and γ and setting the partial derivatives equal to zero will produce maximum likelihood estimates of β and γ, provided that the second-order partial derivatives are negative. But the same estimates will be produced by partially differentiating the logarithm of L with respect to β and γ. Taking logarithms of L produces

$$\ln L = \sum_{i=1}^{n} \delta_i(\ln\beta + x_i\ln\gamma) - \sum_{i=1}^{n} \int_0^{x_i} \beta\gamma^u\, du.$$

Performing the integration in the second term on the right-hand side produces

$$\ln L = \sum_{i=1}^{n} \delta_i(\ln\beta + x_i\ln\gamma) - \sum_{i=1}^{n} \left(\frac{\beta}{\ln\gamma} \right)(\gamma^{x_i} - 1),$$

provided that $\gamma > 0$ and $\gamma \neq 1$. Partially differentiating this expression with respect to β and γ produces

$$\frac{\partial(\ln L)}{\partial\beta} = \frac{\sum_{i=1}^{n}\delta_i}{\beta} - \sum_{i=1}^{n} \left(\frac{1}{\ln\gamma} \right)(\gamma^{x_i} - 1)$$

and

$$\frac{\partial(\ln L)}{\partial\gamma} = \frac{\sum_{i=1}^{n}\delta_i x_i}{\gamma} - \sum_{i=1}^{n} \left(\frac{\beta}{\ln\gamma} \right)\left(x_i\gamma^{(x_i-1)} - \frac{\gamma^{x_i}+1}{\gamma(\ln\gamma)} \right).$$

Therefore, the two equations which, when solved, produce maximum likelihood estimates for β and γ are

$$\frac{\sum_{i=1}^{n}\delta_i}{\beta} = \frac{1}{\ln\gamma} \sum_{i=1}^{n}(\gamma^{x_i} - 1)$$

and

$$\frac{\sum_{i=1}^{n}\delta_i x_i}{\gamma} = \frac{\beta}{\ln\gamma} \sum_{i=1}^{n} \left(x_i\gamma^{(x_i-1)} - \frac{\gamma^{x_i}+1}{\gamma(\ln\gamma)} \right).$$

(That these represent maxima may be checked using second derivatives by readers who like algebra.)

6.5   Since the force of mortality, μ_x, and the hazard rate, $h(x)$, are the same, if $S(x)$ is the probability of surviving to exact age x, then we have

$$S(x) = \exp\left(-\int_0^x \mu_u\, du \right).$$

Thus, for the Gompertz hazard,

$$S(x) = \exp\left(-\int_0^x \beta\gamma^u\, du \right) = \exp\left(-\frac{\beta\gamma^x - \beta}{\ln\gamma} \right).$$

Defining x in years since a person's 60th birthday, we are told that $S(10) = 2S(20)$. Thus, we have

$$\exp\left(-\frac{\beta\gamma^{10} - \beta}{\ln\gamma}\right) = 2\exp\left(-\frac{\beta\gamma^{20} - \beta}{\ln\gamma}\right).$$

Taking logarithms produces

$$-\frac{\beta\gamma^{10} - \beta}{\ln\gamma} = \ln 2 - \frac{\beta\gamma^{20} - \beta}{\ln\gamma},$$

$$\beta - \beta\gamma^{10} = (\ln 2)(\ln\gamma) + \beta - \beta\gamma^{20},$$

$$\beta\gamma^{10} = \beta\gamma^{20} - (\ln 2)(\ln\gamma),$$

$$\beta(\gamma^{20} - \gamma^{10}) = (\ln 2)(\ln\gamma),$$

$$\beta = \frac{(\ln 2)(\ln\gamma)}{\gamma^{10}(\gamma^{10} - 1)}.$$

Chapter 7

7.1 For Mexico in 1986 the crude marriage rate is $579\,895/81\,200\,000 = 0.00714$ (or 7.14 per thousand). For Paraguay in 1987 it is 7.82 per thousand, and for Uruguay in 1987 it is 7.43 per thousand.

7.2 See Table S.24. The age-specific first marriage rates in this table are presented per thousand never married persons. Thus, for males aged 16–19 years, the rate is $(4630/1\,326\,324) \times 1000 = 3.49$ per thousand. These rates are decremental, as, upon marriage, a person leaves the population of never married persons.

7.3 We are only concerned with persons who ultimately marry. Assuming that no one never marries for the first time at an age greater than 50 years, the remaining expectation of life in the never married state at each age x is given by the formula

$$\text{remaining expectation of life in the never married state} = \frac{\substack{\text{person-years lived in the never married state between} \\ \text{exact ages } x \text{ and 50 years by those who marry}}}{\substack{\text{persons never married at exact age } x \text{ years} \\ \text{who will ultimately marry}}}.$$

To calculate the numerator of this formula, we calculate the number of person-years lived in the never married state in each age group given in Table 7.1, excluding the person-years lived by those who will never marry. Since the number of persons who will never marry is, according to

Table S.24

Age group	Age-specific first marriage rate	
	Males	Females
16–19	3.49	14.10
20–24	45.66	72.75
25–29	86.43	93.23
30–34	68.23	62.99
35–39	37.19	34.28
40–44	19.11	17.69
45–49	10.77	11.68

Table S.25

Exact age x	Person-years lived in never married state in age group by those who ultimately marry	Person-years lived between exact ages x and 50 years by those who ultimately marry	Remaining expectation of life in the never married state	
	Males	Males	Males	Females
16	788.0	9316.0	11.8	9.8
17	787.5	8528.0	10.8	8.8
18	783.5	7740.5	9.8	7.9
19	773.0	6597.0	8.9	7.2
20	752.0	6184.0	8.1	6.6
21	715.0	5432.0	7.4	6.2
22	663.0	4717.0	6.8	6.0
23	600.0	4054.0	6.4	5.8
24	531.5	3454.0	6.1	5.7
25	462.0	2922.5	5.9	5.7
26	396.0	2460.5	5.8	5.6
27	337.5	2064.5	5.7	5.6
28	285.5	1727.0	5.6	5.6
29	239.0	1441.5	5.5	5.6
30	760.0	1202.5	5.5	5.7
35	302.5	442.5	5.1	5.3
40	112.5	140.0	4.1	4.2
45	27.5	27.5	2.5	2.5
50	0.0	0.0		

our assumption, l_{50}, the formulae to use are, for $x \leq 29$,

person-years lived in the never married state by those who will ultimately marry $= \frac{1}{2}(l_x + l_{x+1}) - l_{50}$

and, for $x = 30, 35, 40$ and 45,

person-years lived in the never married state by those who will ultimately marry $= 2.5(l_x + l_{x+5}) - 5l_{50}.$

(These formulae assume that first marriages are distributed evenly within each age group.)

Summing the results between exact age x and exact age 50 years for each age x ($16 \leq x \leq 49$) gives us the numerators of the formula for the remaining expectation of life in the never married state. The denominators of this formula are obtained as $l_x - l_{50}$, since this is the number of persons remaining single at exact age x who will marry before age 50 years.

The calculations for males are shown in Table S.25, together with the remaining expectation of life at each age for females.

7.4 (a) See Table S.26. The dependent q-type rates in each age group are obtained simply by dividing the numbers of deaths and persons marrying for the first time in an age group by the number of survivors at the beginning of that age group.

(b) The independent first marriage rates, assuming no mortality, are obtained using equation (5.15), where first marriage is decrement α and death is decrement β. The results are shown in Table S.27

Because mortality in the age range 16–50 years is very low, the dependent and independent first marriage rates are very close.

7.5 Since the percentage never married in the age group 45–49 years exceeds that in the age group 40–44 years, these age groups should be amalgamated. Simple averaging gives a percentage never married in the combined age group 40–49 age group of 0.55.

Table S.26

Exact age x	Dependent q-type death rates		Dependent q-type first marriage rates	
	Males	Females	Males	Females
0	0.015	0.012	0.000	0.000
16	0.001	0.000	0.000	0.003
17	0.001	0.000	0.001	0.009
18	0.000	0.001	0.007	0.030
19	0.001	0.000	0.014	0.050
20	0.001	0.000	0.028	0.072
21	0.001	0.000	0.048	0.095
22	0.001	0.001	0.064	0.103
23	0.000	0.000	0.080	0.111
24	0.001	0.000	0.089	0.111
25	0.001	0.000	0.098	0.110
26	0.000	0.002	0.098	0.103
27	0.002	0.000	0.096	0.099
28	0.000	0.000	0.094	0.094
29	0.002	0.000	0.091	0.089
30	0.007	0.006	0.301	0.282
35	0.014	0.004	0.176	0.156
40	0.017	0.011	0.090	0.085
45	0.033	0.023	0.053	0.053
50				

Table S.27

Exact age x	Independent q-type first marriage rates	
	Males	Females
0	0.000	0.000
16	0.000	0.003
17	0.001	0.009
18	0.007	0.030
19	0.014	0.050
20	0.028	0.072
21	0.048	0.095
22	0.064	0.103
23	0.080	0.111
24	0.089	0.111
25	0.098	0.110
26	0.098	0.103
27	0.096	0.099
28	0.094	0.094
29	0.091	0.089
30	0.303	0.283
35	0.178	0.156
40	0.091	0.085
45	0.054	0.053
50		

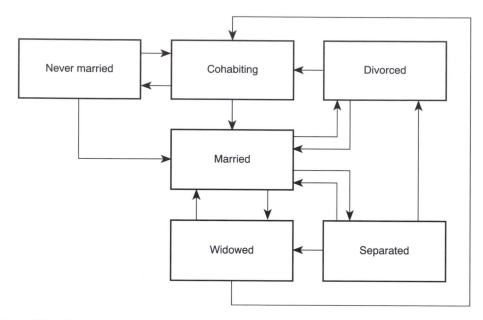

Figure S.7 Solution to Exercise 7.7. The arrows denote possible transitions. Transitions are also possible from every state identified in the figure to the state 'dead', but in the interests of clarity, these are not shown. Depending on the precise definitions of some of the states, minor variations to the solution shown in this figure are possible

This value may also be taken as an estimate of the percentage who never marry. The percentage who ever marry is, therefore, 99.45.

We therefore have

$$\text{number of person-years lived in the single state by those who marry before age 50 years} = (10 \times 100) + (5 \times 95.4) + (5 \times 71.1)$$

$$+ (5 \times 32.1) + (5 \times 17.1) + (5 \times 1.7)$$

$$+ (5 \times 1.7) + (10 \times 0.55) - (50 \times 0.55)$$

$$= 2073.5$$

and

$$\text{SMAM} = \frac{2073.5}{99.45} = 20.85.$$

7.6   The SMAM for males is 29.11, and for females 26.87.

7.7   See Figure S.7.

Chapter 8

8.1   The crude birth rate is obtained by dividing the total births by the total population: thus for 1971 it is $86\,700/5\,236\,000 = 0.0166$ (or 16.6 per thousand). The general fertility rate is obtained by dividing the total births by the population of females aged 15–44 years: thus for 1971 it is $86\,700/1\,011\,000 = 0.0858$ (or 85.8 per thousand). The full set of crude birth rates and general fertility rates is given in Table S.28.

Table S.28

Year	Crude birth rate	General fertility rate
1971	0.0166	0.0858
1981	0.0133	0.0632
1991	0.0131	0.0597
1993	0.0124	0.0574
1994	0.0120	0.0560
1995	0.0117	0.0546

8.2 Fertility is increasing in the population, but the proportion of the population which is composed of women of childbearing age is, simultaneously, falling just enough to cancel out the effect of the increased fertility on the crude birth rate.

8.3 (a) If B_i is the number of births in age group i, and P_i^f is the female population in age group i, then

$$\text{general fertility rate} = \frac{\sum_{15-19}^{40-44} B_i}{\sum_{15-19}^{40-44} P_i^f},$$

P_i^f for each age group is equal to B_i/f_i. Therefore, using the summations in Table S.29,

$$\text{general fertility rate for 1976} = \frac{584.2}{9602.3} = 0.0608 \text{ (or 60.8 per thousand)},$$

and

$$\text{general fertility rate for 1993} = \frac{673.5}{10\,802.4} = 0.0623 \text{ (or 62.3 per thousand)}.$$

(b) See Table S.29. The age-specific fertility rate for age group i, f_i, is equal to B_i/P_i^f. Thus, for example, for age group 15–19 years in 1976 it is $57.9/1809 = 0.032$.

(c) The standardized fertility rate is calculated using the formula:

$$\text{standardized fertility rate} = \frac{\sum_{15-19}^{40-44} P_i^f(1976) \times f_i(1993)}{\sum_{15-19}^{40-44} P_i^f(1976)}.$$

Table S.30 shows the calculations. Thus,

$$\text{standardized fertility rate} = \frac{599.3}{9602.3} = 0.0624 \text{ (or 62.4 per thousand)}.$$

Table S.29

Age group	1976			1993		
	B_i (thousands)	f_i	P_i^f (thousands)	B_i (thousands)	f_i	P_i^f (thousands)
15–19	57.9	0.032	1 809.4	45.1	0.031	1 454.8
20–24	182.2	0.109	1 671.6	152.0	0.083	1 831.3
25–29	220.7	0.119	1 854.6	236.0	0.114	2 070.2
30–34	90.8	0.057	1 593.0	171.1	0.087	1 966.7
35–39	26.1	0.019	1 373.7	58.8	0.034	1 729.4
40–44	6.5	0.005	1 300.0	10.5	0.006	1 750.0
Σ	584.2		9 602.3	673.5		10 802.4

Table S.30

Age group	P_i^f (1976) (thousands)	f_i (1993)	P_i^f (1976) $\times f_i$ (1993)
15–19	1809.4	0.031	56.1
20–24	1671.6	0.083	138.7
25–29	1854.6	0.114	211.4
30–34	1593.0	0.087	138.6
35–39	1373.7	0.034	46.7
40–44	1300.0	0.006	7.8
$\sum$	9602.3		599.3

8.4 First we estimate the number of women in each age group, by dividing the number of births by the corresponding age-specific fertility rate (ASFR). For example, the number of women in age group 15–19 years in Egypt is equal to $43\,600/0.021 = 2\,076\,000$. The results are shown in Table S.31 (all figures are in thousands).

From the data given, there were $1\,773\,900$ births in Egypt in 1988, and $175\,100$ births in Tunisia in 1989.

(a) The GFR for Egypt is $1773.9/10\,776 = 0.165$, and that for Tunisia $175.1/1676 = 0.104$.

(b) The standardized fertility rate for Tunisia, using the female population of Egypt as the standard population, is 0.104.

(c) The formula for the standardized fertility ratio for Tunisia is:

$$\text{standardized fertility ratio} = \frac{\text{total births in Tunisia}}{\sum_i \left(\begin{array}{c}\text{population in age}\\\text{group } i \text{ in Tunisia}\end{array}\right)\left(\begin{array}{c}\text{ASFR for age}\\\text{group } i \text{ in Egypt}\end{array}\right)}.$$

Table S.32 shows the calculation of the denominator of this formula. The standardized fertility ratio in Tunisia is therefore equal to $175.1/275.5 = 0.635$.

(d) Fertility is higher in Egypt than Tunisia.

The general fertility rate and standardized fertility rate for Tunisia are very similar, suggesting that differences in the age structure within the childbearing age ranges between the two countries are small, and that the GFRs give a good impression of the relative levels of fertility.

All these figures are based on the experience of a particular period, and may not represent the experience of any real women.

8.5 The period total fertility rate can be calculated using the formula

$$\text{period total fertility rate} = 5\sum_i ASFR_i,$$

Table S.31

Age group (years)	Egypt 1988	Tunisia 1989
15–19	2076	371
20–24	2076	333
25–29	1826	286
30–34	1500	234
35–39	1269	191
40–44	1064	139
45–49	965	122
$\sum$	10776	1676

Table S.32

Age group	Number of Tunisian women (thousands)	Egyptian ASFR	Expected births to Tunisian women using Egyptian ASFR (thousands)
15–19	371	0.021	7.8
20–24	333	0.194	64.6
25–29	286	0.317	90.6
30–34	234	0.269	62.9
35–39	191	0.191	36.5
40–44	139	0.073	10.1
45–49	122	0.026	3.2
		$\sum$	275.8

where the summation is over the range of five-year age groups. Using the summed ASFRs on the last line of the first table, therefore, we have

$$\text{period TFR for Egypt} = 5(0.021 + 0.194 + 0.317 + 0.269 + 0.191 + 0.073 + 0.026)$$
$$= 5.455.$$

The period TFR for Tunisia is 3.41.

8.6 (a) The calculations are shown in Table S.33. The general fertility rates are, therefore, for urban areas $112/596 = 0.188$ and for rural areas $899/4280 = 0.210$.

(b) Urban areas, 5.51; rural areas, 6.89.

(c) The standardized fertility rate for urban areas is 0.190. This should be compared with the general fertility rate for rural areas.

(d) Fertility is higher in rural areas than in urban areas. Fertility in rural areas exceeds that in urban areas by 25% according to the total fertility rate, but by only about 10% according to the general fertility rate and the standardized fertility rate.

The standardized rate should be compared with the general fertility rate. It appears to be much the same. This suggests that differences in the age structure of the female population between urban and rural areas are not confounding the comparison much.

In fact, there are differences between urban and rural areas in the age structure, so this is surprising. However, the ratios between the age-specific fertility rates in rural and urban areas are not constant, and the relative magnitudes of these ratios may also be influencing the comparison, possibly 'compensating' for the differences in the age structures.

Table S.33

Age group (years)	Number of women in age group		Number of births	
	Urban areas	Rural areas	Urban areas	Rural areas
15–19	129	988	17	163
20–24	135	820	36	239
25–29	120	663	29	181
30–34	84	557	18	145
35–39	63	463	9	94
40–44	40	463	3	57
45–49	25	326	0	20
$\sum$	596	4280	112	899

Table S.34

Age group	Calendar time						
	1960–64	1965–69	1970–74	1975–79	1980–84	1985–89	1990–94
	A	B	C				
15–19	0.040	0.049	0.045	0.030	0.030	0.031	0.033
20–24	0.175	0.165	0.130	0.108	0.095	0.094	0.088
25–29	0.183	0.164	0.135	0.125	0.127	0.125	0.120
30–34	0.105	0.090	0.067	0.065	0.075	0.084	0.087
35–39	0.049	0.040	0.025	0.020	0.024	0.030	0.033
40–44	0.015	0.010	0.007	0.005	0.005	0.005	0.006
45–49	0.000	0.000	0.000	0.000	0.000	0.000	0.000
Period TFR	2.84	2.59	2.05	1.77	1.78	1.85	1.84

8.7 (a) The period total fertility rates are calculated for each five-year period by summing the age-specific fertility rates (ASFRs) for each period and multiplying by 5 (because we have five-year age groups). They are shown in Table S.34.

(b) Cohort fertility rates for women born during the period 1943–47 may be estimated by summing the ASFRs in the diagonal marked A in Table S.34. This is because the women who were 15–19 years of age in 1960–64 were, on average, born in 1945, and the majority of them were born between 1943 and 1947. Similarly, cohort total fertility rates (TFRs) for women born during the periods 1948–52 and 1953–57 can be estimated by summing the ASFRs in the diagonals marked B and C in Table S.34. The women born during the period 1948–52 and 1953–57 have not completed their fertility, but we can assume that fertility in the age groups 40–44 years and 45–49 years during the next ten years will remain the same as it was in 1990–94 (this will not affect the cohort TFRs very much since the ASFRs at ages over 40 years are so small).

Thus we have

$$\text{cohort TFR for women born 1943–47} = 5(0.04 + 0.165 + 0.135 + 0.065 + 0.024 + 0.005)$$
$$= 2.17,$$

$$\text{cohort TFR for women born 1948–52} = 5(0.049 + 0.130 + 0.125 + 0.075 + 0.030 + 0.006)$$
$$= 2.08,$$

$$\text{cohort TFR for women born 1953–57} = 5(0.045 + 0.108 + 0.127 + 0.084 + 0.033 + 0.006)$$
$$= 2.02.$$

(c) Since the ASFR at any age x last birthday, f_x, measures the number of children born per woman-year lived between exact ages x and $x + 1$ years, assuming no mortality among the women, we can work out the proportion of births which take place to women aged under 30 years using the formula

$$\text{proportion of births to woman aged under 30 years} = \frac{\sum_{x=15}^{29} f_x}{\sum_{x=15}^{49} f_x}.$$

Since we have already calculated the TFRs, we can cut down on the calculations by using the formula

$$\text{proportion of births to woman aged under 30 years} = \frac{5 \sum_{i=15-19}^{i=25-29} f_i}{\text{TFR}},$$

Table S.35

Period	Proportion of births to women aged under 30 years
1960–64	0.701
1965–69	0.730
1970–74	0.756
1975–79	0.745
1980–84	0.708
1985–89	0.676
1990–94	0.655

where the subscripts i refer to the five-year age groups. Thus, for the years 1960–64, we have

$$\text{proportion of births to woman aged under 30 years} = \frac{5(0.04 + 0.175 + 0.183)}{2.84} = 0.701.$$

Similar calculations for the other periods produce the results shown in Table S.35.

(d) The fluctuations in period fertility seem to be partly due either to changes in the age pattern of childbearing or to transient period effects, since the cohort TFRs do not seem to have fluctuated by anything like as much. Of course, the problem of censoring means that we cannot make accurate estimates of the cohort TFRs for women born after about 1953–57.

Whereas the decline in period fertility between 1960–64 and 1970–74 seems to have affected all age groups (and is probably, therefore, due to general period effects), the continued decline in fertility between 1970–74 and 1975–79 seems to have been mainly the result of the postponement of childbearing by the women born during the 1950s (who were aged 20–24 years in 1975–79).

The period TFRs for the late 1970s thus seem to have been temporarily depressed by changes in the age pattern of childbearing. Thus, although it may have seemed that fertility in the late 1970s was insufficient to replace the population, the three birth cohorts who were contributing most to the fertility of the population in 1975–79 (cohorts A, B and C in Table S.34) all produced more than two children per woman on average.

8.8 The following points might be made.

1 Period fertility declined by about one birth per woman between the early 1960s and the late 1970s.

2 Since the late 1970s the period TFR has remained roughly constant, at about 1.8 births per woman.

3 Between the late 1970s and the early 1990s, the apparent stability of the period TFRs masked a change in the age pattern of fertility. The fertility of women in their early twenties declined by 19% between the periods 1975–79 and 1990–94, whereas the fertility of women in their early thirties rose by 34% over the same period (and the fertility of women in their late thirties rose by 65%). The percentage of children born to women in their thirties and forties has risen from about 25% to about 35% since the early 1970s.

Chapter 9

9.1 (a) The calculations are shown in Table S.36. Age-order specific fertility rates are required before the order-specific total fertility rates (TFRs) can be calculated.

(b) For first births, we have

$$\text{TFR}_1 = 5(0.0557 + 0.0751 + 0.0251 + 0.0109 + 0.0020)1 = 0.844.$$

Table S.36

Age group	Age-order-specific fertility rates			
	First births	Second births	Third births	Fourth births
15–19	$204/3664 = 0.0557$	0.0074	0.0008	0.0000
20–24	$280/3726 = 0.0751$	0.0580	0.0220	0.0056
25–29	$77/3062 = 0.0251$	0.0474	0.0441	0.0261
30–34	$23/2117 = 0.0109$	0.0179	0.0288	0.0222
35–39	$3/1490 = 0.0020$	0.0134	0.0081	0.0128
40–44	$0/1227 = 0.0000$	0.0016	0.0024	0.0024
45–49	$0/1330 = 0.0000$	0.0000	0.0000	0.0000

The TFRs for the other birth orders are as follows: second births, 0.729; third births, 0.531; fourth births, 0.346.

(c)   The TFR for all birth orders combined is calculated by summing the births of all orders and calculating overall age-specific fertility rates (ASFRs) for each age group, as shown in Table S.37. The TFR is then equal to $5 \sum$ ASFRs:

$$\text{TFR} = 5(0.0639 + 0.1621 + 0.1574 + 0.1105 + 0.0644 + 0.0187 + 0.0015) = 2.89.$$

9.2   The period TFR is a synthetic measure which does not refer to real women. Clearly a real cohort of women cannot, on average, have more than one birth of any order. Because period measures are synthetic, however, it may be that an order-specific period TFR exceeds 1.0.

For example, suppose that the timing of childbearing is changing progressively. Suppose that successive birth cohorts are starting childbearing at ever younger ages. It is possible, if this is happening, that in a particular period, the order-specific TFRs (especially for birth orders 1 and 2) will exceed 1.0. This may happen because the order-specific TFR for, say, first births will be based on high first birth rates for older women (who come from cohorts in which childbearing starts late) and similarly high first birth rates for younger women (who come from cohorts who start childbearing early). The effect is a squeezing of a large number of first births into the period in question.

9.3   (a)   If the parity progression ratio from parity j to parity $j + 1$ is a_j, then the proportion of a cohort who have exactly zero children, n_0, is equal to $1 - a_0$. For $j > 0$, the proportion of a cohort who have exactly j children, n_j, is given by

$$n_j = \left[\prod_{k=0}^{j-1} a_k \right] (1 - a_j).$$

Table S.37

Age group	Total births	ASFR
15–19	$204 + 27 + 3 = 234$	$234/3664 = 0.0639$
20–24	$280 + 216 + 82 + 21 + 5 = 606$	$604/3726 = 0.1621$
25–29	482	0.1574
30–34	234	0.1105
35–39	96	0.0644
40–44	23	0.0187
45–49	2	0.0015

Table S.38

Birth cohort	Proportion of women with exactly				
	0 children	1 child	2 children	3 children	4 children
1931–33	0.139	0.169	0.308	0.185	0.199
1934–36	0.115	0.152	0.326	0.208	0.199
1937–39	0.114	0.136	0.343	0.222	0.185
1940–42	0.110	0.127	0.369	0.230	0.164
1943–45	0.108	0.130	0.413	0.217	0.132
1946–48	0.115	0.134	0.437	0.209	0.105

Thus, for the 1931–33 birth cohort, we have

$$n_0 = (1 - 0.861) = 0.139,$$

$$n_1 = 0.861(1 - 0.804) = 0.169,$$

$$n_2 = 0.861(0.804)(1 - 0.555) = 0.308,$$

$$n_3 = 0.861(0.804)(0.555)(1 - 0.518) = 0.185,$$

and

$$n_4 = 0.861(0.804)(0.555)(0.518) = 0.199.$$

The assumption that no woman has more than four children is equivalent to assuming that $a_j = 0$ for all $j > 3$.

Similar calculations for the other five birth cohorts produce the results given in Table S.38.

(b) The total fertility rate, TFR, is given by the formula

$$\text{TFR} = a_0 + a_0 a_1 + a_0 a_1 a_2 + a_0 a_1 a_2 a_3.$$

For example, for the birth cohort 1934–36, we therefore have

$$\text{TFR} = 0.885 + 0.885(0.828) + 0.885(0.828)(0.555) + 0.885(0.828)(0.555)(0.489)$$

$$= 0.885 + 0.733 + 0.407 + 0.199$$

$$= 2.224.$$

Similar calculations for the other birth cohorts produce the results shown in Table S.39.

9.4 On the basis of these figures, one can say the following things about recent fertility trends in England and Wales.

1 Fertility has been declining among birth cohorts since 1937–39, but only very slowly.

2 There has been a particularly noticeable decline in the proportion of women going on to have more than three children among women born since the late 1930s.

Table S.39

Birth cohort	Total fertility rate
1931–33	2.136
1934–36	2.224
1937–39	2.229
1940–42	2.210
1943–45	2.135
1946–48	2.055

Table S.40

Index year (year of fourth births)	True parity cohort PPR	Year of fifth births	Synthetic parity cohort PPR
1970	0.912		
1971	0.929		
1972	0.976		
1973	0.977		
1974	0.943		
1975	0.964		
1976	0.974		
1977	0.973		
1978	0.822		
1979	0.893		
1980	0.960	1980	0.918
		1981	0.863
		1982	0.957
		1983	0.959
		1984	0.916
		1985	0.916
		1986	0.934
		1987	0.886
		1988	0.893
		1989	0.958
		1990	0.949

3 There has been very little change in the proportion of women remaining childless among birth cohorts since 1934–36.

4 The proportion of women having exactly two children rose steadily among the birth cohorts of the 1930s and 1940s, until, among the birth cohorts of the late 1940s, almost half the women had exactly two children. Possibly a 'two-child norm' has been strengthening.

5 It is not possible to draw any conclusions about short-term period changes in fertility from the figures in the table, as these figures give no indication about the age patterns of fertility among the various birth cohorts. These age patterns may have changed quite a lot, and such changes would affect period TFRs.

9.5 See Table S.40.

9.6 Clearly, true parity cohort PPRs based on small numbers of women will be less reliable than those based on larger numbers. One way of making the results more reliable might be to select an index period longer than one year. Thus, for example, we could take a sample of women who had their fourth births during the period 1970–74, and work out true parity cohort PPRs by following this sample through and observing how many of the women had a fifth child within one year of their fourth birth, within two years of their fourth birth, and so on. A set of q_x^* values could then be calculated based on these data, and used to estimate true parity cohort PPRs.

Chapter 10

10.1 (a) The age-order-specific marital fertility rate for order j is obtained by dividing the number of legitimate births of order j to women in a particular age group by the number of married women in that age group. Thus, for order 1 for women aged under 20 years, the age-order-specific marital fertility rate is equal to $6.7/26.1 = 0.257$. Table S.41 shows the full set of age-order-specific marital fertility rates.

Table S.41

Age group	Age-order-specific marital fertility rate			
	1st births	2nd births	3rd births	4th and higher-order births
Under 20	0.257	0.077	0.008	0.000
20–24	0.119	0.076	0.022	0.005
25–29	0.076	0.067	0.024	0.010
30–34	0.031	0.041	0.021	0.011
35–39	0.008	0.011	0.008	0.007
40–49	0.001	0.001	0.001	0.002

(b) The total marital fertility rate, $TMFR_j$, for birth order j is obtained by summing the figures in the appropriate column of Table S.41 (ignoring the figures in the first row, because these relate to women aged under 20 years) and multiplying the result by 5. Thus, for first births, we have

$$TMFR_1 = 5(0.119 + 0.076 + 0.031 + 0.008 + 0.001 + 0.001) = 1.172.$$

The remaining results are: $TMFR_2 = 0.981$, $TMFR_3 - 0.383$, and $TMFR_4 = 0.178$.

(c) The sum of the $TMFR_j$s over all values of j gives the overall total marital fertility rate (TMFR). Thus,

$$TMFR = 1.172 + 0.981 + 0.383 + 0.178 = 2.713.$$

(d) Marital fertility is very much concentrated on births of orders 1 and 2. Relatively few births take place at orders 3 or higher. The first birth rate for married women aged under 20 years is very high (some of these women may have married because they were pregnant).

10.2 The general legitimate fertility rate is obtained by dividing the births to married women by the mid-year married female population aged 16–44 years. Thus, for 1971 we have

$$\text{general legitimate fertility rate} = \frac{717\,500}{6\,419\,000} = 0.112.$$

The general illegitimate fertility rate is obtained by dividing the births to non-married women by the mid-year never married, widowed and divorced female population aged 16–44 years. Thus, for 1971 we have

$$\text{general illegitimate fertility rate} = \frac{783\,200 - 717\,500}{2\,434\,000 + 200\,000} = 0.025.$$

For 1981, the general legitimate fertility rate is 0.089 and the general illegitimate fertility rate 0.022; while for 1991 the general legitimate fertility rate is 0.084 and the general illegitimate fertility rate is 0.043.

10.3 (a) The main effect of bridal pregnancy is to associate birth more closely with ages at which marriage typically takes place. Assuming that it does not result in any 'extra' births (compared with the number that would occur in the absence of high rates of bridal pregnancy), then bridal pregnancy will be associated with relatively high age-specific fertility rates (ASFRs) at the ages at which women commonly marry, and rather lower rates at other ages.

(b) Since the births taking place within nine months of a marriage will be included in the age-specific marital fertility rates, the latter will be increased at ages at which women commonly marry, and decreased at other (especially older) ages. The effect will be rather greater than that for the ASFRs outlined in the solution to part (a) above.

Table S.42

Method, m	u_m	e_m	$u_m e_m$
Pill	0.05	0.90	0.0450
Female sterilization	0.06	1.00	0.0600
Intra-uterine device	0.15	0.95	0.1425
Periodic abstinence	0.04	0.80	0.0320
Other modern methods	0.01	0.90	0.0090
Other traditional methods	0.08	0.70	0.0560
		$\sum$	0.3445

(c)   The usefulness of the total marital fertility rate (TMFR) as a measure of fertility should not be compromised. Indeed, high rates of bridal pregnancy are often associated with a low proportion of illegitimate births, since one reason why bridal pregnancy is common is the existence of social pressures on couples to 'legitimize' their children. A low proportion of illegitimate births means that the TMFR is calculated using most of the births that take place, which increases its attractiveness as a measure of fertility.

10.4   The index C_i is given by the formula

$$C_i = \frac{20}{18.5 + i}.$$

We are told that $i = 8$, so $C_i = 20/(18.5 + 8) = 0.755$.

The index C_c may be calculated from the formula

$$C_c = 1 - 1.08 \sum u_m e_m.$$

Using the data given, the summation is calculated as shown in Table S.42. So

$$C_c = 1 - 1.08 \times 0.3445 = 0.628.$$

The index C_m is calculated from the data given using the formula

$$C_m = \frac{\sum_{x=15}^{x=49}(\text{ASMFR}_x \times \pi_x)}{\sum_{x=15}^{x=49}(\text{ASMFR}_x)}.$$

So, from Table S.43, $C_m = 1.114/2.006 = 0.555$.

The total marital fertility rate (TMFR) is equal to the sum of the age-specific fertility rates for married women. Using Table S.43 it is clear, therefore, that the

$$\text{TMFR} = 5 \times 2.006 = 10.03.$$

We then have

$$\text{TN} = \text{TMFR}/C_c = 10.03/0.628 = 15.97$$

Table S.43

Age group	ASMFR	π	$ASMFR \times \pi$
15–19	0.471	0.104	0.049
20–24	0.492	0.445	0.219
25–29	0.416	0.712	0.296
30–34	0.307	0.860	0.264
35–39	0.211	0.893	0.188
40–44	0.087	0.908	0.079
45–49	0.022	0.883	0.019
$\sum$	2.006		1.114

and

$$\mathrm{TF} = \mathrm{TN}/C_i = 15.97/0.755 = 21.15.$$

This value of the total fecundity rate is somewhat higher than the normal range of 13 to 17. This may be because the index of contraception, C_c, is underestimated in this example – that is, the fertility-inhibiting effect of contraceptive use is overestimated.

Chapter 11

11.1 (a) See Figure S8.
  (b) Whereas almost every woman in the 1930–39 birth cohort has completed her childbearing by 1980 (since even the youngest woman in that birth cohort is at least 40 years old by then), the same is not true of the other birth cohorts. The histories of these other cohorts are censored.
    Hence the 1950–64 birth cohort contains a large proportion of women (53%) who did not have a second birth before the survey. A large proportion of these women will probably go on to have a second child.
    Censoring is not so serious for the 1940–49 birth cohort as it is for the 1950–64 birth cohort, but the proportion of those who did not have a second birth before the survey is higher for the 1940–49 birth cohort than it is for the 1930–39 birth cohort, and the proportion with an interval between the first and second births of 61 months and over is smaller. The most likely explanation for this is censoring.

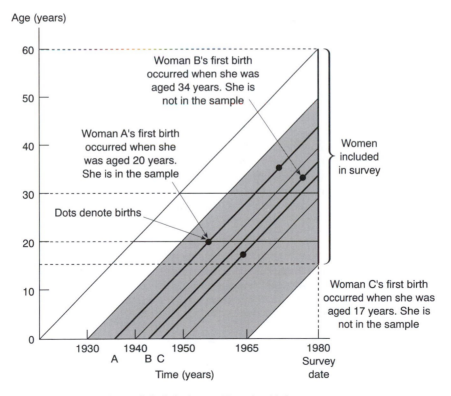

Figure S.8 Solution to Exercise 11.1

The 1950–64 birth cohort is smaller than the others. This is because many women in this birth cohort had not had a first birth before the survey. This creates a potential problem of selection.

11.2   Since $h(x) = \lambda$,

$$S(x) = \exp\left(-\int_0^x h(u)\,\mathrm{d}u\right) = \mathrm{e}^{-\lambda x}.$$

The expected waiting time to conception is analogous to the life expectation, e_x, in the conventional life table. If $l_0 = 1$, l_x and $S(x)$ are equal. Therefore, we can write

$$\frac{\text{expected waiting time to conception}}{\text{after } m \text{ months}} = \frac{\int_m^\infty S(x)\,\mathrm{d}x}{S(m)} = \frac{\int_m^\infty \mathrm{e}^{-\lambda x}\,\mathrm{d}x}{\mathrm{e}^{-\lambda m}}.$$

Thus, integrating, we have

$$\frac{\text{expected waiting time to conception}}{\text{after } m \text{ months}} = \left[\frac{\mathrm{e}^{-\lambda x}}{\ln \mathrm{e}^{-\lambda}}\right]_m^\infty \cdot \frac{1}{\mathrm{e}^{-\lambda m}} = \frac{-\mathrm{e}^{-\lambda m}}{-\lambda} \cdot \frac{1}{\mathrm{e}^{-\lambda m}} = \frac{1}{\lambda}.$$

This applies for all m, including $m = 3$. So, if the expected remaining waiting time to conception after three months is 5 months, $\lambda = 0.2$.

11.3   (a)   Since many birth intervals will be censored by the survey, survival analysis methods are appropriate.

A proportional hazards model is a convenient way of incorporating the effects of covariates on the hazard of a birth interval ending (or, better, on the hazard of conception of a subsequent child). To apply such a model, the following steps are necessary.

1   Prepare a set of data which has one case per birth interval, and which includes the following variables: the duration between the index birth and the conception of the next child (or a point nine months before the survey date, whichever is the shorter); a 0–1 censoring indicator (δ_i); and the covariates measuring the relevant social and economic factors (which may be continuous or categorical).

2   The model is

$$h(t) = h_0(t)\exp(\beta_1 x_1 + \beta_2 x_2 + \cdots),$$

where $h_0(t)$ is some baseline hazard, $x_1, x_2, \ldots$ are the covariates and $\beta_1, \beta_2, \ldots$ are parameters to be estimated.

3   Estimate the parameters of the model using maximum likelihood.

4   Exponentiate the estimates of the βs to obtain relative risks which measure the effect of social and economic factors on the hazard – and hence on birth interval length.

5   If desired, estimated survivor functions, and so on, can be presented to illustrate the effects further.

(b)   The method is as described in part (a), save that instead of the social and economic covariates, a time-varying covariate representing breastfeeding should be incorporated. This covariate will typically be a 0–1 variable, which takes the value 1 at all durations since the index birth (the birth which starts the birth interval) during which the woman is breastfeeding, and 0 at all durations after the end of breastfeeding.

11.4   (a)   We have

$$h(x) = -\frac{\mathrm{d}}{\mathrm{d}x}\ln S(x).$$

Since

$$S(x) = \exp[-(\lambda x)^\beta],$$

then

$$\ln S(x) = -(\lambda x)^\beta,$$

and, therefore,

$$h(x) = -\frac{d}{dx} - (\lambda x)^\beta = \frac{d}{dx}(\lambda x)^\beta,$$

so

$$h(x) = \lambda^\beta \beta x^{\beta-1}. \tag{1}$$

(b)   First, note that since $h(x)$ cannot be negative, equation (1) implies that β cannot be negative.

   If $\beta = 0$, then $\lambda^\beta \beta x^{\beta-1} = 0$ for all durations x, and if $\beta = 1$, then $\lambda^\beta \beta x^{\beta-1} = \lambda$ for all durations x. Both these values of β, therefore, lead to constant hazards.

   In addition, it is clear that if $\beta > 1$, then the fact that $h(x)$ is a function of $x^{\beta-1}$ means that $h(x)$ will increase with increasing duration x.

   The only values of β which will lead to a hazard which decreases monotonically with duration are $0 < \beta < 1$.

Chapter 12

12.1 (a)   The gross reproduction rate (GRR) is equal to $\sum_x f_x^d$ where f_x^d is the age-specific fertility rate at age x last birthday for daughters only. Since $f_x^d = (100/205)f_x$, the GRR may be calculated from the data given using the formula

$$GRR = 5\frac{100}{205}\sum_i f_i,$$

where the i are five-year age groups. Thus

$$GRR = \frac{500}{205}(0.033 + 0.09 + 0.12 + 0.087 + 0.032 + 0.006) = 0.898.$$

The net reproduction rate (NRR) may be calculated using five-year age groups for fertility and mortality according to the formula

$$NRR = \frac{100}{205}\sum f_x \cdot {}_5L_x,$$

where ${}_5L_x$ is the number of women alive in each age group using a relevant life table in which l_0 is equal to 1.

   We can use the approximate formula

$${}_5L_x = 5l_{x+2.5}$$

in the calculation. This assumes that deaths are distributed uniformly within each age group.

   Using the data given, the calculations are summarized in Table S.44. We obtain

$$NRR = \frac{100}{205}(1.8157) = 0.886.$$

(b)   A net reproduction rate of 0.886 suggests that the next generation will only be 89% of the size of the present one. Therefore the population will decrease by about 11% per generation.

   However, this will only happen if the fertility and mortality rates used in the calculation apply to a whole birth cohort. Since the actual figures used in part (a) relate to a single calendar year, we cannot conclude that the population of England and Wales will decline in the long run. In fact, we can say almost nothing about the long-run evolution of the population of England and Wales from the result of part (a) above.

Table S.44

Age group	Exact age x	f_x	$l_{x+2.5}$	$_5L_x$	$f_x \cdot {_5L_x}$
15–19	15	0.033	0.9903	4.9515	0.1634
20–24	20	0.090	0.9890	4.9450	0.4451
25–29	25	0.120	0.9871	4.9355	0.5923
30–34	30	0.087	0.9850	4.9250	0.4285
35–39	35	0.032	0.9817	4.9085	0.1571
40–44	40	0.006	0.9766	4.8830	0.0293
45–49	45	0.000	0.9685	4.8425	0.0000
				$\sum$	1.8157

12.2 (a) Using the exponential growth equation, with an annual rate of growth r, we can relate the populations, P_{t_1} and P_{t_2}, in any two years t_1 and t_2, by the equation

$$P_{t_2} = P_{t_1}\, e^{r(t_2 - t_1)}.$$

Taking logarithms of this equation produces

$$\ln P_{t_2} = \ln P_{t_1} + r(t_2 - t_1),$$

and solving for r produces

$$r = \frac{\ln P_{t_2} - \ln P_{t_1}}{t_2 - t_1}. \qquad (1)$$

Thus, for growth between 1982 and 1990, we can use equation (1) to write

$$
\begin{aligned}
r &= \frac{\ln P_{1990} - \ln P_{1982}}{1990 - 1982} \\
&= \frac{\ln 26\,000\,000 - \ln 18\,000\,000}{8} \\
&= \frac{17.074 - 16.706}{8} \\
&= 0.046 \ (\text{or } 4.6\%).
\end{aligned}
$$

For growth between 1990 and 2000, we use equation (1) to write

$$
\begin{aligned}
r &= \frac{\ln P_{2000} - \ln P_{1990}}{2000 - 1990} \\
&= \frac{\ln 40\,000\,000 - \ln 26\,000\,000}{10} \\
&= \frac{17.074 - 17.074}{10} \\
&= 0.043 \ (\text{or } 4.3\%).
\end{aligned}
$$

(b) Solving equation (1) for $t_2 - t_1$ produces

$$t_2 - t_1 = \frac{\ln P_{t_2} - \ln P_{t_1}}{r}.$$

Thus the number of years it will take for the population to reach 80 million, if it is 40 million in the year 2000, and continues growing at an annual rate of 0.043, is given by the equation

$$t = \frac{\ln 80\,000\,000 - \ln 40\,000\,000}{0.043} = 16.1 \text{ years.}$$

Therefore, the population will reach 80 million 16.1 years after 30 June 2000, which is sometime in August 2016.

12.3 Suppose a population growing at an annual rate r doubles in t years. We have

$$P_t = P_0 e^{rt} = 2P_0.$$

This implies that

$$e^{rt} = 2$$

$$\ln 2 = rt$$

whence

$$t = \frac{\ln 2}{r} = \frac{0.693}{r}.$$

The annual growth rate as a percentage is equal to $100r$. We have, therefore,

$$t = \frac{69.3}{100r} \cong \frac{70}{100r}.$$

Chapter 13

13.1 See Figure 13.3.

13.2 (a) The proportion of the population alive aged between exact ages x and $x + n$ in a *stationary* population, $_n c_x$, is given by the formula

$$_n c_x = b \cdot {_n L_x},$$

where b is the birth rate in the stationary population and $_n L_x$ is derived from the life table of a stationary population in which $l_0 = 1.0$. If $l_0 \neq 1.0$, we have

$$_n c_x = \frac{b \cdot {_n L_x}}{l_0}.$$

In a stationary population with l_0 births per year, the constant total population is equal to the value T_0 in the life table describing the mortality of that population. Thus the birth rate, b, is equal to l_0/T_0, and we can write

$$_n c_x = \frac{l_0}{T_0} \cdot \frac{_n L_x}{l_0} = \frac{_n L_x}{T_0}.$$

The calculation of the $_n c_x$ is shown in Table S.45. We make the assumption that deaths are uniformly distributed within each age group (even though this is a bad assumption for the age group 0 years), so we can use the formula

$$_n L_x = \frac{n}{2}(l_x + l_{x+n}).$$

To calculate $_\infty L_{80}$, we note that the population aged 80 years and over is 1.5% of the total. Thus we can write, using standard life table notation,

$$_\infty L_{80} = 0.015 T_0.$$

Since the total population, T_0, is equal to the sum of those aged under 80 years and those aged 80 years and over, we have

$$_\infty L_{80} = 0.015(_\infty L_{80} + {_{80} L_0}),$$

which may be rearranged to obtain

$$_\infty L_{80} = \frac{0.015 \cdot {_{80} L_0}}{0.985}.$$

Table S.45

Age x	n	l_x	$_nL_x$	$_nc_x$
0	1	1 000	975	0.019
1	4	950	3 700	0.070
5	5	900	4 450	0.085
10	10	880	8 650	0.165
20	10	850	8 200	0.156
30	10	790	7 450	0.142
40	10	700	6 400	0.122
50	10	580	5 150	0.098
60	10	450	3 950	0.075
70	10	340	2 850	0.054
80		230	788	0.015
			Σ 52 563	1.000

The quantity $_{80}L_0$ can be calculated from the data given, so we can use this equation to evaluate $_\infty L_{80}$.

(b) For a stable population growing at annual rate r (or $100r\%$), the proportion of the population at each age x last birthday is given by the expression $(b\,e^{-r(x+1/2)}L_x)$. The proportion of the population in an age group n years wide, $_nc_x^*$, is given by the expression

$$_nc_x^* = b^*\,e^{-r(x+n/2)}\,_nL_x.$$

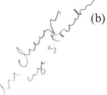

In this equation, $_nL_x$ refers to a life table in which l_0 is equal to 1.0, and b^* is the constant birth rate in the stable population. Although we do not know b^*, we note that since it is a constant, and is the same for all age groups, we can write

$$_nc_x^* = \frac{e^{-r(x+n/2)}\,_nL_x}{\sum_x e^{-r(x+n/2)}\,_nL_x}. \tag{1}$$

We apply equation (1) to the data we are given, taking r to be 0.02. For the oldest age group, we have assumed that $x + n/2$ is 85, since we are told that the average age of the people in this age group is 85 years. The denominator of equation (1) calculated using our data is 29 534. The calculations are shown in Table S.46.

(c) The stable population age structure is considerably younger than the age structure of the stationary population. The growing stable population has a larger proportion of its members in age groups below age 30 years than does the stationary population. For the age groups above 30 years, the reverse is the case.

Table S.46

Age x	n	$_nL_x$	$e^{-r(x+n/2)}$	$e^{-r(x+n/2)}\,_nL_x$	$_nc_x^*$
0	1	975	0.990	965	0.033
1	4	3 700	0.942	3 485	0.118
5	5	4 450	0.861	3 830	0.130
10	10	8 650	0.741	6 408	0.217
20	10	8 200	0.607	4 974	0.168
30	10	7 450	0.497	3 700	0.125
40	10	6 400	0.407	2 602	0.088
50	10	5 150	0.333	1 714	0.058
60	10	3 950	0.273	1 077	0.036
70	10	2 850	0.223	636	0.022
80		788	0.183	144	0.005
		Σ 52 563	1.000	29 534	1.000

13.3 In a declining stable population, the number of births is decreasing each year. Therefore, older age groups are survivors of a larger original number of births than younger age groups. However, as each cohort of births grows older, it is depleted by mortality. The age pattern of mortality in most populations is such as to lead to an increasing rate of depletion at older ages. The consequence of these two factors acting together is that, at ages up to about 40–45 years, the larger original number of births outweighs the effect of mortality in reducing survivorship. At older ages, the depleting effect of mortality gradually becomes dominant, leading to a smaller number of persons alive at those ages.

(Mathematically, the number of persons alive between exact ages x and $x + 1$ is proportional to $e^{r(x+1/2)}L_x$, where r is the annual rate of decline, and L_x measures survivorship. As x increases, $e^{r(x+1/2)}$ increases and L_x decreases. At ages up to about 40–45 years, the combined effect of this is to increase $e^{r(x+1/2)}L_x$, but at older ages the decrease in L_x accelerates, so that $e^{r(x+1/2)}L_x$ declines.)

13.4 (a) The population's growth rate will increase, since mortality will fall. The increase in numbers will be experienced at ages over 45 years. Eventually, after $\omega - 45$ years, the population will become stable once more (ω is the limiting age). A suitable sketch is shown in Figure S.9(a).

(b) The population's growth rate will decrease, since fertility falls. The number of children born will decrease gradually for more than 20 years, because even after age-specific fertility rates have stopped falling, the number of women exposed to the risk of child-bearing will continue to fall, so the number of children being born will continue to decline. Stability will not be achieved for several generations: indeed, if the initial rate of growth was slow, a fall in fertility of 50% might be enough to turn it negative, resulting in the population eventually dying out. There will be 'knock-on' effects on the number of children born in subsequent generations. See Figure S.9(b).

(c) Since the fall in the number of children born is temporary, it will be reflected in a gash in the side of the age pyramid, which will gradually move up the pyramid over time. It may be that parents attempt to compensate for the one-year fall in fertility by having extra children in subsequent years. In that case, a small bulge in the age pyramid might be seen below the gash. Any 'knock-on' effects in the next generation will be very small. See Figure S.9(c).

(d) The growth rate of the population will increase. Changes in infant mortality will ultimately affect the shape of age pyramid only at the base. The proportion of the population aged under 1 year will decrease, and the proportion at all other ages will increase. However, during the transition from one stable population to the other the shape will change, such that after x years, the proportion at ages below x years will increase, and the proportion at ages above x years will decrease. See Figure S.9(d).

13.5 The number of persons alive between 40 and $40\frac{1}{2}$ years of age, $_{\frac{1}{2}}A_{40}$, is given by

$$_{\frac{1}{2}}A_{40} = bP e^{-40.25r} {}_{\frac{1}{2}}L_{40}.$$

Similarly,

$$_{\frac{1}{2}}A_{40\frac{1}{2}} = bP e^{-40.75r} {}_{\frac{1}{2}}L_{40\frac{1}{2}}.$$

Therefore, if these two quantities are equal, we have

$$bP e^{-40.25r} {}_{\frac{1}{2}}L_{40} = bP e^{-40.75r} {}_{\frac{1}{2}}L_{40\frac{1}{2}}.$$

Dividing both sides of this equation by $bP e^{-40.25r}$ produces

$$_{\frac{1}{2}}L_{40} = e^{-0.5r} {}_{\frac{1}{2}}L_{40\frac{1}{2}}.$$

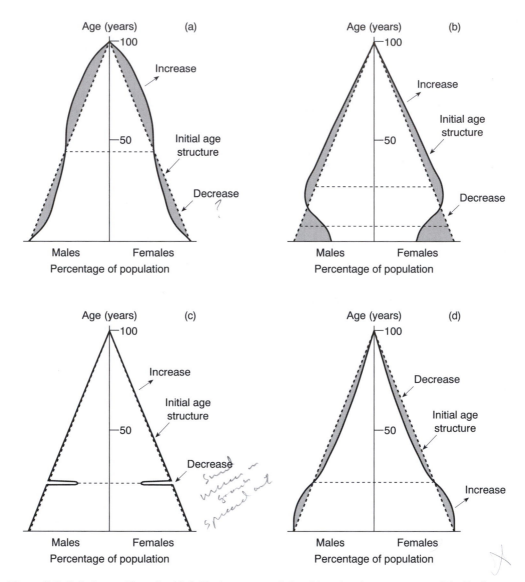

Figure S.9 Solution to Exercise 13.4. Various ways of sketching the changes are possible. In the ones illustrated here, the shape of the age pyramid is shown in proportional terms – that is, the area of the pyramid remains the same as before the change took place

Taking logarithms gives

$$\ln \tfrac{1}{2}L_{40} = \ln \tfrac{1}{2}L_{40\frac{1}{2}} - 0.5r.$$

Therefore,

$$r = -2(\ln \tfrac{1}{2}L_{40} - \ln \tfrac{1}{2}L_{40\frac{1}{2}})$$

$$= -2\ln \left(\frac{\tfrac{1}{2}L_{40}}{\tfrac{1}{2}L_{40\frac{1}{2}}} \right). \tag{1}$$

Table S.47

Age group	n	$_nL_x$	Percentage in age group
0	1	99 945	1.26
1–4	4	397 496	5.01
5–9	5	496 408	6.26
10–14	5	496 073	6.26
15–19	5	495 552	6.25
20–24	5	494 797	6.24
25–29	5	493 969	6.23
30–34	5	492 909	6.22
35–39	5	491 273	6.20
40–44	5	488 653	6.16
45–49	5	485 521	6.11
50–54	5	477 785	6.03
55–59	5	466 812	5.89
60–64	5	448 812	5.66
65–69	5	419 962	5.30
70–74	5	377 302	4.76
75–79	5	317 206	4.00
80–84	5	238 404	3.01
85–89	5	149 669	1.89
90 and over		99 424	1.25

Assuming an even distribution of deaths within each half year of age between exact ages 40 and 41, we have

$$_{\frac{1}{2}}L_{40} = \tfrac{1}{4}(l_{40} + l_{40\frac{1}{2}}),$$ (2)

and

$$_{\frac{1}{2}}L_{40\frac{1}{2}} = \tfrac{1}{4}(l_{40\frac{1}{2}} + l_{41}).$$ (3)

Substituting from equations (2) and (3) into equation (1) produces

$$r = -2\ln\left(\frac{l_{40} + l_{40\frac{1}{2}}}{l_{40\frac{1}{2}} + l_{41}}\right).$$

a repeat

13.6 (a) See Table S.47. Note that a_0 was taken as 0.2, but that deaths were assumed evenly distributed within each of the remaining age groups; $l_0 = 100\,000$.
(b) See Table S.47.
(c) See Table S.48. Note that the average age of persons alive in the oldest age group was assumed to be 95 years.
(d) See Table S.48.

13.7 The equation

$$\text{mean age at childbearing} = \frac{\sum_x xf_xL_x}{\sum_x f_xL_x},$$

shows that the mean age at childbearing is a weighted average of the ages x, where the weights are the f_xL_x values. Using these weights means that the number of women at each age is represented purely by L_x. This will only reflect the case in a real population if the population is stationary, for in a stationary population the number of women at each age is proportional to the values of L_x.

Table S.48

Age group	Percentage in age group			
	$r = 0.01$	$r = 0.02$	$r = 0.03$	$r = -0.01$
0	1.83	2.52	3.30	0.81
1–4	7.12	9.56	12.21	3.32
5–9	8.50	10.91	13.32	4.34
10–14	8.08	9.86	11.46	4.56
15–19	7.68	8.92	9.85	4.79
20–24	7.29	8.05	8.47	5.03
25–29	6.93	7.28	7.27	5.28
30–34	6.57	6.57	6.25	5.54
35–39	6.23	5.92	5.36	5.80
40–44	5.90	5.33	4.59	6.07
45–49	5.56	4.78	3.92	6.32
50–54	5.22	4.27	3.32	6.56
55–59	4.85	3.77	2.80	6.73
60–64	4.43	3.28	2.31	6.81
65–69	3.95	2.78	1.86	6.70
70–74	3.37	2.26	1.44	6.32
75–79	2.70	1.72	1.04	5.59
80–84	1.93	1.17	0.67	4.42
85–89	1.15	0.66	0.36	2.91
90 and over	0.71	0.38	0.19	2.09

In a growing population, there will be a greater proportion of women at a younger age, and a smaller proportion of women at older ages, than there will be in a stationary population. Thus, in a growing population, the weights attached to the younger ages will be increased, and the weights attached to the older ages will be decreased, relative to their values in the stationary population. The result is that the mean length of a generation will be less than the mean age at childbearing. The opposite is true of a declining population.

However, since the extent to which the age structure within the childbearing age range varies with the rate of population growth is quite small, the difference between the mean length of a generation and the mean age at childbearing is also usually small.

13.8   The mean age at childbearing is calculated using the formula

$$\text{mean age at childbearing} = \frac{\sum_x x f_x L_x}{\sum_x f_x L_x},$$

where f_x is the age-specific fertility rate at age x last birthday, and L_x is the number of woman-years lived between exact ages x and $x + 1$ years. For five-year age groups we can modify this formula to read

$$\text{mean age at childbearing} = \frac{\sum_i (a_i + 2.5) f_i \cdot {}_5L_{a_i}}{\sum_i f_i \cdot {}_5L_{a_i}}, \tag{1}$$

where a_i is the youngest age in age group i (so $a_i + 2.5$ is the mid-point of age group i), f_i is the age-specific fertility rate in age group i, and ${}_5L_{a_i}$ is the number of woman-years lived in age group i.

Assuming that deaths are evenly distributed within each age group, we can write

$$_5L_{a_i} = 5l_{a_i + 2.5}. \tag{2}$$

Table S.49

Age group	f_i	$l_{a_i+2.5}$	$a_i + 2.5$	$f_i l_{a_i+2.5}$	$(a_i + 2.5)f_i l_{a_i+2.5}$
15–19	0.033	9 903	17.5	326.80	5 718.98
20–24	0.090	9 890	22.5	890.10	20 027.25
25–29	0.120	9 871	27.5	1 184.52	32 574.30
30–34	0.087	9 850	32.5	856.95	27 850.88
35–39	0.032	9 817	37.5	314.14	11 780.40
40–44	0.006	9 766	42.5	58.60	2 490.33
45–49	0.000	9 685	47.5	0.00	0.00
				Σ 3 631.11	100 442.10

Substituting from equation (2) into equation (1) produces

$$\text{mean age at childbearing} = \frac{\sum_i (a_i + 2.5)f_i \cdot 5l_{a_i+2.5}}{\sum_i f_i \cdot 5l_{a_i+2.5}}$$

$$= \frac{\sum_i (a_i + 2.5)f_i \cdot l_{a_i+2.5}}{\sum_i f_i \cdot l_{a_i+2.5}}.$$

The relevant calculations are shown in Table S.49. The mean age at childbearing is therefore equal to 100 442.10/3631.11, which is 27.66 years.

Chapter 14

14.1 (a) Since the population of computers is constant, and the rate of breakdown is constant over time, we can assume that the population of computers is stationary.

Suppose that the number of computers the university must buy each year is l_0. The proportion surviving to exact age x years ($0 < x \le 5$) is given by

$$\frac{l_x}{l_0} = \exp\left[-\int_0^x \lambda \, du\right] = e^{-\lambda x}. \qquad (1)$$

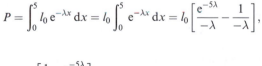

Since the population is stationary, we can write

$$P = \int_0^\omega l_x \, dx,$$

where ω is the limiting age. In this case, we are told that $\omega = 5$, so

$$P = \int_0^5 l_x \, dx. \qquad (2)$$

From equation (1) above, we know that $l_x = l_0 e^{-\lambda x}$, and substituting this expression into equation (2) produces

$$P = \int_0^5 l_0 e^{-\lambda x} \, dx = l_0 \int_0^5 e^{-\lambda x} \, dx = l_0 \left[\frac{e^{-5\lambda}}{-\lambda} - \frac{1}{-\lambda}\right],$$

so

$$P = l_0 \left[\frac{1 - e^{-5\lambda}}{\lambda}\right]. \qquad (3)$$

Rearranging equation (3) produces the required expression for l_0:

$$l_0 = \frac{P\lambda}{1 - e^{-5\lambda}}.$$

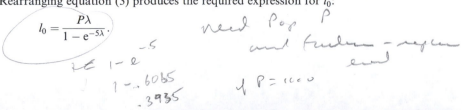

(b)   Since the population of microcomputers is stationary, its age structure is constant, and is proportional to the value of L_x in the life table representing the mortality (that is, break-down) experience. Since

$$_nL_x = \int_x^{x+n} l_u \, du,$$

we have

$$_2L_3 = \int_3^5 l_0 \, e^{-\lambda x} \, dx = l_0 \left[\frac{e^{-3\lambda} - e^{-5\lambda}}{\lambda} \right].$$

The proportion of computers which will not run the new package is $_2L_3/P$, and, using equation (3)

$$\frac{_2L_3}{P} = \frac{l_0 \left[\dfrac{e^{-3\lambda} - e^{-5\lambda}}{\lambda} \right]}{l_0 \left[\dfrac{1 - e^{-5\lambda}}{\lambda} \right]} = \frac{e^{-3\lambda} - e^{-5\lambda}}{1 - e^{-5\lambda}}.$$

14.2   (a)   The population of students on the course can be regarded as stable, provided that we assume there are no transfers into or out of the course. If we make this assumption, then the ratio of new recruits to the number of students on the course will be constant (this quantity is analogous to the birth rate in a stable population). Thus the number of students admitted to the degree course will increase at a constant annual rate.

Let this rate be r. If the number of recruits in year t is R_t, we have, in order for student numbers to double in 20 years,

$$R_{t+20} = 2R_t.$$

But

$$R_{t+20} = R_t \, e^{20r}.$$

Thus

$$2R_t = R_t \, e^{20r};$$

$$2 = e^{20r}.$$

Taking logarithms and rearranging this equation produces

$$r = \frac{\ln 2}{20} = 0.03466.$$

So the number of students admitted to the degree course must increase by 3.47% per year.

(b)   Suppose that, in the year before the freeze (which we shall call year t), the number of students in their first year is N_1, the number of students in their second year is N_2, the number of students in their third year is N_3, and the number of students in their fourth year is N_4.

In general, the number of students in the jth year of their studies depends on the number of students recruited, and on the failure rate at the end of all years from the first to the $(j-1)$th. We are told that the failure rate at the end of all years is 0.1.

Since there are no students dropping out during their first year, then N_1 is simply R_t, the number of new students admitted in the last year before the freeze.

We can also write down the following expressions for N_2, N_3 and N_4:

$$N_2 = 0.9R_{t-1}, \tag{2}$$

$$N_3 = (0.9)^2 R_{t-2}, \tag{3}$$

$$N_4 = (0.9)^3 R_{t-3}. \tag{4}$$

But, we know from (a) that

$$R_{t-n} = R_t e^{-rn}, \tag{5}$$

so, substituting from equation (5) into equations (2), (3) and (4), and replacing R_t by N_1 (since these two quantities are identical), we have

$$N_2 = 0.9 N_1 e^{-r}, \tag{6}$$

$$N_3 = (0.9)^2 N_1 e^{-2r}, \tag{7}$$

$$N_4 = (0.9)^3 N_1 e^{-3r}. \tag{8}$$

The total number of students on the course in the last year prior to the freeze is, of course, $N_1 + N_2 + N_3 + N_4$. Substituting from equations (6), (7) and (8) for N_2, N_3 and N_4, we have

$$\frac{\text{Total number of students}}{\text{before freeze}} = N_1 + 0.9 N_1 e^{-r} + (0.9)^2 N_1 e^{-2r} + (0.9)^3 N_1 e^{-3r}. \tag{9}$$

We can now work out the number of students there will be in each year of the study in year $t+1$, the year of the freeze.

The number of admissions must continue to increase by an annual rate r, so

$$\text{number of students in first year in year } t+1 = N_1 e^r. \tag{10}$$

The number of students in their second year in year $t+1$ is equal to the number of students in their first year in year t multiplied by 1 minus the probability of failure at the end of the first year. Call the failure rate ϕ. We can write

$$\text{number of students in second year in year } t+1 = N_1(1-\phi). \tag{11}$$

The number of students in their third year in year $t+1$ is equal to the number of students in their second year in year t multiplied by 1 minus the probability of failure at the end of the second year:

$$\text{number of students in third year in year } t+1 = N_2(1-\phi)$$
$$= 0.9 N_1 e^{-r}(1-\phi) \tag{12}$$

(using equation (6)).

Finally, we can write, in a similar manner,

$$\text{number of students in fourth year in year } t+1 = N_3(1-\phi)$$
$$= (0.9)^2 N_1 e^{-2r}(1-\phi) \tag{13}$$

(using equation (7)).

The total number of students on the course in the year of the freeze is equal to the sum of the number in each year of study. From equations (10), (11), (12) and (13), this sum is given by the equation

$$\frac{\text{total number of students}}{\text{in year } t+1} = N_1 e^r + N_1(1-\phi) + 0.9 N_1 e^{-r}(1-\phi)$$
$$+ (0.9)^2 N_1 e^{-2r}(1-\phi). \tag{14}$$

The freeze involves making sure that the total numbers of students in years t and $t+1$ are the same. Thus the two sums in equations (9) and (14) must be equal. This implies the following equation:

$$N_1 + 0.9 N_1 e^{-r} + (0.9)^2 N_1 e^{-2r} + (0.9)^3 N_1 e^{-3r}$$
$$= N_1 e^r + N_1(1-\phi) + 0.9 N_1 e^{-r}(1-\phi) + (0.9)^2 N_1 e^{-2r}(1-\phi).$$

To solve this equation for ϕ, first divide by N_1 to obtain

$$1 + 0.9\,e^{-r} + (0.9)^2\,e^{-2r} + (0.9)^3\,e^{-3r}$$
$$= e^r + (1 - \phi) + 0.9\,e^{-r}(1 - \phi) + (0.9)^2\,e^{-2r}(1 - \phi).$$

Now it is probably easiest to substitute into this equation the known value of r from the solution to part (a). This value is 0.03466. The result is

$$1 + 0.9\,e^{-0.03466} + (0.9)^2\,e^{-(2 \times 0.03466)} + (0.9)^3\,e^{-(3 \times 0.03466)}$$
$$= e^{0.03466} + (1 - \phi) + 0.9\,e^{-0.03466}(1 - \phi) + (0.9)^2\,e^{-(2 \times 0.03466)}(1 - \phi).$$

Doing the calculations produces

$$1 + 0.8693 + 0.7558 + 0.6570 = 1.0353 + (1 - \phi) + 0.8693(1 - \phi) + 0.7558(1 - \phi),$$

which may be simplified and solved for ϕ as follows:

$$1 + 0.8693 + 0.7558 + 0.6570$$
$$= 1.0353 + 1 - \phi + 0.8693 - 0.8693\phi + 0.7558 - 0.7558\phi$$
$$0.6570 = 1.0353 - \phi - 0.8693\phi - 0.7558\phi$$
$$0.6570 = 1.0353 - 2.6251\phi$$
$$2.6251\phi = 0.3783$$
$$\phi = \frac{0.3783}{2.6251} = 0.144.$$

So the failure rate will have to be 14.4% in order to maintain the total number of students. This is an increase from the previous failure rate (10%) of 44 per cent.

14.3   To correct for age misreporting is not easy. In order to do this, it is necessary to specify some kind of model which describes the misreporting mechanism. For example, if one could be sure that all the heaping on, say, age 30 years arose because some people who were really 29 or 31 years last birthday said that they were 30 years last birthday; and, further, that all the people who really were 30 years last birthday reported this fact accurately; then it might be possible to redistribute some people to the categories 29 years and 31 years last birthday from the category 30 years last birthday.

This could be done, for example, by assuming that the numbers of people in each of these three categories were really equal (or, that the true numbers at each age were those of some stable population which is believed to have fertility and mortality rates similar to those in the developing country).

It is clear from the above that some assumption about the true age distribution is always going to be necessary if we are to correct the reported age structure. Such an assumption is potentially rather unsafe, since the results of any demographic reconstruction based on the corrected age structure will to some degree reflect this assumption.

However, we can avoid making any correction if we are prepared to assume that all the people who reported their age as, say, 30 last birthday had true ages between 28 years and 32 years last birthday (that is, no one misreported his/her age by more than two years). Similarly, we might assume that all those who reported their age as 35 years last birthday had true ages between 33 and 37 years last birthday.

If we make these assumptions, then we can avoid having to correct the age structure by using, in our analysis, the following age groups: 0–2 years, 3–7 years, 8–12 years, 13–17 years, 18–22 years, and so on. This is because, if the assumptions of the previous paragraph are true, the reported number of people in each of these age groups will be accurate.

14.4 (a) The total fertility rate (TFR) is related to the net reproduction rate (NRR) by the following equation:

$$\text{NRR} \cong \text{TFR} \cdot \frac{100}{205} \cdot \frac{l_{\bar{m}}}{l_0},\tag{1}$$

where $l_{\bar{m}}$ is the number of women (out of l_0 births) who survive to the mean age at childbearing. Rearranging equation (1), we have

$$\text{TFR} \cong \text{NRR} \cdot \frac{205}{100} \cdot \frac{l_0}{l_{\bar{m}}}.\tag{2}$$

We also know that

$$\text{NRR} = e^{rg},\tag{3}$$

where g is the mean length of a generation, and r is the annual rate of growth of the population.

Substituting from equation (3) into equation (2) gives

$$\text{TFR} \cong e^{rg} \cdot \frac{205}{100} \cdot \frac{l_0}{l_{\bar{m}}}.\tag{4}$$

$$e^{rg} = \left(\frac{100}{205}\right)\left(\frac{l_{\bar{m}}}{l_0}\right) \quad TFR$$

Now let us assume that the mean length of a generation is equal to the mean age at childbearing, and that both are equal to 28 years (remember that this is not too bad an assumption for the vast majority of human populations).

From the life table, we have $l_0 = 1000$, $l_{20} = 850$ and $l_{30} = 790$. Interpolating linearly between l_{20} and l_{30} to find l_{28} gives

$$l_{28} = 850 - 0.8(850 - 790) = 802.$$

We are also told that r is equal to 2%, or 0.02, per year. Substituting all these values into equation (4) produces

$$\text{TFR} \cong e^{(0.02 \times 28)} \cdot \frac{205}{100} \cdot \frac{1000}{802} \cong 1.7507 \times 2.05 \times 1.2469 \cong 4.47.$$

(b) Solving equation (4) above for r gives

$$r \cong \frac{\ln\left[\dfrac{\text{TFR} \cdot 100 \cdot l_{\bar{m}}}{205 \cdot l_0}\right]}{g}.$$

$$rg =$$

Assuming that $l_{\bar{m}}$, l_0 and g have the same values as they did in (a), then, if the TFR is equal to 6.0, we have

$$r \cong \frac{\ln\left[\dfrac{6.0 \times 100 \times 802}{205 \times 1000}\right]}{28} \cong \frac{\ln 2.3473}{28} \cong 0.03047.$$

Thus the population will grow at a rate of 3.05% per year.

14.5 (a) The information given means that we can assume that the population of furry animals is stable. The calculations are shown in Table S.50.

(b) We have

$$\text{net reproduction rate} = e^{rg} = \sum f_x^{\text{d}} L_x,$$

where g is the length of a generation, f_x^{d} is the age-specific fertility rate for female babies only and L_x is survivorship (assuming that $l_0 = 1$).

The length of a generation, g, can be approximated as the mean age at childbearing. Since female animals bear children between the ages of 1 and 3 years, but their age-specific

Table S.50

Age x	l_x	L_x	Average age y	$e^{-0.02y}L_x$	Percentage at each age
0	1000	900	0.5	891	34.2
1	800	750	1.5	728	27.9
2	700	600	2.5	571	21.9
3	500	350	3.5	326	12.5
4	200	100	4.5	91	3.5
				$\sum$ 2607	

fertility is twice as high between exact ages 1 and 2 years as it is between exact ages 2 and 3 years, we can write

$$\text{mean age at childbearing} \cong 0.667 \times 1.5 + 0.333 \times 2.5 \cong 1.83.$$

Thus

$$\text{NRR} = e^{(0.02 \times 1.83)} = 1.037.$$

If the age-specific fertility rate at age 1 last birthday is f_1 and equal numbers of males and females are born, then $f_1^d = 0.5f_1$, and $f_2^d = 0.25f_1$.

From above, we have

$$\text{NRR} = \sum f_x^d L_x = f_1^d L_1 + f_2^d L_2 = 0.5f_1 L_1 + 0.25f_1 L_2.$$

Inserting the value we have calculated for the NRR, and values for L_1/l_0 and L_2/l_0 from the table in the solution to (a), we obtain

$$1.037 = (0.5f_1 \times 0.75) + (0.25f_1 \times 0.6) = 0.375f_1 + 0.15f_1 = 0.525f_1,$$

whence $f_1 = 1.975$. Thus the age-specific fertility rate at age 1 last birthday is 1.975, and the age-specific fertility rate at age 2 last birthday is 0.9875.

Chapter 16

16.1 (a) See Figure S.10.

(b) See Figure S.11. Since the logarithms of the population, when plotted against time, produce lines which are nearly straight, the exponential model seems to be a good description of the growth of the populations of these two countries between 1950 and 1985.

(c) Linear regression using equation (16.8) produces the following estimated annual rates of growth: Algeria, 0.02617; Jordan, 0.02879. Using the exponential growth formula with the estimated rates of growth we have, for Algeria in 1995,

$$P_{1995} = P_{1950}\, e^{(45 \times 0.02617)}$$

$$= 8\,753\,000\, e^{(45 \times 0.02617)}$$

$$= 28\,419\,000.$$

For Algeria in 2005 we have

$$P_{2005} = P_{1950}\, e^{(55 \times 0.02617)}$$

$$= 8\,753\,000\, e^{(55 \times 0.02617)}$$

$$= 36\,920\,000.$$

For Jordan we obtain projections of 4 519 000 for 1995 and 6 026 000 for 2005.

Population
(thousands)
25000
20000
15000
10000
5000
0
Algeria
Jordan
1950
1955
1960
1965
1970
1975
1980
1985
Year

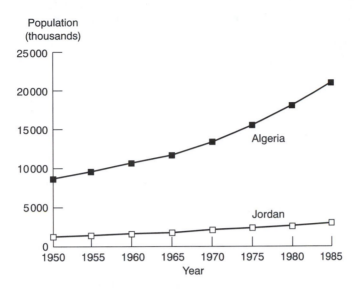

Figure S.10 Solution to Exercise 16.1(a)

Chapter 17

17.1 Denote the population aged between x and $x + n$ years last birthday in year t by the symbol ${}_nP_{x,t}$. Assume that mortality does not vary over time. Assume, further, that migration is negligible (this is not usually too bad an assumption for a national population, like that of England and Wales).

Then the proportion of those from age group 0–19 years in 1841 surviving and in age group 20–39 years in 1861 will be the same as the proportion of those from age group 0–19 years in 1861 surviving and in age group 20–39 years in 1881, or, in symbols,

$$\frac{{}_{20}P_{20,1881}}{{}_{20}P_{0,1861}} = \frac{{}_{20}P_{20,1861}}{{}_{20}P_{0,1841}}.$$

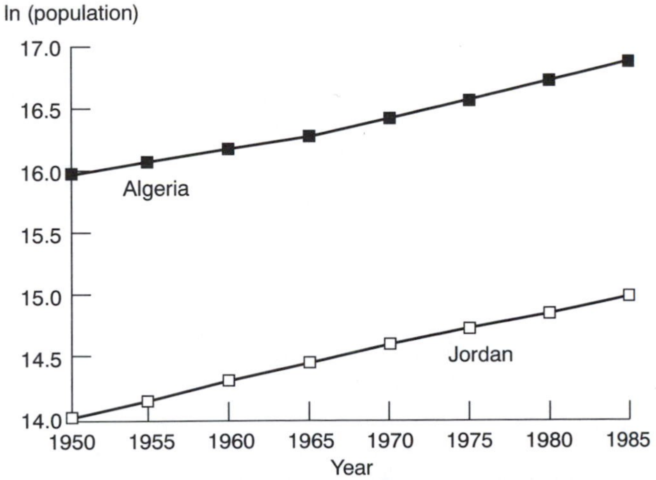

ln (population)
17.0
16.5
16.0
15.5
15.0
14.5
14.0
Algeria
Jordan
1950
1955
1960
1965
1970
1975
1980
1985
Year

Figure S.11 Solution to Exercise 16.1(b)

Substituting into the right-hand side of this equation the relevant data, we have

$$\frac{_{20}P_{20,1881}}{_{20}P_{0,1861}} = \frac{3164}{3667} = 0.8628.$$

So,

$$\text{population aged 20–39 years in 1881} = 0.8628 \times \text{population aged 0–19 years in 1861}$$

$$= 0.8628 \times 4\,537\,000$$

$$= 3\,915\,000.$$

If mortality is constant, and migration is negligible, then we can also say that the proportion of those from age group 20–39 years in 1841 surviving and in age group 40–59 years in 1861 will be the same as the proportion of those from age group 20–39 years in 1861 surviving and in age group 40–59 years in 1881. However, in this exercise, we are not dealing with the population aged 40–59 years, but the population aged 40 years and over.

To project the population aged 40 years and over in 1881, then, we need to make some other assumption. Two possible approaches are outlined here. One approach is to work out the ratio between the population aged 40 and over, and the population aged 20–39 years, in 1841 and 1861, and use the revealed trend to project this ratio in 1881. The required calculations are as follows:

$$\frac{\text{population aged 40 and over in 1841}}{\text{population aged 20–39 in 1841}} = \frac{1919}{2551} = 0.7523,$$

$$\frac{\text{population aged 40 and over in 1861}}{\text{population aged 20–39 in 1861}} = \frac{2589}{3164} = 0.8183.$$

Between 1841 and 1861 the ratio increased by $0.8183 - 0.7523 = 0.0660$. Assuming it increases by the same amount between 1861 and 1881, the ratio in 1881 will be $0.8183 + 0.0660 = 0.8843$.

Therefore, using the figure already calculated for the population aged 20–39 years in 1881, the population aged 40 years and over in 1881 will be equal to $0.8843 \times 3\,915\,000$, which is $3\,462\,000$.

The other approach is to use an assumption about survivorship which may be written in symbols as follows:

$$\frac{_{\omega-40}P_{40,1881}}{_{20}P_{20,1861}} = \frac{_{\omega-40}P_{40,1861}}{_{20}P_{20,1841}}, \tag{1}$$

where ω is the limiting age. (This assumption implies that the population aged 40–59 years last birthday is a constant proportion of the population aged 40 years and over: in other words, that the age structure is constant in proportional terms.)

Thus, substituting the known values from the data given in equation (1), we have

$$\frac{_{\omega-40}P_{40,1881}}{3164} = \frac{2589}{2551},$$

and so

$$\text{population aged 40 years and over in 1881} = \frac{2589}{2551} \times 3\,164\,000$$

$$= 3\,211\,000.$$

(I prefer this second approach, since a constant age structure is consistent with the assumption of constant mortality, which we have already made, and constant fertility, which we shall be making below.)

Table S.51

Age group	Female population, 1881 (thousands)
0–19	5620
20–39	3915
40 and over	3211

Now for the population aged 0–19 years in 1881. Those alive aged 0–19 years in 1881 are the survivors of the births between 1861 and 1881. Thus,

$$_{20}P_{0,1881} = \text{births between 1861 and 1881} \times \frac{_{20}L_0}{_{20}l_0},$$

where $_{20}L_0/_{20}l_0$ is the fraction of those born between 1861 and 1881 surviving until 1881 according to the life table representing constant mortality, using standard life table notation.

Now the number of births between 1861 and 1881 will be equal to the average population aged 20–39 years during this period multiplied by some 'birth rate' (this birth rate is not the same as any of the birth rates we defined in Chapter 8). Thus, we have

$$_{20}P_{0,1881} = \frac{\text{average population aged 20–39 years}}{\text{between 1861 and 1881}} \times \text{birth rate} \times \frac{_{20}l_0}{_{20}l_0}. \tag{2}$$

Now, assuming that fertility is constant over time, we can say that

$$_{20}P_{0,1861} = \frac{\text{average population aged 20–39 years}}{\text{between 1841 and 1861}} \times \text{birth rate} \times \frac{_{20}L_0}{_{20}l_0}. \tag{3}$$

Thus, substituting into equation (3) the appropriate values from the data given, we have

$$4537 = \tfrac{1}{2}(2551 + 3164) \times \text{birth rate} \times \frac{_{20}L_0}{_{20}l_0}$$

so

$$\text{birth rate} \times \frac{_{20}L_0}{_{20}l_0} = \frac{4537}{\tfrac{1}{2}(2551 + 3164)} = 1.5878.$$

Therefore, using the projected value of $_{20}P_{20,1881}$, we can apply equation (2) as follows:

$$_{20}P_{0,1881} = \tfrac{1}{2}(3\,164\,000 + 3\,915\,000) \times 1.5878 = 5\,620\,000.$$

The completed projection is shown in Table S.51.

17.2 The following data would be required.

1 A base year population (for example, the mid-year population in 1997) of males by single years of age from age 45 last birthday to the limiting age, which we denote by the symbol ω.
2 Age-specific death rates for males by single years of age from age 45 last birthday to the limiting age for each of the next 20 years (strictly speaking, we can get away with death rates for ages 45 to ω for the first year of the projection period, ages 46 to ω for the second year of the projection period, ages 47 to ω for the third year of the projection period, and so on).
3 Age-specific net migration rates for males by single years of age from age 45 last birthday to the limiting age for each of the next 20 years (again, we can get away with data for ages 45 to ω for the first year of the projection period, ages 46 to ω for the second year of the projection period, ages 47 to ω for the third year of the projection period, and so on).

17.3 (a) Assume the projection is being made from the present day. The data requirements are as follows.

1   Age structure data. The latest available mid-year population estimate will be for 30 June 1997. Since all the children aged under 5 years on 30 June 2002 and those aged under 4 years on 30 June 2001 will be born after 30 June 1997, to project the number of these children we only need data for females aged over about 10 years and under 50 years on 30 June 1997 (since these women will be the mothers of the relevant children). We need these by single years of age.

Children aged 3 and 4 years last birthday on 30 June 2000 will already have been born on 30 June 1997, at which point they will be aged under 2 years. We therefore need to know how many children were aged under 2 years on 30 June 1997.

2   A set of age-specific fertility rates by single years of age for the period 30 June 1997 to 30 June 1999 (all children aged 3 years and over on 30 June 2002 will have been born on or before 30 June 1999).

3   A set of age-specific death rates for males and females from ages 0 to 5 years for the period 30 June 1997 to 30 June 2002 (we need both sexes because males have different mortality levels than females).

4   A set of age-specific death rates for females from about age 10 years to age 50 years for the period 30 June 1997 to 30 June 1999.

5   Data on migration by single years of age for males and females aged 0–5 years and for females aged 10–50 years from 30 June 1997 to 30 June 2002.

(b)   The component method of projection should be adopted. The first stage is to project the number of females in the fertile age groups on 30 June 1998 and 30 June 1999. This can be done for the first period using the worksheet in Table S.52. A similar worksheet can be used for the following year.

Using the notation in Table S.52, we can say that the average female population at age x last birthday during the period 1997–98 is $\frac{1}{2}[^{1997}P_x^f + {}^{1998}P_x^f]$. A similar expression can be used for subsequent years. The number of children born during the period from 30 June 1997 to 30 June 1998 to women aged x last birthday is then equal to $\frac{1}{2}[^{1997}P_x^f + {}^{1998}P_x^f] \cdot {}^{1998}f_x$, where ${}^{1998}f_x$ is the age-specific fertility rate for the period. Summing this quantity over all the fertile ages gives the total number of children born between 30 June 1997 and 30 June 1998. If we denote this by ${}^{1998}N$, then the number of boys born is equal to $s \cdot {}^{1998}N$, and the number of girls born is equal to $(1 - s) \cdot {}^{1998}N$, where s is the assumed proportion of babies which are boys. Similar formulae can be used for the period from 30 June 1998 to 30 June 1999.

The children must then be 'survived forward' in time, to take account of infant and child mortality between 30 June 1997 and 30 June 2002. We need to include in the analysis those aged under 2 years on 30 June 1997, together with those projected to be born during the period between 30 June 1997 and 30 June 1999. The analysis requires another worksheet, like that shown in Table S.53. The one shown is for males; a similar one can be used

Table S.52

Age x	Females aged x birthday on 30 June 1997	Proportion surviving from age x to age $x + 1$	Females surviving to be $x + 1$ last birthday on 30 June 1998	Adjustment for net migration between 30 June 1997 and 30 June 1998	Females aged x last birthday on 30 June 1998
10	${}^{1997}P_{10}^f$	${}^{1997}S_{10}^f$			
11	${}^{1997}P_{11}^f$	${}^{1997}S_{11}^f$	${}^{1997}P_{10}^f \cdot {}^{1997}S_{10}^f$	${}^{1998}M_{11}^f$	${}^{1997}P_{10}^f \cdot {}^{1997}S_{10}^f + {}^{1998}M_{11}^f$
12	${}^{1997}P_{12}^f$	${}^{1997}S_{12}^f$	${}^{1997}P_{11}^f \cdot {}^{1997}S_{11}^f$	${}^{1998}M_{12}^f$	${}^{1997}P_{11}^f \cdot {}^{1997}S_{11}^f + {}^{1998}M_{12}^f$
⋮	⋮				
49	${}^{1997}P_{49}^f$				

Table S.53

Age x last birthday	Number alive on 30 June 1997	Survivorship factor for the period from 30 June 1997 to 30 June 1998	Number alive on 30 June 1998	Survivorship factor for the period from 30 June 1998 to 30 June 1999	Number alive on 30 June 1999	Survivorship factor for the period from 30 June 1999 to 30 June 2000	Number alive on 30 June 2000
0	$^{1997}P_0^m$	$^{1997}S_0^m$	$s \cdot {}^{1997}N \dfrac{^{1997}L_0^m}{l_0}$	$^{1998}S_0^m$	$s \cdot {}^{1998}N \dfrac{^{1998}L_0^m}{l_0}$	$^{1999}S_0^m$	$s \cdot {}^{1999}N \dfrac{^{1999}L_0^m}{l_0}$
1	$^{1997}P_1^m$	$^{1997}S_1^m$	$^{1997}P_0^m \cdot {}^{1997}S_0^m$	$^{1998}S_1^m$	$s \cdot {}^{1997}N \dfrac{^{1997}L_0^m}{l_0} \cdot {}^{1998}S_0^m$	$^{1999}S_1^m$	$s \cdot {}^{1998}N \dfrac{^{1998}L_0^m}{l_0} \cdot {}^{1999}S_0^m$
2			$^{1997}P_1^m \cdot {}^{1997}S_1^m$	$^{1998}S_2^m$	$^{1997}P_0^m \cdot {}^{1997}S_0^m \cdot {}^{1998}S_1^m$	$^{1999}S_2^m$	.
3					$^{1997}P_1^m \cdot {}^{1997}S_1^m \cdot {}^{1998}S_2^m$	$^{1999}S_3^m$	.
4					.	.	.

for females. (Only part of the worksheet is shown: it can be extended to the year 2002 to complete the calculations.)

Adjustments for net migration can also be made, although these are likely to be very small.

Repeated application of the worksheets will yield the number of males and females alive aged 3 and 4 years last birthday on each of the dates 30 June 2000, 30 June 2001 and 30 June 2002. The figures for males and females can then be summed to give the total number of children of these ages alive on each of these dates.

17.4 (a) The factors which may cause the outcome to differ from the projections are as follows:

1   errors in the age structure data;
2   uncertainty about the projection of fertility rates;
3   uncertainty about the projection of mortality rates;
4   uncertainty about migration;
5   incorrect assumptions about the sex ratio at birth.

(b) Of these factors, the most important is certainly going to be uncertainty about future fertility rates. This uncertainty will affect the projections of the population aged 3 and 4 years last birthday on 30 June 2002 and also that of the children aged 3 years last birthday on 30 June 2001. Fertility is subject to considerable fluctuations from year to year, depending on social, economic and attitude changes. These are very hard to predict.

International migration among women in the fertile age groups and among young children is relatively small. Mortality will probably decline slowly over the projection period, and the sex ratio at birth is almost constant at 105 males per 100 females. It is also likely that the base population age structure is accurate to within 1–2%.

17.5 Projections of the population by marital status might be required under the following circumstances: where estimates of the number of people in a component of the population explicitly defined on the basis of marital status are needed (for example, the number of war widows); and where marital status is an important intervening variable which will affect the results (for example, in projections of the number of children born in each of the next ten years, which may well be influenced by the marriage behaviour of the population during that time).

One common application of projections by marital status is in forecasting the demand for housing. The number of housing units which will be required in the future depends on the number of households there will be, and this is related to the distribution of the adult population by marital status.

17.6 (a) We use the component method. We assume that the life table we have describes mortality during the next five years (that is, that mortality remains constant).

Since we have five-year age groups, we need five-year survivorship probabilities, $_5L_{x+5}/_5L_x$, for values of x from 45 to 75 years. Given the mortality data we have, we can estimate $_5L_x$ either by assuming an even distribution of deaths within each five-year age group, and applying the formula

$$_5L_x = \tfrac{5}{2}(l_x + l_{x+5}),$$

or by working out L_x for each single year of age, and using the formula

$$_5L_x = \sum_{u=x}^{x+4} L_u.$$

Over the age range 45–75 years, the first of these methods is satisfactory. The calculations are shown in Table S.54. The age structure on 30 June 2000 is then obtained by

Table S.54

Exact age x	l_x	$_5L_x$	$\dfrac{_5L_{x+5}}{_5L_x}$
45	96 573	479 542.5	0.98155
50	95 244	470 695	0.96983
55	93 034	456 495	0.95263
60	89 564	434 870	0.92699
65	84 384	403 120	0.88528
70	76 864	356 875	0.81393
75	65 886	290 465	
80	50 300		

multiplying the base year age structure by the survivorship probabilities, as shown in Table S.55.

(b) The migration will only affect the projected numbers in the age group 65–69 years. Those moving in between 30 June 1995 and 30 June 1996 will be aged 69 years last birthday on 30 June 2000; those moving in between 30 June 1996 and 30 June 1997 will be aged 68 years last birthday on 30 June 2000, and so on.

Of course, some of the in-migrants will die before 30 June 2000. Consider those who move in between 30 June 1995 and 30 June 1996. They are, on average, aged exactly 65 years when they move in. Therefore, on 30 June 2000, they will be aged, on average, $69\frac{1}{2}$ years. The proportion who survive will be given by $1 - {}_{4\frac{1}{2}}q_{65}$. A value for this proportion can be obtained from the life table given using the formula

$$1 - {}_{4\frac{1}{2}}q_{65} \cong \frac{\frac{1}{2}(l_{69} + l_{70})}{l_{65}}.$$

Similar survival proportions can be derived for in-migrants in subsequent years. Applying these to the number of in-migrants in each year, and summing the results, will produce the expected number of in-migrants who survive until 30 June 2000. This figure can be added to the projected figure of 908 survivors from the population aged 60–64 years on 30 June 1995 to produce a revised estimate of the population aged 65–69 years on 30 June 2000.

Table S.55

Age group	Population on 30 June 1995	Population on 30 June 2000
45–49	1500	
50–54	1468	$1500 \times 0.98155 = 1472$
55–59	1370	$1468 \times 0.96983 = 1424$
60–64	980	$1370 \times 0.95263 = 1305$
65–69	940	$980 \times 0.92699 = 908$
70–74	800	$940 \times 0.88528 = 832$
75–79	730	$800 \times 0.81393 = 651$

References

Allison, P.D. (1984) *Event History Analysis: Regression for Longitudinal Survey Data*. London: Sage.

Benjamin, B. and Overton, E. (1981) Prospects for mortality decline in England and Wales. *Population Trends*, **23**, 22–28.

Benjamin, B. and Pollard, J.H. (1993) *The Analysis of Mortality and Other Actuarial Statistics*. London: Institute of Actuaries and Faculty of Actuaries in Scotland.

Blossfeld, H.-P., Hamerle, A. and Meyer, K.U. (1989) *Event History Analysis: Statistical Theory and Applications in the Social Sciences*. London: Lawrence Erlbaum Associates.

Boden, P., Stillwell, J. and Rees, P. (1992) How good are the NHSCR data? In J. Stillwell, P. Rees and P. Boden (eds), *Migration Processes and Patterns, Volume 2: Population Redistribution in the United Kingdom*. London: Belhaven, pp. 13–27.

Bogue, D., Hinze, K. and White, M.J. (1982) *Techniques of Estimating Net Migration*. Chicago: Community and Family Study Center, University of Chicago.

Bongaarts, J. (1978) A framework for analysing the proximate determinants of fertility. *Population and Development Review*, **4**, 105–132.

Bongaarts J. and Potter, R.G. (1983) *Fertility, Biology and Behaviour: An Analysis of the Proximate Determinants*. New York: Academic Press.

Brass, W. (1975) *Methods for Estimating Fertility and Mortality from Limited and Defective Data*. Chapel Hill, NC: Poplab, University of North Carolina.

Brass, W. (1989) Is Britain facing the twilight of parenthood? In H. Joshi (ed.), *The Changing Population of Britain*. Oxford: Blackwell, pp. 12–26.

Brass, W. (1996) Demographic data analysis in less developed countries, 1946–1996. *Population Studies*, **50**, 451–468.

Brass, W. and Jolly, C.L. (1993) *Population Dynamics of Kenya*. Washington, DC: National Academy Press.

Britton, M. (1989) Mortality and geography. *Population Trends*, **56**, 16–23.

Bulatao, R.A. and Lee, R.D. (eds) (1983) *Determinants of Fertility in Developing Countries* (2 volumes). London: Academic Press.

Bulusu, L. (1985) Area mortality comparisons and institutional deaths. *Population Trends*, **42**, 36–41.

Bulusu, L. (1991) A review of migration data sources. *Population Trends*, **66**, 45–47.

Bureau of Statistics (1972) *Japan Statistical Yearbook 1971*. Tokyo: Bureau of Statistics, Office of the Prime Minister.

Census of England and Wales, 1841 (1843) *Age Abstract, England and Wales and Islands in the British Seas*. London: Her Majesty's Stationery Office, pp. 458–459.

Census of England and Wales, 1861 (1863) *Population Tables Vol. II. Ages, Civil Condition, Occupations and Birth-Places of the People*. London: Her Majesty's Stationery Office, pp. x–xi.

Champion, T. and Fielding, A. (eds) (1992) *Migration Processes and Patterns, Volume 1: Research and Prospects*. London: Belhaven.

Coale, A.J. (1967) Factors associated with the development of low fertility: an historic summary. In United Nations, *Proceedings of the World Population Conference, 1965, Volume 2*. New York: United Nations Department of Economic and Social Affairs, pp. 205–209.

Coale, A.J. (1972) *The Growth and Structure of Human Populations: A Mathematical Investigation*. Princeton, NJ: Princeton University Press.

Coale, A.J. (1987) Stable populations. In J. Eatwell, M. Milgate and P.K. Newman (eds), *The New Palgrave: A Dictionary of Economics*, Volume 4. London: Macmillan, pp. 466–469.

Coale, A.J. and Demeny, P. (1983) *Regional Model Life Tables and Stable Populations* (2nd edition). New York: Academic Press.

Coale, A.J. and Watkins, S.C. (eds) (1986) *The Decline of Fertility in Europe*. Princeton, NJ: Princeton University Press.

Coulibaly, S., Dicko, F., Traoré, S.M., Sidibé, O., Seroussi, M. and Barrère, B. (1996) *Enquête Démographique et de Santé, Mali 1995–96*. Bamako, Mali: Cellule de Planification et de Statistique, Ministère de la Santé, de la Solidarité et des Personnes Agées; Direction Nationale de la Statistique et de l'Informatique; and Calverton, MD: Macro International Inc.

Cox, P.R. (1975a) Life tables: the measure of mortality. *Population Trends*, **1**, 13–15.

Cox, P.R. (1975b) Life tables (2): wider applications. *Population Trends*, **2**, 19–21.

Cox, P.R. (1976) *Demography* (5th edition). Cambridge: Cambridge University Press.

Dale, A. and Marsh, C. (eds) (1993) *The 1991 Census User's Guide*. London: Her Majesty's Stationery Office.

Daugherty, H.G. and Kammeyer, K.C.W. (1995) *An Introduction to Population* (2nd edition). New York: Guilford Press.

Davis, K. and Blake, J. (1956) Social structure and fertility: an analytic framework. *Economic Development and Cultural Change*, **4**, 211–235.

Daykin, C. (1986) Projecting the population of the United Kingdom. *Population Trends*, **44**, 28–33.

DeGroot, M.H. (1986) *Probability and Statistics* (2nd edition). Reading, MA: Addison-Wesley.

Diamond, I.D. and McDonald, J.W. (1992) Analysis of current status data. In J. Trussell, R. Hankinson and J. Tilton (eds), *Demographic Applications of Event History Analysis*. Oxford: Clarendon Press, pp. 231–252.

Elandt-Johnson, R.C. and Johnson, N.L. (1980) *Survival Models and Data Analysis*. New York: Wiley.

Feeney, G. and Yu, J. (1987) Period parity progression measures of fertility in China. *Population Studies*, **41**, 77–102.

Flowerdew, R. and Green, A. (1993) Migration, transport and workplace statistics from the 1991 census. In A. Dale and C. Marsh (eds), *The 1991 Census User's Guide*. London: Her Majesty's Stationery Office, pp. 269–294.

Fox, J., Jones, D., Moser, K. and Goldblatt, P. (1985) Socio-demographic differentials in mortality. *Population Trends*, **40**, 10–16.

Hajnal, J. (1953) Age at marriage and proportions marrying. *Population Studies*, **7**, 111–136.

Hajnal, J. (1965) European marriage patterns in perspective. In D.V. Glass and D.E.C. Eversley (eds), *Population in History: Essays in Historical Demography*. London: Edward Arnold, pp. 101–143.

Harding, S. (1995) Social class differences in mortality of men: recent evidence from the OPCS Longitudinal Study. *Population Trends*, **80**, 31–37.

Hinde, P.R.A. (1987) The population of a Wiltshire village in the nineteenth century: a reconstitution study of Berwick St James 1841–71. *Annals of Human Biology*, **14**, 475–485.

Hinde, P.R.A. and Mturi, A.J. (1996) Social and economic factors related to breast-feeding durations in Tanzania. *Journal of Biosocial Science*, **28**, 347–354.

Hobcraft, J. and Rodríguez, G. (1980) *Illustrative Analysis: Life Table Analysis of Birth Intervals in Colombia,* World Fertility Survey Scientific Report No. 16. Voorburg: International Statistical Institute; and London: World Fertility Survey.

Institut National d'Études Demographiques (1992) Vingt et unième rapport sur la situation demographique de la France. *Population*, 47, 1113–1186.

Kalbfleisch, J.D. and Prentice, R.L. (1980) *The Statistical Analysis of Failure Time Data*. New York: Wiley.

Keyfitz, N. (1977) *Introduction to the Mathematics of Population*. Reading, MA: Addison-Wesley.

Keyfitz, N. (1985) *Applied Mathematical Demography*. Reading, MA: Addison-Wesley.

Kleinbaum, D.G. (1996) *Survival Analysis: A Self-learning Text*. New York: Springer-Verlag.

Larson, H.J. (1982) *Introduction to Probability Theory and Statistical Inference* (3rd edition). New York: Wiley.

Lee, E.T. (1980) *Statistical Methods for Survival Data Analysis*. Belmont, CA: Lifetime Learning.

Lopez, A. (1967) Asymptotic properties of a human age distribution under a continuous net maternity function. *Demography*, **4**, 680–687.

Lotka, A.J. and Sharpe, F.R. (1911) A problem in age distribution. *Philosophical Magazine*, **21**, 435–438.

Lucas, D. and Meyer, P. (1994) *Beginning Population Studies* (2nd edition). Canberra: National Centre for Development Studies, Australian National University.

Lutz, W. (1993) Nuptiality rates. In D.J. Bogue, E.E. Ariaga and D.L. Anderton (eds), *Readings in Population Research Methodology, Volume 4: Nuptiality, Migration, Household, and Family Research*. Chicago, IL: Social Development Center, pp. 13.3–13.6.

Mood, A.M., Graybill, F.A. and Boes, D.C. (1974) *Introduction to the Theory of Statistics* (3rd edition). New York: McGraw-Hill.

Mturi, A.J. and Hinde, P.R.A. (1994) Fertility decline in Tanzania. *Journal of Biosocial Science*, **26**, 529–538.

National Council for Population and Development (1989) *Kenyan Demographic and Health Survey, 1989*. Nairobi: Ministry of Home Affairs and National Heritage; Columbia, MD: Institute for Resource Development/Macro Systems, Inc.

Newell, C. (1988) *Methods and Models in Demography*. London: Belhaven.

Ngallaba, S., Kapiga, S.H., Ruyobya, I. and Boerma, T. (1993) *Tanzania Demographic and Health Survey 1991/92*. Dar es Salaam: Bureau of Statistics; and Calverton, MD: Macro International.

Ní Bhrolcháin, M. (1987) Period parity progression ratios and birth intervals in England and Wales, 1941–1971: a synthetic life table analysis. *Population Studies*, **41**, 103–125.

Ní Bhrolcháin, M. (1992) Period paramount: a critique of the cohort approach to fertility. *Population and Development Review*, **18**, 599–629.

Office for National Statistics (1997a) *English Life Table No. 15*, series DS, no. 14. London: The Stationery Office.

Office for National Statistics (1997b) *Mortality Statistics: Cause*, series DH2, no. 22. London: The Stationery Office.

Office for National Statistics (1997c) *Birth Statistics*, series FMI, no. 24. London: The Stationery Office.

Office of Population Censuses and Surveys (1987a) *English Life Table No. 14: The Report Prepared by the Government Actuary for the Registrar General for England and Wales*. London: Her Majesty's Stationery Office.

Office of Population Censuses and Surveys (1987b) *Birth Statistics*, series FM1, no. 13. London: Her Majesty's Stationery Office.

Office of Population Censuses and Surveys (1989) *Marriage and Divorce Statistics*, series FM2, no. 14. London: Her Majesty's Stationery Office.

Office of Population Censuses and Surveys (1992) *1991 Census County Report: East Sussex*, Part 1. London: Her Majesty's Stationery Office.

Office of Population Censuses and Surveys (1993a) *1991 Census Report for Great Britain*, Part 1, Vol. 1. London: Her Majesty's Stationery Office.

Office of Population Censuses and Surveys (1993b) *Marriage and Divorce Statistics*, series FM2, no. 19. London: Her Majesty's Stationery Office.

Office of Population Censuses and Surveys (1994) *Mortality Statistics: General 1992*, series DH1, no. 27. London: Her Majesty's Stationery Office.

Political and Economic Planning (1948) *Population Policy in Great Britain: A Report by PEP*. London: Political and Economic Planning.

Preston, S. (1975) Estimating the proportion of American marriages that end in divorce. *Sociological Methods and Research*, **4**, 435–460.

Rees, P.H. and Wilson, A.G. (1977) *Spatial Population Analysis*. London: Edward Arnold.

Republic of Malawi (1994) *Malawi Demographic and Health Survey 1992*. Zomba, Malawi: National Statistical Office; Calverton, MD: Macro International Inc.

Republic of Seychelles (1991) *1987 Census Report*. Victoria: Statistics and Database Administration, Management Information Systems Division.

Rogers, A. (1975) *Introduction to Multiregional Mathematical Demography*. London: Wiley.

Shaw, C. (1993) 1991-based national population projections for the United Kingdom and constituent countries. *Population Trends*, **72**, 45–50.

Shaw, C. (1994) Accuracy and uncertainty of the national population projections for the United Kingdom. *Population Trends*, **77**, 24–32.

Shryock, H.S. and Siegel, J.S. (1975) *The Methods and Materials of Demography* (2 volumes). Washington, DC: US Bureau of the Census.

Statistics Sweden (1997) *Statistical Yearbook of Sweden*. Stockholm: Statistics Sweden.

United Nations (1970) *Methods of Measuring Internal Migration*. Population Studies 47. New York: United Nations Department of Economic and Social Affairs.

United Nations (1982) *Model Life Tables for Developing Countries*. New York: United Nations Department of International Economic and Social Affairs.

United Nations (1983) *Indirect Techniques for Demographic Estimation*. Manual X, Population Studies 81. New York: United Nations.

United Nations (1991) *World Population Prospects 1990*. New York: United Nations Department of International Economic and Social Affairs.

United Nations (1993) *Demographic Yearbook 1991*. New York: United Nations.

United Nations (1995a) *World Population Prospects 1994*. New York: United Nations Department of International Economic and Social Affairs.

United Nations (1995b) *Demographic Yearbook 1993*. New York: United Nations,

Weeks, J.R. (1989) *Population: An Introduction to Concepts and Issues* (4th edition). Belmont, CA: Wadsworth.

Werner, B. (1988a) Birth intervals: results from the OPCS Longitudinal Study 1972–84. *Population Trends*, **51**, 25–29.

Werner, B. (1988b) Spacing of births to women born in 1935–1959: evidence from the OPCS Longitudinal Study. *Population Trends*, **52**, 20–25.

Wilkie, J.W., Komisaruk, C. and Ortega, J.G. (1996) (eds), *Statistical Abstract of Latin America, Vol. 32*. Los Angeles: UCLA Latin American Centre.

Woods, R. (1982) *Theoretical Population Geography*. London: Longman.

World Bank (1984) *Population Change and Economic Development*. Oxford: Oxford University Press.

Wrigley, E.A. and Schofield, R.S. (1989) *The Population History of England 1541–1871: A reconstruction*. Cambridge: Cambridge University Press.

Yamaguchi, K. (1991) *Event History Analysis*. London: Sage.

Yaukey, D. (1990) *Demography: The Study of Human Population*. Prospect Heights, IL: Waveland.

Zou'bi, A.A., Poedjastoeti, S. and Ayad, M. (1992) *Jordan Population and Family Health Survey 1990*. Amman: Department of Statistics and Ministry of Health; and Columbia, MD: Institute for Resource Development/Macro International Inc.

Index

Page numbers in *italic* refer to figures and tables.